Fluids and Electrolytes:
A Practical Approach

Fluids and Electrolytes:
A Practical Approach
Edition 4

Carla A. Bouska Lee, PhD, RNC, ARNP, FAAN
Associate Professor and Coordinator of
Postgraduate Studies in Nursing
Department of Nursing
Clarkson College
Omaha, Nebraska

Vice President of Development and Director
Associated Learning Centers, Inc.
GRC Industries, Inc.
Kansas City, Missouri

C. Ann Barrett, MA
Assistant Director
Research and Planning
Mesa Community College
Mesa, Arizona

Principal
Barrett Communications
Phoenix, Arizona

Donna D. Ignatavicius, MS, RNC
Instructor, Charles County Community College
School of Nursing
Instructor, University of Maryland School of Nursing
Consultant and Owner, DI Associates
La Plata, Maryland

F. A. DAVIS COMPANY · Philadelphia

F. A. Davis Company
1915 Arch Street
Philadelphia, PA 19103

Printed in the United States of America

Last digit indicates print number: 10 9 8 7 6 5 4 3 2 1

Publisher: Robert G. Martone
Nursing Editor: Alan Sorkowitz
Production Editor: Jessica Howie Martin
Cover Designer: Louis J. Forgione

As new scientific information becomes available through basic and clinical research, recommended treatments and drug therapies undergo changes. The authors and publisher have done everything possible to make this book accurate, up to date, and in accord with accepted standards at the time of publication. The authors, editors, and publisher are not responsible for errors or omissions or for consequences from application of the book, and make no warranty, expressed or implied, in regard to the contents of the book. Any practice described in this book should be applied by the reader in accordance with professional standards of care used in regard to the unique circumstances that may apply in each situation. The reader is advised always to check product information (package inserts) for changes and new information regarding dose and contraindications before administering any drug. Caution is especially urged when using new or infrequently ordered drugs.

Library of Congress Cataloging in Publication Data

Lee, Carla A. B
    Fluids and electrolytes : a practical approach / Carla A. B. Lee, C. Ann Barrett,
Donna D. Ignatavicius.—Ed. 4
       p.   cm.
    Rev. ed. of : Fluids and electrolytes / Violet R. Stroot, Carla A. B. Lee, C. Ann Barrett,
Ed. 3. c1984.
    Includes bibliographical references and index.
    ISBN 0-8036-5531-2. — ISBN 0-8036-8531-2. (pbk.)
    1. Body fluid disorders—Nursing.   2. Fluid-electrolyte imbalances—Nursing.
I. Barrett, C. Ann, 1942– .
    II. Ignatavicius, Donna D.   III. Stroot, Violet R. Fluids and electrolytes.   IV. Title.
    [DNLM: 1. Water-Electrolyte Imbalance—nurses' instruction.   2. Water-Electrolyte
Balance—nurses' instruction.   WD 220 L477f 1996]
RC630.S77   1996
616.3'99—dc20
DNLM/DLC
for Library of Congress
                                                96-11892
                                                   CIP

Dedicated to Thomas J. Luellen, MD, and Violet R. Stroot, RN, BS

Preface

Over the past 10 years, the acuity and complexity of health care have increased dramatically. Nurses are expected to make rapid, accurate assessments and implement appropriate, cost-effective interventions in a variety of settings.

With these changes in mind, we made major revisions in the fourth edition of *Fluids and Electrolytes: A Practical Approach* to provide the reader with accessible, vital information. Although the book continues to attract the nursing student population, clinicians and case managers will also find this edition a valuable resource for professional practice in the 21st century in multiple settings with application across the lifespan.

This edition is divided into four parts. Part One, Fluid, Electrolyte, and Acid-Base Balance and Imbalance, is composed of three chapters that provide an overview of basic principles related to fluids and electrolytes, including the key fluid disorders.

Part Two, Electrolytes and Other Elements: Balance and Imbalance, contains 10 chapters. Five of these chapters are new to this edition: Chapter 9, Magnesium; Chapter 10, Copper; Chapter 11, Iron; Chapter 12, Zinc; and Chapter 13, Trace Minerals. These new chapters reflect the increase in knowledge during the last few years concerning the importance of additional electrolytes and elements in the maintenance of health and the pathophysiology of illness.

Part Three, Techniques and Procedures for Maintenance of Fluid and Electrolyte Balance, has also been revised. The last edition's chapters on venipuncture and airways have been omitted. The chapter on intravenous therapy (Chapter 15), now includes expanded content on parenteral nutrition. Each chapter in this part of the book focuses on the nursing assessments and interventions related to each selected treatment modality.

The last part of the book, Part Four, is entitled Selected Clinical Disorders that Affect Fluid, Electrolyte, and Acid-Base Balance. Although not intended to replace information found in medical-surgical nursing textbooks, the material in this part enhances and supplements their content. The discussion in each chapter illustrates the relationship of each disorder to fluid, electrolyte, and acid-base balance.

In addition to the revisions in each part of the book, several new features are incorporated throughout. First, where appropriate, additional lifespan considerations are identified with run-in headings for elderly persons, children, and pregnant women. Second, in view of increasing cultural diversity, transcultural content is clearly delineated where applicable.

New tables that focus on nursing assessment and key interventions help students and practitioners identify the most important content. Critical thinking exercises have been added to many chapters, as well as 40 NCLEX-style test questions with answers. All of the reference and reading lists have been updated and expanded.

Finally, in this edition the reader will find a more friendly, concise reading style with multiple headings to make material easy to find. The intent of this book is to provide the information that nurses *need* to build a solid foundation for practice.

Carla A. B. Lee
C. Ann Barrett
Donna D. Ignatavicius

Acknowledgments

To produce and maintain a book through four editions requires the efforts of many people, and we wish to acknowledge some of them here. First we would like to thank the consultants who have reviewed and commented upon the manuscript for this new fourth edition, and we would also like to acknowledge the work of the contributors and consultants to the previous editions.

This edition is dedicated to both Thomas J. Luellen, MD, and Violet R. Stroot, RN, BS. Dr. Luellen served as project director of the Kansas Regional Medical Program's Fluid and Electrolytes Project, which conducted fluid and electrolyte workshops throughout the state. Violet Stroot was project coordinator. We wish to acknowledge them both for establishing these workshops, which led ultimately to this book. Violet served as senior author through the first three editions.

The staff at F. A. Davis Company have all been excellent to work with. Their skill, kindness, and professionalism have improved this edition in many ways. We owe a special note of thanks to Alan Sorkowitz, Nursing Editor; Robert G. Martone, Nursing Publisher; Robert Butler, Production Manager; Jessica Howie Martin, Production Editor; and Louis J. Forgione, Cover Designer.

Our families have been a tremendous source of support, and we thank them for the patience they have shown during this project: Gordon Larry Lee; Kristin and Scott Schaper; Orville and Charline Barrett; and Charles and Stephanie Ignatavicius.

We thank the instructors who have chosen our book for their classes, and especially those who, over the years, have provided us with comments and suggestions for each new edition. Finally, we thank our students, who over the years have taught us as much as we have taught them.

Carla A. B. Lee
C. Ann Barrett
Donna D. Ignatavicius

Consultants

Sheri L. Bremer, RN, MSN
Community College of Denver
Denver, Colorado

Yvonne Chipman, RN, MSN
School of Nursing
St. Francis Medical Center
Pittsburgh, Pennsylvania

Peggy Craik, MBA, MS, RN
Associate Professor
School of Nursing
University of Mary Hardin-Baylor
Belton, Texas

Lona W. Crane, MSN, RNC
Instructor, Department of Nursing
Xavier University
Cincinnati, Ohio

Peggy Dease, BS, MS
Director, Associate Degree Nursing
Pearl River Community College
Poplarville, Mississippi

Carol A. Fanutti, EdD, RN
Assistant Director of Nursing
Trocaire College
Buffalo, New York

Joan Fregoe, RN, BSN, MEd
Associate Professor of Nursing
State University of New York
  at Canton
Canton, New York

Jeri Jaquis, RN, MSN
Galveston College
Galveston, Texas

Sherri Kembel, RN, MSN
Nursing Instructor
Morgan Community College
Ft. Morgan, Colorado

Barbara Rexilius, RN, MSN
Associate Professor of Nursing
Director of Nursing/Chair of Allied
  Health
North Country Community College
Saranac Lake, New York

Contents

• • • • • • • • •

PART ONE

FLUID, ELECTROLYTE, AND ACID-BASE BALANCE AND IMBALANCE

CHAPTER 1

. . . .
. . . .
. . . .
. . . .
. . . .
. . . .
. . . .
. . . .
. . . .
. . . .
. . . .

Concepts of Fluid and Electrolyte Balance

FLUID BALANCE

Solutions are made up of fluid (the solvent) and particles dissolved in the fluid (the solute). In the body, solutes are both electrolytes and nonelectrolytes; fluid is primarily water. When dissolved in water, electrolytes, such as potassium and sodium, dissociate into electrically charged atoms or groups of atoms known as ions. Nonelectrolytes, such as urea, do not have this electrical property and disperse as electrically neutral (uncharged) molecules.

Fluid Distribution

Fluids and electrolytes are found throughout the body and are necessary for normal cellular function. Approximately 60% of an adult's body weight is water.

Body fluids are located in two major compartments—intracellular fluid (ICF) and extracellular fluid (ECF). As seen in Figure 1–1, ICF is contained within the cells and accounts for about two thirds of all body fluids in adults. In infants, less than half of all fluid is intracellular. ECF can be further divided into three compartments: interstitial fluid, intravascular fluid, and transcellular fluid.

Interstitial fluid (ISF) is the fluid around the cells and includes lymph. Fluids and electrolytes can move freely between this compartment and the

3

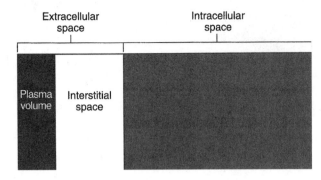

FIGURE 1–1
Normal distribution of total body water. (From Ignatavicius, DD, Workman, ML, and Mishler, MA: Medical-Surgical Nursing: A Nursing Process Approach, ed 2. WB Saunders, Philadelphia, 1995, p 253. © 1992 M. Linda Workman. Reprinted with permission.)

intravascular fluid (IVF), which is the plasma component of the blood. The average blood volume for an adult is 5 L; 3 L are plasma and the remaining 2 L are blood cells. Transcellular fluid (TCF) is secreted by epithelial cells and, therefore, is found in many places within the body. Cerebrospinal fluid, digestive fluids, and synovial fluid are examples of TCF.

Transcultural Considerations

A person's age, sex, racial/ethnic origin, and weight can influence the amount and distribution of body fluid. Infants have as much as 70% body water, whereas the elderly have less than 50%. Men have more body water than women because they have fewer fat cells. Fat cells contain little water. African Americans often have larger numbers of fat cells compared to other groups and, therefore, have less body water (Giger & Davidhizar, 1991).

Implications for Infants and Children

Infants and children have a higher percentage of total body water than their adult counterparts. The largest percentage of an infant's body is extracellular. However, as the individual grows and matures, more fluid moves into the intracellular compartment.

Implications for Pregnant Women

When a woman becomes pregnant, the IVF compartment (blood volume) increases to between 45% and 50% of total body fluids. The excess fluid causes a physiologic anemia (hemodilution) evidenced by a decreased hematocrit. ISF also increases as a result of the sodium retention associated with pregnancy.

Homeostasis

Numerous physical and biologic processes work together to maintain the proper balance of fluids and electrolytes in the body; this balance is referred to as homeostasis. Homeostasis is regulated by fluid and electrolyte transport systems and by regulatory mechanisms, such as hormones and body organs.

Fluid and Electrolyte Transport Systems

Two types of transport systems move fluids and electrolytes within the body: passive and active. In *passive* transport systems, no energy is expended. By contrast, *active* transport requires energy.

Passive Transport Systems

Body fluid and electrolyte movement takes place by means of three passive transport systems: diffusion, filtration, and osmosis.

Diffusion

Diffusion is based on the principle that molecules and ions flow freely, but randomly, from an area of higher concentration of solute across a biologic membrane to an area of lower concentration to establish an equilibrium (Fig. 1–2). This process is due to the molecular movement referred to as Brownian motion. The speed of diffusion is related to the concentration difference between the two sides of the membrane (concentration gradient). The greater the difference in concentration, the more rapidly diffusion occurs.

In the body, diffusion is the process by which gases (oxygen and carbon dioxide), electrolytes, and molecules move across a capillary or cell membrane. In some cases, the substance diffuses with the assistance of another. For example, insulin promotes the movement of glucose and potassium into the cell. This type of diffusion is called facilitated diffusion.

Although most capillary membranes allow indiscriminate diffusion, cell membranes are more selective, permitting movement of some solutes but preventing movement of others. When the cell allows movement through its membrane, the membrane is referred to as permeable. When it does not, the membrane is described as impermeable. Diffusion, then, is a passive transport system that depends on the concentration gradients of the solute, and is permitted or enhanced by the permeability of the cell membrane.

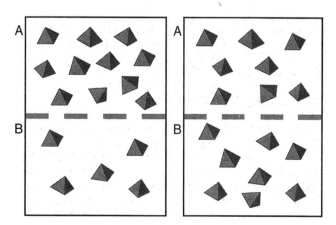

FIGURE 1–2
Diffusion of a solute. (From Ignatavicius, DD, Workman, ML, and Mishler, MA: Medical-Surgical Nursing: A Nursing Process Approach, ed 2. WB Saunders, Philadelphia, 1995, p 245. © 1992 M. Linda Workman. Reprinted with permission.)

Filtration

Filtration is the movement of the solute *and* solvent (fluid) from one side of a membrane to another caused by hydrostatic pressure differences between the two sides of the membrane. Hydrostatic pressure is the force per unit area exerted by a liquid at rest. In a column of liquid, for example, molecules exert force because of the weight of the solution. Thus, hydrostatic pressure may be called the "water pushing" pressure.

In the body, filtration is important for the movement of nutrients, water, and waste products when blood arrives at the tissue capillaries. Because the capillaries are lined by a single layer of cells and large pores are present between those cells, water can filter freely through capillary membranes in either direction if a hydrostatic pressure gradient (difference in pressures) is present.

If the normal pressure gradient changes, edema can develop. In patients with right-sided congestive heart failure, for example, blood backs up into the systemic venous circulation causing the venous and capillary hydrostatic pressures to increase. As a result, water moves from the capillaries into the interstitial spaces in the periphery causing visible edema—a process called "third spacing."

Osmosis

Osmosis is the movement of a *solvent* from a solution of *lower* solute concentration to one of *higher* concentration through a semipermeable membrane that separates the two solutions. It is the process by which fluids move to and from intracellular and extracellular areas (Fig. 1–3). The concentration of solutes dissolved in body fluids, therefore, affects the movement of water. This effect is called osmotic pressure or "water pulling pressure."

The osmotic pressures of electrolytes and nonelectrolytes in a solution differ. There are two basic types of osmotic pressure: crystalloid, which is due to dissolved ions that can pass freely through a membrane, and colloid, due to substances with larger molecules, that, in healthy persons, do not pass through a membrane. Examples of colloid substances are plasma proteins, such as albumin.

The concentration of substances in body fluids is expressed as milliosmoles per liter (mOsm/L) or milliequivalents per liter (mEq/L). Most laboratories in the United States measure electrolytes in mEq/L.

The number of milliosmoles in a liter of solution is called osmolarity. A similar term, osmolality, refers to the number of milliosmoles in a kilogram (kg) of solution. The normal osmolarity of plasma and other body fluids [extracellular fluid (ECF)] is between 270 and 300 mOsm/L (Guyton, 1995). Sodium, an important body electrolyte, is primarily responsible for extracellular osmolarity and helps to hold water in that fluid compartment.

Some solutes, such as sodium and glucose, are very effective in moving fluids across a biologic membrane. Tonicity is the term used to describe effective osmolarity.

Fluids or solutions can be classified as isotonic, hypertonic, or hypotonic. A fluid into which cells can be placed without affecting the cell size is described as *isotonic* with the cells. For example, a 0.9% saline solution [nor-

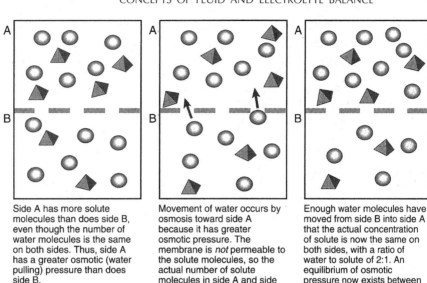

Side A has more solute molecules than does side B, even though the number of water molecules is the same on both sides. Thus, side A has a greater osmotic (water pulling) pressure than does side B.

DISEQUILIBRIUM
side A 1.5:1 ratio of water to solute
side B   3:1 ratio of water to solute

Movement of water occurs by osmosis toward side A because it has greater osmotic pressure. The membrane is *not* permeable to the solute molecules, so the actual number of solute molecules in side A and side B does not change. *Only the water molecules move because the membrane is not permeable to the solute molecules.*

Enough water molecules have moved from side B into side A that the actual concentration of solute is now the same on both sides, with a ratio of water to solute of 2:1. An equilibrium of osmotic pressure now exists between the two compartments, and no further *net* movement of water molecules or solute molecules will occur.

EQUILIBRIUM
side A 2:1 ratio of water to solute
side B 2:1 ratio of water to solute

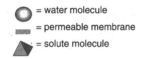

○ = water molecule

▬ = permeable membrane

◆ = solute molecule

FIGURE 1–3
The osmosis process. (From Ignatavicius, DD, Workman, ML, Mishler, MA: Medical-Surgical Nursing: A Nursing Process Approach, ed 2. WB Saunders, Philadelphia, 1995, p 247. © 1992 M. Linda Workman. Reprinted with permission.)

mal saline (NS)] is isotonic with the plasma and other body fluids. Isotonic solutions are frequently used for intravenous infusion because they do not affect the plasma osmolarity.

A solution that can make cells shrink is referred to as *hypertonic*. A hypertonic solution exerts *greater* osmotic pressure (water pulling pressure) than the solution with which it is compared. A 3% saline solution, for instance, would cause water to leave the cells (intracellular compartment) and enter the ECF compartment.

Conversely, a solution that can make cells swell is called a *hypotonic* solution. A hypotonic solution exerts *less* osmotic pressure than the solution with which it is compared. For example, a 0.2% saline solution causes water to leave the ECF compartment and enter the cells.

Active Transport System

The second major type of transport system for fluids and electrolytes is active transport. The principle of active transport is that *solutes* can be moved from an area of *lower* concentration of solute to an area of *higher* concentration, that is, against a concentration gradient. This process is also referred to as "pumping."

The most common example of active transport is the sodium-potassium pump. Located in the cell membrane, this pump allows the movement, when needed, of sodium out of the cell and of potassium into the cell. Because both electrolytes move against their respective concentration gradients, energy is required for the process to occur. Therefore, the optimum function of active transport pumps depends on the presence of adequate amounts of cellular adenosine triphosphate (ATP).

Fluid Gain and Loss

Water is vital for body temperature regulation, nutrient and electrolyte transport, blood volume maintenance, and cellular metabolism. Although people can live without food for several weeks, they can survive for only a few days without water.

Fluid Gain

Water is gained and lost from the body every day, but the net amount usually remains fairly constant in a healthy person. Fluid intake is regulated by the thirst mechanism in the brain. In a healthy person, most of the fluid maintenance is achieved through the oral intake of liquids. In individuals who are ill, parenteral fluids and irrigations may be sources of fluid ingestion. Nonmeasurable sources of fluid are solid foods and fluid resulting from cellular metabolism.

Fluid Loss

Sensible and insensible obligatory losses of approximately 2500 mL (2.5 L) of fluid occur daily in an adult. *Sensible* losses are those of which a person is aware, such as urination, which accounts for the loss of approximately 1500 mL of extracellular fluid during each 24-hour period.

Insensible losses, such as perspiration, may occur without a person's awareness. In an adult, approximately 1000 mL of extracellular fluid (ECF) are lost every day through the skin (600 mL) and lungs (400 mL). Electrolytes can be lost if this amount increases, for example, when profuse perspiration occurs with a fever or excessive exercise.

Not all secretions are lost from the body. Digestive juices are secreted and reabsorbed at the rate of approximately 10.0 L every 24 hours in an adult. This gastrointestinal (GI) fluid contains sodium, chloride, and potassium. All the food and fluid ingested into the body, excluding undigestible waste byproducts, are absorbed into the ECF. During normal body function, GI fluids are reabsorbed at 9.5 L through the small intestine and 0.5 L through the large intestine. However, during illness, the loss of GI fluids from the body as a result of vomiting and diarrhea can cause fluid and electrolyte imbalance. Upper GI fluid contains 5 to 10 mEq/L of potassium; lower GI fluid has 30 to 50 mEq/L.

Additional fluid losses from the body can occur through excessive salivation, GI suction, and fistula and wound drainage.

Implications for Infants and Children

The daily intake and output for an infant is about half the volume of the ECF. Therefore, in an infant fluid loss can equal the total ECF in 2 days. By contrast, equivalent fluid loss in an adult takes 6 days.

ELECTROLYTE BALANCE

The human body makes use of substances contained in foodstuffs, which, through the processes of digestion, hydrolysis, and absorption, are broken down into more basic elements: carbohydrates, fats, proteins, vitamins, minerals, water, and fiber. Hydrolysis further converts carbohydrates, simple and complex, to glucose; fats to fatty acids and glycerol; and proteins to amino acids. Vitamins, minerals, and water obtained from the foodstuffs enter the bloodstream to function in key metabolic actions.

Natural minerals found in foodstuffs become dissociated or otherwise transformed through digestion or metabolic processes to form electrolytes, that is, charged particles in solution. These electrolytes play a key role in bodily functions. Their concentrations are usually expressed as milliequivalents per liter (mEq/L) or milligrams per deciliter (mg/dL).

Electrolyte Distribution

Electrolytes can be divided into two types—cations, which are positively charged, and anions, which are negatively charged. The key ions for the body are the cations sodium (Na^+), potassium (K^+), calcium (Ca^{2+}), and magnesium (Mg^{2+}); and the anions chloride (Cl^-), bicarbonate (HCO_3^-), phosphate (PO_4^{3-}), and sulfate (SO_4^{2-}). Phosphate at its physiological pH is actually hydrogen phosphate (HPO_4^{2-}).

The major electrolytes of the intracellular fluid (ICF) are potassium, magnesium, and phosphate. Potassium is the primary cation; phosphate is the primary anion. Extracellular fluid (ECF) contains the cation sodium and the anions chloride and bicarbonate. The primary electrolytes and the major regulators of fluid and electrolyte balance in the ECF are sodium and chloride, or saline. Part II of this book describes each of the major electrolytes in detail.

Electrolyte Functions

Electrolytes maintain cell structure by their osmotic effect and are an integral part of metabolic and cellular processes. Examples of electrolyte functions in cellular processes are the sodium-potassium pump, in which sodium and potassium interact to create the electrical charges innervating every nerve in the body, the interaction of calcium and phosphorous in skeletal stability, and the role of potassium in myocardial contractility and conduction.

The amount of an electrolyte considered to be within the normal range in the body can vary in two ways. First, there can be a variation in total body electrolyte content. Second, there can be a variation in the specific proportions maintained in the two major fluid compartments. Understanding these two concepts of fluid and electrolyte variation is important to comprehend the differences between true electrolyte readings and false electrolyte readings.

ELECTROLYTE IMBALANCE

Imbalances may be caused by inadequate intake, actual loss from the body, or dislocation of the electrolyte plus fluid to a "third space" where it is trapped and not physiologically accessible. The two most common imbalances of electrolytes are excess or deficit. The location of many electrolytes in the body may change to assist in maintaining equilibrium of the total body electrolytes at the expense of maintaining the correct ratio between the fluid compartments. The classic example of this process is potassium concentration. When a person is experiencing acidosis (i.e., the serum is acidic), the compensatory mechanism that is activated causes hydrogen ions (acid) to move into the cell and potassium ions to be released into the extracellular fluid. The movement of the hydrogen ions into the cells reduces the acidotic state; however, a "false" elevated serum potassium reading results. This reading is called "false" because the actual total body potassium amount has not changed, just the location (from the cell into the serum) and the concentration of potassium in the ECF. The fact that this is a "false" elevation, however, does not reduce the serious consequences if the potassium concentration level in the serum reaches 6.0 mEq/L or greater. Extremely high potassium levels can lead to death.

Imbalances are specifically labeled by the Greek prefixes "hyper" (over) and "hypo" (under). The prefix is followed by the English or Latin name of the electrolyte, which is then followed by the Latin suffix "emia" meaning blood. For example, the term for sodium excess uses the prefix "hyper"; the Latin term for sodium, "natrium"; and "emia," which means blood. Thus a sodium excess is hypernatremia. The terms describing magnesium follow the same pattern but imbalances use the English name for magnesium. Thus, a magnesium excess in the blood is hypermagnesemia. Table 1–1 lists terms used to describe common fluid and electrolyte imbalances.

Electrolyte Concentrations

Although the quantities of electrolytes vary in both total body amounts and the specific proportions of each electrolyte maintained in the two fluid compartments (ICF and ECF), the actual concentration in the serum must remain within a very narrow range. For instance, the concentration of potassium in the serum (ECF) normally ranges between 3.5 and 5.5 mEq/L. A normal range of potassium for the ICF is 30 times higher at 105 to 165 mEq/L.

Regulatory Mechanisms

Maintenance of fluid and electrolyte balance is a complex process based on the interaction of regulatory mechanisms in the body. There are four major

TABLE 1-1
TERMS FOR COMMON FLUID AND ELECTROLYTE IMBALANCES

Hypovolemia	Fluid deficit in the IVF compartment
Hypervolemia	Fluid excess in the IVF compartment
Hypoalbuminemia	Albumin deficit
Hypoproteinemia	Protein deficit
Hyponatremia	Sodium deficit
Hypernatremia	Sodium excess
Hypokalemia	Potassium deficit
Hyperkalemia	Potassium excess
Hypocalcemia	Calcium deficit
Hypercalcemia	Calcium excess
Hypophosphatemia	Phosphorus/phosphate deficit
Hyperphosphatemia	Phosphorus/phosphate excess
Hypomagnesemia	Magnesium deficit
Hypermagnesemia	Magnesium excess

IVF = intravascular fluid

control mechanisms to regulate ECF volume: the baroreceptor reflex, volume receptors, the renin-angiotensin-aldosterone mechanism, and antidiuretic hormone. *Baroreceptors* in the atrial walls, vena cava, aortic arch, and carotid sinus react to a fluid deficit because they respond to low arterial blood pressure. They work to retain ECF by bringing about constriction of the kidneys' afferent arterioles.

The *volume receptors* respond to a fluid excess in the atria of the heart and great veins. The stimulation of the volume receptors causes a strong renal response that increases urine output, thus reducing ECF.

The *renin-angiotensin-aldosterone* mechanism is more complex. Renin is an enzyme secreted by the kidneys when arterial pressure or ECF volume is decreased. Conversely, when renal perfusion is increased, the release of renin is reduced. Renin interacts with angiotensinogen produced by the healthy liver to form a vasopressor (blood vessel constrictor) called angiotensin. Angiotensin I, an inactive form of angiotensin, is then converted in the lungs, with the assistance of the angiotensin-converting enzyme (ACE), to the active angiotensin II.

Angiotensin II is not found in the plasma of normotensive people, but is present in patients with essential hypertension. It provides an immediate response to a decrease in arterial pressure by inducing vasoconstriction of arterioles and veins, thus increasing peripheral resistance, which elevates arterial blood pressure.

Angiotensin II also stimulates the adrenal cortex to secrete aldosterone, an active mineralocorticoid that is responsible for controlling sodium and potassium levels in the blood. It also increases chloride and bicarbonate concentrations and ECF volume.

The aldosterone negative feedback mechanism can respond in another way to an extracellular volume excess or deficit. This mechanism involves aldosterone, adrenocorticotrophic hormone (ACTH, secreted by the anterior pituitary gland), and sodium. As the sodium level declines, so does the ECF level. The reduced sodium causes the secretion of ACTH, which stimulates the adrenal cortex to release aldosterone. ACTH plays a minor role in aldosterone release. Angiotensin stimulates the glomerulosa zone of the adrenal cortex to secrete aldosterone. Aldosterone, in turn, causes the retention of sodium and, therefore, water. As the fluid and sodium levels return to normal, the stimulation of the pituitary decreases and the release of aldosterone is reduced. This cycle, which involves both the nervous system and endocrine system, is repeated continuously to maintain the correct extracellular fluid level.

Antidiuretic hormone (ADH, also called vasopressin) is released by the posterior pituitary gland when there is a need for water to restore intravascular fluid volume. The pituitary gland responds to the osmoreceptors located in the thirst center of the hypothalamus. When the osmolarity of the blood is reduced, as with excessive fluid intake, the amount of ADH released is decreased. Conversely, a volume deficit leads to increased ADH secretion.

Antidiuretic hormone acts on the distal tubules of the kidney to promote retention of water in the body. While ADH causes the distal tubules to passively reabsorb water, the active pumping of sodium into the capillaries controls the water reabsorption in the proximal tubules and the loops of Henle. The proximal and ascending loops respond to circulating blood volume largely by retaining sodium and chloride.

TABLE 1–2

MAJOR PHYSIOLOGIC CHANGES OF AGING THAT AFFECT FLUID AND ELECTROLYTE BALANCE

Change	Consequences
Decreased total body fluid	Prone to dehydration
Dry, inelastic skin	Unreliable as an indicator of fluid status
Diminished thirst mechanism	Decreased fluid intake that can result in dehydration
Decreased glomerular filtration and concentrating capacity	Decreased ability to excrete wastes and regulate water and electrolytes
Atrophy of endocrine glands	Decreased ability to regulate sodium and potassium (adrenal atrophy); decreased ability to respond to fluid and electrolyte changes
Decreased cardiovascular function	Decreased cardiac output
	Increased peripheral resistance
	Decreased renal blood flow
Decreased respiratory function	Decreased elasticity of lung tissue
	Decreased strength of expiratory muscles
	Decreased partial pressure of oxygen
Diminished gastrointestinal function	Decreased gastric juices
	Decreased intestinal absorption

Implications for Elderly Persons

In addition to having less body water, the elderly individual has decreased neuroendocrine and renal function. As a result, regulatory control and compensatory mechanisms for maintaining fluid and electrolyte balance are less efficient in an elderly person than in a younger person. Therefore, fluid and electrolyte imbalances are very common in elderly persons. Table 1–2 summarizes the major physiologic changes associated with aging that can affect fluid and electrolyte balance in the elderly.

REFERENCES AND READINGS

Booker, MF, & Ignatavicius, DD (1996). Infusion Therapy: Techniques and Procedures. Philadelphia: WB Saunders.

Cook-Fuller, CC (1991). Nutrition. Guilford, CT: Dishkin.

DeAngelis, R (1991). Hypokalemia. Critical Care Nurse 11, 71–72; 74–75.

Ebersole, P, & Hess, P (1994). Toward Healthy Aging: Human Needs and Nursing Response. St. Louis: Mosby.

Gahart, BL (1995). Intravenous Medications. St. Louis: Mosby.

Giger, JN, & Davidhizar, RE (1991). Transcultural Nursing: Assessment and Intervention. St. Louis: Mosby–Year Book.

Guyton, AC (1995). Textbook of Medical Physiology. Philadelphia: WB Saunders.

Ignatavicius, DD, Workman, ML, & Mishler, MA (1995). Medical-Surgical Nursing: A Nursing Process Approach, ed 2. Philadelphia: WB Saunders.

Kokko, JP, & Tannen, RL (1995). Fluids and Electrolytes. Philadelphia: WB Saunders.

National Research Council (1989). Diet and Health: Implications for Reducing Chronic Disease Risk. Washington, DC: National Academy Press.

National Research Council (1989). Recommended Dietary Allowances. Washington, DC: National Academy Press.

Quillman, SM (1990). Nutrition and Diet Therapy: A Study and Learning Tool. Springhouse, PA: Springhouse.

Stater, MB (1992). Fluids and electrolytes in infants and children. Seminars in Pediatric Surgery 1(3), 208–211.

Theunissen, IM, & Parer, JT (1994). Fluid and electrolytes in pregnancy. Clinical Obstetrics and Gynecology 37(1), 3–15.

CHAPTER 2

. . . .
. . . .
. . . .
. . . .
. . . .
. . . .
. . . .
. . . .
. . . .
. . . .

Fluid Imbalance

The two broad classifications of fluid imbalance are fluid volume deficit, commonly called dehydration, and fluid volume excess, referred to as over-hydration or fluid overload. These problems result from a variety of clinical situations and disease states.

FLUID VOLUME DEFICIT

Fluid volume deficit, or dehydration, is a fairly common fluid imbalance. Dehydration may be an *actual* decrease in body water, caused by excessive fluid loss or inadequate fluid intake, or a *relative* decrease in which fluid (plasma) shifts from the intravascular compartment to the interstitial space, a process called "third spacing."

Fluid deficit can be further divided into three types—isotonic dehydration, hypertonic dehydration, and hypotonic dehydration (Fig. 2–1).

Isotonic Dehydration

Isotonic dehydration, also called hypovolemia, is the most common type of fluid volume deficit. Clinical manifestations are the result of a decreased volume of circulating plasma. Isotonic dehydration occurs when isotonic fluids are lost from the extracellular fluid (ECF), both the plasma and inter-stitial fluid. In this type of dehydration, plasma osmolarity (concentration) is not affected because there is no shift between fluid compartments. When isotonic dehydration begins, the body attempts to retain fluid and maintain tissue perfusion by compensatory mechanisms. Two main compensatory

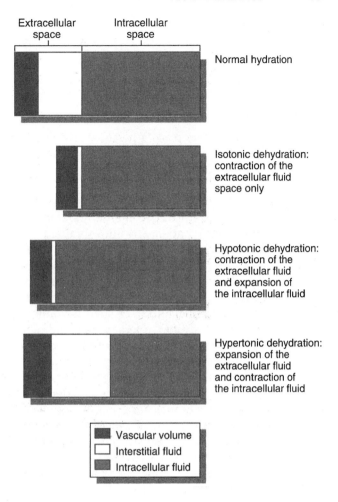

Extracellular space   Intracellular space

Normal hydration

Isotonic dehydration: contraction of the extracellular fluid space only

Hypotonic dehydration: contraction of the extracellular fluid and expansion of the intracellular fluid

Hypertonic dehydration: expansion of the extracellular fluid and contraction of the intracellular fluid

■ Vascular volume
□ Interstitial fluid
▦ Intracellular fluid

FIGURE 2–1
Three types of dehydration. (From Ignatavicius, DD, Workman, ML, and Mishler, MA: Medical-Surgical Nursing: A Nursing Process Approach, ed 2. WB Saunders, Philadelphia, 1995, p 264. © 1992 M. Linda Workman. Reprinted with permission.)

mechanisms, sympathetic and hormonal, are triggered when plasma volume decreases.

Compensatory Mechanisms

When 3% of the plasma volume is lost, the *sympathetic* nervous system is activated (Workman, 1995). The catecholamines epinephrine and norepinephrine are released by the adrenal medulla and increase cardiac output by increasing myocardial contractility and heart rate. Constriction of small vessels shunts blood away from the skin and periphery to make more blood available to major organs. This compensation is usually effective in maintaining perfusion until about 25% of plasma volume is lost. At this point, the patient can experience hypovolemic shock, discussed in Chapter 23.

In addition to the sympathetic response, a *hormonal* response also functions to maintain tissue perfusion. As discussed in Chapter 1, when the kidneys sense a decreased fluid volume, they release the hormone renin. Renin activates a process that results in the formation of angiotensin II in the lungs. Angiotensin II is a powerful vasoconstrictor that increases blood pressure and stimulates aldosterone release from the adrenal cortex; aldosterone signals the kidney to retain water and sodium. This mechanism does not operate over as long a period as the sympathetic response. Figure 2–2 illustrates the compensatory mechanisms associated with isotonic dehydration.

Causative Factors

Many factors place a person at risk for isotonic dehydration, including inadequate fluid intake, excessive fluid loss, or third spacing. *Decreased intake* is common in patients who are unable to eat or drink because of a pathologic condition, such as coma, or who are allowed nothing by mouth (NPO), as occurs immediately before and after surgery. Patients with infection or fever can become dehydrated very quickly, especially very young and elderly patients, if fluid intake does not meet the body's demand for fluid. Hyperthermia from heat exposure can rapidly result in heat exhaustion or heat stroke, which can be fatal.

Excessive losses of isotonic fluids can result from profuse bleeding, diuresis, diaphoresis, copious wound or fistula drainage, and losses from the gastrointestinal (GI) tract, such as vomiting, diarrhea, and GI suction.

Third spacing refers to a fluid shift from the intravascular space (plasma) into another part of the body where the fluid is not functional. For example, fluid can move from the plasma into body cavities such as the peritoneal, pleural, or pericardial spaces; or into the interstitial fluid compartment, causing edema. Such edema is likely to develop after burns and other types of major trauma. In some cases, fluid can become trapped: for example, in the bowel, especially after an obstruction; or in inflamed tissues or organs, as in pancreatitis or peritonitis.

In most cases of third spacing, the fluid leaves the bloodstream because capillaries are damaged and allow leakage of fluids out of the intravascular space. This occurs, for example, with burns and trauma. In other cases, however, the fluid moves from the intravascular to the interstitial fluid compartment because there are inadequate amounts of serum protein and other substances that exert osmotic (pulling) pressure. An example is the development of ascites in patients with liver disease.

Table 2–1 summarizes the major causes of isotonic dehydration.

Implications for Elderly Persons

In an elderly person, compensatory mechanisms lose their effectiveness as a result of the physiologic changes of aging. For example, renal function decreases by as much as 50%, and therefore the kidneys are not able to conserve water and sodium as efficiently as those of a younger person. This problem is confounded by the fact that elderly people have less total body water than their younger counterparts and therefore can more readily experience signs and symptoms of dehydration. Elderly individuals are also at a higher risk for illnesses that lead to dehydration, such as intestinal obstruction, infection (especially of the urinary tract), and hyperthermia. In addition, drugs that cause diuresis are more commonly prescribed for older adults to treat a variety of chronic health problems. Table 1–2 in Chapter 1 summarizes the physiologic changes associated with aging that affect fluid and electrolyte status.

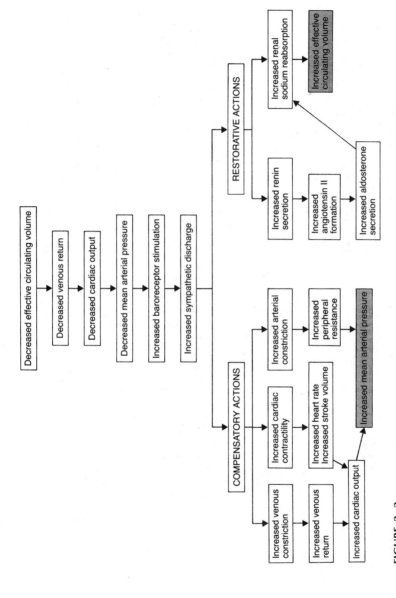

FIGURE 2-2

Compensatory mechanisms associated with isotonic dehydration. (From Ignatavicius, DD, Workman, ML, and Mishler, MA: Medical-Surgical Nursing: A Nursing Process Approach, ed 2. WB Saunders, Philadelphia, 1995, p 265. © 1992 M. Linda Workman. Reprinted with permission.)

TABLE 2–1
COMMON CAUSES OF ISOTONIC DEHYDRATION

Cause	Altered Function
GASTROINTESTINAL	
Vomiting	Reverse peristalsis causes loss of extracellular secretions, which contain large amounts of sodium and chloride
Diarrhea	Increased forward peristalsis shortens absorption period, preventing ECF absorption
Fistulous drainage	Tubelike passageway between the gastrointestinal (GI) tract and an *external* surface, allowing loss of ECF
Gastrointestinal suction	Evacuation of ECF by mechanical means, such as Levin or Cantor tubes
Excessive tap water enemas	Hypotonic solution draws electrolytes from ECF, causing saline loss when expelled
RENAL	
Overzealous use of diuretics	Excessive loss of Na, Cl, and H_2O
Salt-losing nephritis	Tubules, damaged by either chemical or physical means, may allow saline to flow out
DIAPHORESIS	
	Saline loss due to hyperthermia or a thermoregulatory response triggered by stress
THIRD SPACE	
Peritonitis	Fluid accumulation in the peritoneal cavity due to an infectious and/or inflammatory process
Intestinal obstruction	Fluid accumulation within intestines due to decreased or absent peristalsis
Postoperative condition	Fluid moves into traumatized operative site due to local inflammatory stress response
Thrombophlebitis	Blood clot(s) with inflammation in a vein with resulting venous obstruction
Acute pancreatitis	Inflammatory response with fluid pulled into inflamed site
Ascites	Serous fluid accumulation in the peritoneal cavity related to interference with portal circulation
Fistulous drainage	Tubelike passageway between the GI tract and *internal* adjacent cavity allowing loss of ECF
Burns	Fluid accumulation in blisters or interstitial space (Fluid may be lost due to evaporation [white bleeding], which is not a third space accumulation.)

Implications for Infants and Children

Infants and small children are especially prone to isotonic dehydration. In addition to having a relatively low percentage of total body water, infants are not able to control their fluid intake. Crying, restlessness, and irritability may be signs of hypovolemia. Skin turgor is not a reliable sign in infants and sunken fontanels are not present until dehydration becomes severe.

Hypertonic Dehydration

Hypertonic dehydration is the second most common type of fluid volume deficit. In this type of dehydration, the loss of extracellular fluid (ECF) is increased relative to the loss of solute, including electrolytes, so that plasma osmolarity increases. Compensatory mechanisms respond to the hyperosmolar (hypertonic) plasma.

Compensatory Mechanisms

Three mechanisms work together to increase ECF volume. First, the hyperosmolar plasma, which has an increased osmotic pressure, pulls fluid rapidly from the intracellular fluid and causes the cells to shrink. This process results in cellular dehydration. Second, the posterior pituitary gland releases more antidiuretic hormone (ADH) to increase water reabsorption by the kidneys and diminish urinary output. Third, the thirst center in the hypothalamus is activated to increase oral fluid intake.

Causative Factors

Hypertonic dehydration can result from loss of hypotonic fluids, an excessive intake of hypertonic fluids, or an inability to respond to the thirst reflex. Hypotonic fluid loss usually occurs in conditions that increase loss of insensible fluids, such as diaphoresis, diabetic ketoacidosis (DKA), hyperglycemic, hyperosmolar nonketotic syndrome (HHNS), prolonged fever, hyperventilation, watery diarrhea, diabetes insipidus, and renal insufficiency. Excessive intake of intravenous hypertonic solutions, particularly those containing sodium, and of high-osmolarity enteral feeding formulas with inadequate water supplementation can also lead to hypertonic dehydration. If patients are unable to swallow or not aware of the thirst reflex, they are prone to fluid volume deficit as well.

Implications for Elderly Persons

As a result of the aging process, the elderly person has a decreased thirst reflex and therefore may not be aware of additional water needs. Elderly people are also at higher risk for many of health problems that cause hypertonic dehydration, such as diabetes mellitus and gastroenteritis, which causes diarrhea. Enteral feedings via a feeding tube are also more common in elderly patients.

Hypotonic Dehydration

The least common type of fluid volume deficit is hypotonic dehydration. The situation is the opposite of that with hypertonic dehydration, in which relatively more water is lost than solutes. Most commonly, in hypotonic dehydration, sodium and potassium losses increase, so that plasma osmolarity *decreases*. As a result, fluid moves from the plasma and interstitial fluid into the cells, causing them to swell. Consequently, extracellular volume depletion, or hypovolemia, occurs.

Compensatory Mechanisms

Because the plasma volume decreases in hypotonic dehydration, the compensatory mechanisms are the same as those discussed under isotonic dehydration earlier in this chapter.

Causative Factors

Chronic illness, renal insufficiency (in which sodium is lost by the kidney), chronic malnutrition, and excessive use of intravenous hypotonic solutions are the most common causes of hypotonic dehydration.

Assessment

Physical Assessment

Signs and symptoms are exhibited by patients with fluid volume deficit and electrolyte imbalances.

Cardiovascular Assessment

Cardiovascular assessment is the most important part of the assessment to determine plasma volume changes. In patients who are hypovolemic (isotonic and hypotonic dehydration), the heart rate increases, the blood pressure decreases, and the peripheral pulses are weak. Postural (orthostatic) hypotension is common. Postural hypotension is a decrease in blood pressure (usually more than 20 mmHg systolic) when the patient changes from a lying to a sitting or standing position.

In addition to monitoring blood pressure changes, the nurse should assess neck and hand vein filling. In dehydration, the neck and hand veins remain flat, regardless of the patient's position.

Renal Assessment

When plasma volume decreases, the kidneys reabsorb water to conserve plasma volume. Therefore, urine output decreases. However, at least 30 mL/h should be excreted to demonstrate adequate kidney function. Prolonged dehydration can cause renal tubular damage. Intake and output measurements are an important part of nursing assessment, but daily weights are the most reliable indicator of fluid loss or gain. One liter (about a quart) of water weighs about 1 kg (2.2 lb). Therefore, a decrease in weight of 1 lb is equivalent to a fluid loss of about 500 mL. Patients with dehydration experience weight loss because they have lost body fluid.

Skin Assessment

The skin and mucous membranes may also be assessed for moisture and elasticity. The skin of a dehydrated person is typically dry and warm. Body temperature may be elevated, although the elderly often do not have an elevation until severe dehydration occurs. Mucous membranes are also dry and sticky, and may be covered with a pasty coating (Table 2–2).

Assessment for skin turgor, in which the nurse pinches the skin for "tenting," is not a very reliable indicator of hydration, especially in the elderly. Theoretically, the skin of a dehydrated person stays in the "tent" posi-

TABLE 2–2
NURSING ASSESSMENT OF FLUID BALANCE

Observation	Significance
Change in daily weight (loss in dehydration, gain in overload)	Most accurate indicator of fluid balance; 1 L water weighs about 1 kg. • 2% weight change = mild deficit or overload • 5% weight change = moderate deficit or overload • 8% weight change = severe deficit or overload
Intake and output comparison	Excessive intake compared to output can indicate overload Excessive output compared to intake can indicate dehydration
Poor (decreased) skin turgor	Can indicate dehydration
Edema (dependent)	Can indicate overload
Increased temperature	Can indicate dehydration; prolonged hyperthermia can lead to dehydration
Dry conjunctiva of eyes	Along with "soft" eyeballs, can indicate dehydration
Periorbital edema	Can indicate overload
Tachycardia, decreased blood pressure, decreased pulse pressure, diminished pulses, flat neck veins, decreased CVP	Can indicate dehydration
Increased blood pressure, increased pulse pressure, bounding pulses, distended neck veins (jugular vein distention [JVD]) and hand veins, increased CVP	Can indicate overload
Moist crackles, dyspnea, increased respirations	Can indicate overload (pulmonary edema)
Oliguria (<30 mL/h)	Can indicate dehydration
Decreased level of consciousness	Can indicate dehydration

tion after testing. However, an elderly person who has lost skin elasticity and tissue fluids with aging may have poor skin turgor (tenting) even when well hydrated.

IMPLICATIONS FOR ELDERLY PERSONS • For an elderly person, the skin over the forehead or sternum should be used to assess skin turgor. These areas tend to be more elastic and not as dry as others.

Other Assessment

When acidosis accompanies dehydration, respirations are deep and rapid. This is known as **Kussmaul's breathing.** The body attempts to get rid of excess acid by "blowing off" extra carbon dioxide. (See Chap. 3 on acid-base imbalances.) An increased respiratory rate further depletes fluid volume by increasing insensible loss.

The patient with dehydration also becomes constipated and thirsty, and has diminished bowel sounds. In hypovolemia, dysfunction of brain cells caused by decreased perfusion leads to decreased levels of consciousness.

TABLE 2–3
LAB PROFILE: DEHYDRATION

Values	Isotonic Dehydration	Hypotonic Dehydration	Hypertonic Dehydration
BLOOD VALUES			
BUN	• Normal or increased	• Increased	• Increased
Creatinine	• Normal or increased	• Increased	• Increased
Sodium	• Normal	• <120 mEq/L (mmol)	• >150 mEq/L (mmol)
Osmolarity	• Normal	• Decreased	• Increased
Hematocrit	• Increased	• Increased	• Normal or increased
Hemoglobin	• Increased	• Increased	• Normal or increased
WBC	• Increased	• Increased	• Normal or increased
Protein	• Increased	• Increased	• Increased
URINE VALUES			
Specific gravity	• >1.010	• <1.010	• >1.010
Osmolarity	• Increased	• Decreased	• Increased
Volume	• Decreased	• Increased	• Decreased

*All values reflect dehydration states alone and not the underlying pathologic changes or disease states contributing to the dehydration.
Source: Ignatavicius, DD, Workman, ML, and Mishler, MA: Medical-Surgical Nursing: A Nursing Process Approach, ed 2. WB Saunders, Philadelphia, 1995. Reprinted with permission.

Diagnostic Assessment

A group of laboratory tests combined with assessment findings are needed to confirm a diagnosis of fluid volume deficit. Laboratory values depend partly on the type of dehydration present. Table 2–3 displays the typical laboratory findings in patients with dehydration.

Nursing Diagnoses

The common nursing diagnoses typically seen in patients with dehydration include:

- Fluid Volume deficit, related to inadequate fluid intake or excessive fluid loss
- Skin Integrity, impaired, potential, related to dry skin
- Hyperthermia, related to loss of fluid
- Constipation, related to inadequate water intake

Collaborative Management

Medical Treatment

The goals of treatment are to restore fluid volume, replace electrolytes as needed, and eliminate the individual causes of dehydration. For mild dehydration in a patient who can swallow, oral fluid replacement can restore fluid volume. If the sodium level is decreased, as in hypotonic dehydration, adding salt to the diet or providing high-sodium liquids such as broth and tomato juice is also helpful in replacing sodium. For patients with hypertonic dehydration in which sodium is elevated, foods and fluids with a high sodium content should be avoided.

The cause of the dehydration must be treated. Drugs to control diarrhea and vomiting, for example, can prevent further progression of dehydration. For noninfectious diarrhea, drugs for symptomatic relief, such as Kaopectate, are appropriate. For infectious diarrhea such as that caused by *Clostridium difficile* in the elderly, however, fluid replacement and antibiotic agents are the appropriate treatment. Antidiarrheal drugs should not be administered to these patients because the pathogen needs to be eliminated from the body.

For more severe dehydration, especially that caused by diarrhea, oral rehydration therapy (ORT) is an option, particularly for children and the elderly. ORT is a cost-effective way to replace water, electrolytes, and glucose. In the elderly patient with adequate cardiovascular and renal function, as much as 3 L of liquid can be given per day (Workman, 1995).

For patients who have severe dehydration and are not able to take oral fluids, large quantities of intravenous (IV) fluids may be required. In general, isotonic dehydration is treated with isotonic fluids, hypertonic dehydration is treated with hypotonic fluids, and hypotonic dehydration is treated with hypertonic fluids. Table 15–2 in Chapter 15 lists common IV solutions and their tonicity.

To monitor the patient's fluid status more accurately, a central venous catheter may be inserted to monitor the pressure within the right atrium or vena cava. Normal central venous pressure (CVP) is between 4 and 11 mmH_2O. A patient with dehydration has a low pressure, below 4 mmH_2O. If the patient is in the critical care unit, hemodynamic monitoring may be used to assess both right and left heart pressures. Both CVP and hemodynamic monitoring are discussed in more detail in Chapter 20.

If a patient with severe dehydration has a urinary output of less than 30 mL/h (oliguria), a fluid challenge test may be given to determine if the kidneys have been damaged. In this test, the patient receives a high volume of IV fluid (usually 200 to 300 mL in a 5- to 10-minute period). If the CVP is not affected, additional fluid may be given. If the patient's urinary output does not increase, renal damage is suspected.

Nursing Interventions

In large part, nursing interventions consist of ongoing nursing assessment of fluid and electrolyte status. Table 2–3 outlines the critical components of the nursing assessment for any patient with a suspected fluid imbalance. In addition to the physical assessment, the nurse is responsible for providing and measuring oral and IV fluid replacement as ordered. Chapter 15 describes nursing interventions associated with IV therapy.

Comfort measures are another essential element of nursing care. Patients with dehydration have very dry skin and oral mucous membranes. These tissues can easily break open to create fissures or other lesions. Therefore, meticulous skin and mouth care are vital.

For patients with copious wound drainage, such as that which occurs from fistulas, special care to protect the skin is required. Placing an ostomy bag with a skin barrier over the fistula is better than using traditional dressings that hold drainage against the skin.

Patients with diarrhea also need special skin care, especially in the peri-

TABLE 2–4

NURSING INTERVENTIONS FOR PROMOTING
COMFORT IN PATIENTS WITH DEHYDRATION

- Provide meticulous and frequent mouth care
  - Do not use alcohol-based mouthwashes.
  - Clean mouth, including tongue, with a soft-bristled brush or sponge.
  - Use a solution of one-half peroxide and saline.
  - Pay special attention to upper palate, where secretions can accumulate and lesions can occur.
  - Provide mouth care often, especially after each episode of vomiting.
- Provide frequent skin care.
  - Use moisturizing lotions after cleansing.
  - Do not use soap on dry skin areas.
  - Pat skin dry, rather than rubbing.
  - Apply protective barrier cream topically to perineal/perianal areas as needed.
  - Observe for skin infections, especially monilial infections in the perineal/perianal areas.
- Provide special care for draining wounds/fistulas.
  - For copiously draining wounds, use absorbent dressing material.
  - For fistulas, apply ostomy bag over protective barrier.
  - Change dressing or bag frequently.

anal area. The area must be kept dry and clean. A protective barrier cream should be applied topically after each cleansing.

Table 2–4 lists the primary nursing interventions for promoting comfort in a patient with dehydration.

FLUID VOLUME EXCESS

Fluid Overload

Fluid overload, also called overhydration, may be an actual excess of total body fluid or a relative fluid excess in one or more fluid compartments. Like dehydration, fluid volume excess can be divided into three types—isotonic, hypotonic, and hypertonic. Isotonic and hypotonic fluid overload are most common and are discussed here.

Isotonic Fluid Overload

Isotonic fluid overload is also called hypervolemia because there is excessive fluid in the extracellular fluid (ECF) compartment, that is, in both the intravascular and interstitial spaces. This fluid imbalance is less common than fluid deficit because the body is able to compensate for fluid excess until it becomes severe.

Compensatory Mechanisms

Unless renal function is impaired, the elevated blood pressure resulting from hypervolemia results in an increased venous return to the heart (increased preload) and stretching of the myocardium, thus increasing cardiac output (Starling's law). Renal blood flow and glomerular filtration rate also increase, so that there is a corresponding increase in both water and sodium excretion.

IMPLICATIONS FOR ELDERLY PERSONS • If a person has poor cardiac function, fluid overload can rapidly occur and lead to congestive heart failure or pulmonary edema. This problem is typical in elderly patients, who often have poor cardiac reserve and a history of cardiovascular disease.

Causative Factors

Conditions that can cause isotonic overhydration include excessive administration of oral or IV fluids, excessive irrigation of body cavities or organs (such as overirrigation of the bladder in patients with continuous bladder irrigation, or excessive use of enemas), and the use of hypotonic fluids to replace isotonic fluid losses.

Hypotonic Fluid Overload

Hypotonic fluid overload is also called water intoxication. In this fluid imbalance, the excess fluid is hypotonic to the rest of the body, so that the osmolarity of the plasma is decreased. Consequently, extracellular fluid shifts into the intracellular fluid compartment. The result is overexpansion of all fluid compartments and dilutional electrolyte deficits (low serum electrolytes). The body has difficulty adequately compensating for this fluid shift.

Conditions that cause hypotonic overload are the syndrome of inappropriate antidiuretic hormone (SIADH), excessive water intake, and congestive heart failure. Congestive heart failure is very common in elderly persons.

Assessment

Physical Assessment

The patient with fluid overload has marked cardiovascular changes, including increased, bounding pulses; increased blood pressure; elevated central venous pressure; and distended neck and hand veins (Table 2–5). Respirations are increased and shallow. If pulmonary edema is present, the nurse

TABLE 2–5

ASSESSMENT FINDINGS ASSOCIATED WITH ISOTONIC FLUID OVERLOAD

Bounding pulses
Increased blood pressure
Widened pulse pressure
Dyspnea
Moist crackles
Edema (dependent) in lower extremities and sacrum
Periorbital edema
Increased weight
Decreased hemoglobin and hematocrit values

observes dyspnea with exertion or in the supine position and auscultates moist crackles. Visible edema is also noted in dependent areas such as the feet, ankles, and sacrum (for patients in bed). The nurse assesses whether the edema is pitting or nonpitting by depressing the skin over the edematous area (see Table 2–2).

In addition to these very common assessment findings, some patients also experience confusion or altered level of consciousness (lack of oxygenated blood to the brain), skeletal muscle weakness, and increased bowel sounds. These changes may be related to fluid excess or electrolyte deficits, especially of sodium and potassium (see also Chaps. 4 and 5).

Diagnostic Assessment

In patients with *isotonic* fluid excess, serum electrolytes tend to be normal, but hemoglobin level and hematocrit are low because of hemodilution. If renal failure is the cause of fluid overload, electrolytes, blood urea nitrogen (BUN), and serum creatinine levels are increased because the kidneys are not able to excrete them. *Hypotonic* fluid overload results in decreased blood levels of electrolytes (dilutional effect), protein, hemoglobin, and a decreased hematocrit (hemodilution).

Collaborative Management

Medical Treatment

In hypotonic fluid overload, the goal of treatment is to restore fluid balance, correct electrolyte imbalances if present, and eliminate or control the underlying cause of fluid overload. If renal function is not impaired, diuretics may be used to promote excretion of water in patients with isotonic overload. Osmotic diuretics, such as mannitol, permit renal excretion of water rather than sodium. If these diuretics are not effective or if the patient has congestive heart failure or pulmonary edema, high-ceiling (loop) diuretics, such as furosemide (Lasix), may be given by IV push or orally. In addition, a fluid- or sodium-restricted diet may be required.

To accurately monitor patients with pulmonary edema, a serious complication of left-sided congestive heart failure, a central venous catheter may be inserted to measure pressure in the right atrium or vena cava. When the reading is above 11 mmH$_2$O, the diagnosis of fluid overload is confirmed. In the critical care setting, hemodynamic monitoring, which measures left- and right-sided heart pressures, may be used. Elevated pressures confirm fluid overload. Chapter 20 describes these invasive monitoring procedures in more detail.

Nursing Interventions

As for patients with dehydration, a major part of nursing care involves ongoing assessment of fluid and electrolyte status (see Table 2–3).

If edema is present, the nurse checks for pitting (depression in the skin made by finger pressure). Serial measurements of the depth of the depression made in pitting edema help to determine whether the edema is improving or worsening. The location of the edema is also assessed and documented. Circumferential measurements of the leg or ankle may be helpful, but other conditions such as thrombophlebitis can cause swelling. This swelling may not be an indication of fluid overload.

Because change in weight is the best indicator of fluid gain resulting in fluid overload, the nurse takes daily weights before breakfast using the same scale each time. Liquid intake and output should also be recorded. The relationship of weight to intake and output measurements is evaluated to determine the effectiveness of diuretic therapy.

Prevention of fluid overload or of an increase in overload is a major nursing responsibility. Intravenous fluid infusion rates must be carefully monitored, preferably by an electronic infusion device. Infants, children, and elderly people are at especially high risk for fluid overload and must receive special attention when receiving IV therapy.

CRITICAL THINKING ACTIVITIES

An elderly man is hospitalized with a diagnosis of congestive heart failure. Nursing assessment findings included: 2+ pitting edema in both feet, patient alert and oriented, vital signs within baseline limits, orthopnea, dyspnea, weakness, loss of appetite, weight gain of 5 lb in 1 week.

QUESTIONS

1. *What signs and symptoms suggest a fluid imbalance? With what fluid imbalance are these assessment findings associated?*

2. *What pathophysiologic changes led to this fluid imbalance?*

3. *The initial laboratory results were hematocrit (Hct) 31%, BUN 26 mg/dL, sodium 136 mEq/L, potassium 3.1 mEq/L. Which of these values are abnormal and why? (See normal laboratory values in Appendix.)*

4. *Four days later the Hct was 35%, potassium was 3.6 mEq/L, and the patient had lost 3 lb. Why might the Hct be higher at this time?*

REFERENCES AND READINGS

Bove, LA (1994). How fluids and electrolytes shift after surgery. Nursing94 24(8), 34–40.

Burns, D (1992). Working up a thirst. Nursing Times 88(26), 44–45.

Byers, JF, & Goshorn, J (1995). How to manage diuretic therapy. American Journal of Nursing 95(2), 38–44.

Dennison, R, & Blevins, B (1992). Myths and facts about fluid imbalance. Nursing92 22(3), 22.

Ebersole, P, & Hess, P (1994). Toward Healthy Aging: Human Needs and Nursing Response. St. Louis: Mosby.

Fowler, JP (1994). From chronic to acute . . . when CHF turns deadly. Nursing95 25(1), 54–55.

Gallagher, HG, & Phelan, DM (1992). Oral rehydration therapy: A third world solution applied to intensive care. Intensive Care Medicine 18, 53–55.

Gershan, J, et al (1990). Fluid volume deficit: Validating the indicators. Heart and Lung 19(2), 152–156.

Guyton, A (1995). Textbook of Medical Physiology. Philadelphia: WB Saunders.

Hirschorn, N, & Greenough, W (1991). Progress in oral rehydration therapy. Scientific American 264(5), 50–56.

Jones, A, et al (1990). Fluid volume dynamics. Critical Care Nurse 11(4), 74–76.

Millam, D (1990). Electronic infusion devices: Controlling the flow. Nursing90, 20(8), 65–68.
Porth, C, & Erickson, M (1992). Physiology of thirst and drinking. Heart and Lung 21(3), 273–284.
Price, SA, & Wilson, LM (1992). Pathophysiology: Clinical Concepts of Disease Processes. St. Louis: Mosby–Year Book.
Watt, S (1991). Quenching the body's thirst. New Zealand Nursing Journal 84(10), 18–19.
Workman, ML (1995). Interventions for clients with fluid imbalances. In DD Ignatavicius, ML Workman, & MA Mishler (eds), Medical-Surgical Nursing: A Nursing Process Approach (2nd ed, pp 263–290). Philadelphia: WB Saunders.

CHAPTER 3

. . . .
. . . .
. . . .
. . . .
. . . .
. . . .
. . . .
. . . .

Acid-Base Balance and Imbalance

A solution is identified as acidic or basic on the basis of the concentration of hydrogen ions in the solution, and the pH scale is used to describe the hydrogen ion concentration. The pH scale is a logarithmic scale with values from 0.00 to 14.00. A neutral solution (neither acidic or basic) has a pH of 7.00. Acid solutions have pH values between 0.00 and 7.00, and basic solutions have pH values between 7.00 and 14.00. The further the pH value is from the neutral value of 7.00, the more acidic or basic the solution. Thus, a solution with a pH of 1 is more acidic than a solution with a pH of 2. In fact, a solution with a pH of 1 has 10 times the concentration of hydrogen ions as a solution with a pH of 2. Because the pH scale is logarithmic, a change of one unit in pH corresponds to a 10-fold change in hydrogen ion concentration.

ACID-BASE BALANCE

Acid-base homeostasis means that the hydrogen ion concentration of various biologic fluids is maintained within narrow physiologic limits. For example, the pH of blood normally falls in the range of 7.35 to 7.45; that is, blood is slightly alkaline. Table 3–1 gives normal pH values of various body fluids.

TABLE 3–1
pH VALUES OF BODY FLUIDS

Gastric juices	1.0–5.0	Acid
Urine	5.5–6.5	
Water	7.00	Neutral
Blood	7.35–7.45	
Bile	7.5	Base
Pancreatic juice	8.4–8.9	

Definitions of Acids and Bases

Acids are hydrogen ion (proton) *donors* (Fig. 3–1). This means that acids have hydrogen ions that are given up to neutralize or decrease the strength of a base. For example, hydrochloric acid, when mixed with sodium bicarbonate, yields sodium chloride and carbonic acid, as shown in the equation below:

$$HCl + NaHCO_3 \longrightarrow NaCl + H_2CO_3 \text{ (carbonic acid)}$$

Hydrochloric acid donates its proton to the sodium bicarbonate solution; this results in the neutralization of sodium bicarbonate to a salt, sodium chloride, and the formation of a weaker acid, carbonic acid.

Bases are proton or hydrogen ion *acceptors* (Fig. 3–1). A basic solution is one whose hydrogen ion concentration is low (as evidenced by pH values between 7.00 and 14.00) and that contains bicarbonate or hydroxyl (OH−) groups. In the equation shown above, the base (bicarbonate ion) accepted a proton from the hydrochloric acid, thus converting the strong acid, hydrochloric acid, to the weaker acid, carbonic acid.

Fixed and Nonfixed Acids

The metabolism of all foodstuffs results in the manufacture of acids. There are two types of acids to be excreted: fixed or nonvolatile and nonfixed or

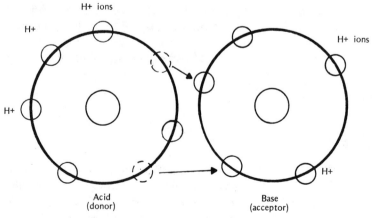

FIGURE 3–1
Schematic hydrogen ion donor-acceptor concept.

volatile. An understanding of how these two types of acids are formed and how they are normally excreted facilitates the comprehension of acid-base regulatory systems.

Fixed Acids

Fixed acids are the products of incomplete metabolism of organic compounds, the metabolism of sulfur-containing amino acids, and the hydrolysis of phosphoprotein. Incomplete metabolism of organic compounds results in such acids as lactic acid, which forms when metabolism of carbohydrates occurs without sufficient oxygen, as in the anaerobic phase of the Krebs cycle. This is typically present in the metabolic acidosis that accompanies shock or other hypoxic states. Sulfur-containing amino acids form sulfuric acid as they are metabolized. Proteins are the origin of nitrogenous waste products in the catabolic phase. These types of acids are called fixed because they are not volatile (not easily vaporized), so they are not normally exhaled through the lungs but must be excreted through the kidneys.

Nonfixed Acids

Nonfixed or volatile acids are the result of metabolism of most carbon-containing compounds, essentially carbohydrates and fats. Carbohydrates and fats, with sufficient oxygen, are catabolized into carbon dioxide (CO_2) and water. Carbon dioxide and water exist in equilibrium with carbonic acid, H_2CO_3.

$$CO_2 + H_2O \rightleftharpoons H_2CO_3$$

The carbon dioxide carried by the hemoglobin in the bloodstream diffuses into the lungs and is exhaled as a gas and eliminated. The water eliminated via the lungs is considered an insensible loss. These nonfixed acids are considered volatile because they are primarily exhaled through the lungs. In general, then, fixed acids are excreted by the kidneys and nonfixed acids are exhaled through the respiratory system.

Regulatory Systems

Regulation of acid-base balance is an intricate process involving the interaction of four basic physiologic systems, which are listed in order of the rapidity of their response to acid-base changes: (i) chemical buffers, (ii) the respiratory system, (iii) cells, and (iv) the kidneys.

Buffer Systems

The buffer systems are the fastest-acting defenses, providing immediate protection against changes in the hydrogen ion concentration of the extracellular fluid. The buffers also serve as transport mechanisms that carry excess hydrogen ions to the lungs. The other regulatory systems are slower to react but provide more thorough protection.

A buffer is a substance that reacts to minimize pH changes when either acid or base is released into the system. It consists of weakly ionized acid (or base) in equilibrium with its fully ionized salt form. For example, carbonic acid and bicarbonate form the bicarbonate buffer system. If a strong acid is added to the system, its protons combine with bicarbonate to form carbonic acid, a weak acid whose free hydrogen ion concentration is low. Thus, there is little change in pH. If a strong base is added to the bicarbonate buffer, the base combines with the protons from carbonic acid and generates bicarbonate. The base is neutralized but there is little change in pH because the neutralizing protons came from carbonic acid, not from free protons in solution. In effect, the buffer acts as a chemical sponge, absorbing or releasing hydrogen ions as needed.

There are three primary buffer systems in the extracellular fluid: the hemoglobin system, the plasma protein system, and the bicarbonate system. These substances react immediately with acids or bases to minimize changes in pH. The capacity of a buffer is limited, so that once the components of a buffer system have reacted, they must be replenished before the body can respond to further stress.

The hemoglobin and deoxyhemoglobin found in red blood cells, together with their potassium salts, act as buffer pairs. In addition, *hemoglobin* helps maintain the acid-base balance by a process called chloride shift. The electrolyte chloride shifts in and out of the red blood cells according to the level of oxygen in the blood plasma. For each chloride ion that leaves the red blood cell, a bicarbonate ion enters the cell; for each chloride ion that enters the red blood cell, a bicarbonate ion is released.

Plasma proteins are large molecules that contain both the acid (or base) and salt form of a buffer. Thus, they have the ability to bind or release hydrogen ions.

The **bicarbonate** buffer system maintains the blood pH in the range of 7.35 to 7.45, with a ratio of 20 parts bicarbonate to 1 part carbonic acid. If a strong acid is added to the body, the ratio is upset because some of the bicarbonate will be lost in the process of neutralizing the strong acid to the weaker carbonic acid. In this acid imbalance, the largest amount of CO_2 diffuses in the plasma to the red blood cells; CO_2 then combines with plasma protein. Lastly, CO_2, which is dissolved in the blood, combines with water to form carbonic acid. This process is called hydration of CO_2 and is a means of buffering the excess acid in the blood.

Respiratory System

In healthy individuals, the lungs form a second line of defense in maintaining acid-base balance and interact with the buffer system to maintain acid-base balance. The carbonic acid that was created by the neutralizing action of the bicarbonate can be carried to the lungs, where it is reduced to carbon dioxide and water, both of which are exhaled. In this way, the hydrogen ions are inactivated and excreted. To excrete excess acid, respirations usually increase and become deeper.

When there is a bicarbonate excess, the speed and depth of respirations are reduced to allow a buildup of carbonic acid. The carbonic acid then serves to neutralize the excess bicarbonate. The action of the lungs is reversible, making it possible for them to control either an excess or deficit.

This means that the lungs can either hold the hydrogen ions until the deficit is corrected or inactivate the hydrogen ions into water molecules to be exhaled with the carbon dioxide as vapor, thereby correcting the excess. It takes from 10 to 30 minutes for the lungs to inactivate the hydrogen molecules by converting them to water molecules. The lungs are capable of inactivating only the hydrogen ions carried by carbonic acid, which are termed volatile acids. Excess hydrogen ions created by other problems must be excreted through the kidneys.

Cellular Systems

The cells can also serve as regulators or buffers. They react in 2 to 4 hours, and they have the ability to absorb or release extra hydrogen ions. Potassium plays a role in temporarily helping to correct the hydrogen ion level. For example, when the hydrogen ion level is too high in the extracellular fluid, potassium leaves the cell in exchange for hydrogen, which goes into the cell (see Chap. 4). The result is an increased serum potassium, as seen in metabolic acidosis.

Renal System

The ultimate correction of acid-base disturbances and the restoration of buffers are dependent on the kidneys, even though renal excretion of acids, and to a lesser degree alkali, restores acid-base balance more slowly than the other regulatory systems. Compensation through the action of kidneys requires a period of a few hours to several days. However, renal action is more thorough, permanent, and selective than that of other acid-base regulators.

The primary mechanism of the kidneys in acid-base balance is the selective regulation of bicarbonate. The kidneys restore bicarbonate by releasing hydrogen ions to the urine and sending bicarbonate ions back to the blood capillaries. Although the kidneys regulate the amount of bicarbonate to be reabsorbed, the amount available for reabsorption is affected by the use of diuretics, stress, body concentration of acids, and loss of chloride or bicarbonate in gastrointestinal fluid and other metabolic conditions.

The secondary and tertiary kidney mechanisms involve the acids secreted by the renal tubular cells. In the second mechanism, the buffer, primarily phosphate, becomes acidic, and the extra hydrogen ions are excreted in the urine in the form of phosphoric acid. The third mechanism, which also involves a chemical change within the renal tubules, is called the ammonia mechanism. The alteration of certain amino acids in the renal tubules results in a diffusion of ammonia into the kidneys, where it picks up excess hydrogen ions and is excreted as ammonium ions. Both phosphoric acid and ammonium ion are fixed acids.

Because fixed acids cannot be destroyed by metabolic processes or eliminated through the lungs, they must instead be excreted through the kidneys. Thus, the kidneys are very important in the regulation of pH balance. Urine can sustain a hydrogen concentration a thousand times greater than that sustained by blood. The maximum urine acidity is pH 4.0, compared with a blood pH of 7.4. However, very little acid can be excreted by the kidneys as free hydrogen ions but instead must be excreted as fixed acids and ammonium in approximately equal amounts.

Thus, in comparing the regulatory action of the lungs and kidneys, it is essential to understand that the action of carbonic acid is identified with the maintenance of respiratory balances, whereas the action of bicarbonate is correlated with maintenance of metabolic balances. The kidneys can also affect acid-base balance by excreting potassium ions as needed. The relationship of potassium and hydrogen is explained in Chapter 4.

Implications for Infants and Children

The kidneys of infants and small children have a decreased ability to remove acids from the body. As a result, their blood pH tends to be lower than that of adults, in the range of 7.30 to 7.35 (Ichikawa, 1990). Infants and small children are also more likely than adults to compensate for an increased acid level in the body.

ACID-BASE IMBALANCE

Acid-base imbalances are classified as either acidosis or alkalosis, and as having a respiratory or metabolic etiology. These disorders are further classified depending on how well the buffering mechanisms are working. A disorder is classified as uncompensated (buffers not working), compensating (some buffering intact, but pH not fully corrected), compensated (pH fully corrected), and decompensating (worsening degenerating state).

Analysis of Acid-Base Imbalance

Acid-base analysis consists of the clinical assessment of the patient's history and physical findings, as well as careful attention to laboratory results including arterial blood gases (ABGs), serum and urinary electrolytes, and the buffering patterns or compensations.

Patient History

When assessing patient history, it is essential to consider three areas of consideration in sequence. The first area focuses on acid-base balance. Is acidosis or alkalosis present? Acidosis is present when the serum pH falls below 7.35; alkalosis occurs when the pH is above 7.45. Is the contributory etiology essentially respiratory or metabolic?

The second area of consideration explores the nature of the respiratory or metabolic problem. If the problem is respiratory, is the situation acute or chronic? An *acute* respiratory problem usually involves hyperventilation leading to respiratory alkalosis. A *chronic* respiratory condition usually causes retention of acids, thus leading to respiratory acidosis. If the situation is one of metabolic origin, further diagnostic input is crucial. If a metabolic acidosis is confirmed, then the determination of the anion gap is needed to determine the source of the acidosis (refer to discussion of anion gap in Chapter 6). If metabolic alkalosis is present, then assessing the urinary chloride is important. Examples of situations that might indicate a metabolic problem are hypoxia, diabetes, and renal failure.

Finally, as the third area of consideration, the level of compensation needs to be assessed. In respiratory disorders, the kidneys need to compensate. In metabolic disorders, the lungs will compensate. If there is not appropriate compensation for the primary disorder, then a combined or mixed acid-base imbalance is present.

Diagnostic Assessment

Arterial Blood Gases

Arterial blood gas assessments are essential to ascertaining the pH level. The formula used to calculate pH is known as the Henderson-Hasselbalch equation:

$$pH = 6.10 + \log \frac{[HCO_3^-]}{0.301 \times P_{CO_2}}$$

where the concentration of bicarbonate ion, $[HCO_3^-]$, is measured in milliequivalents per liter (mEq/L) and the partial pressure of CO_2, P_{CO_2} is measured in millimeters of mercury (mmHg). This equation shows that the pH of blood is dependent on the ratio between bicarbonate ion concentration, which is regulated by the kidneys, and the partial pressure of carbon dioxide, which is regulated by the lungs, or

$$pH = \frac{Kidneys\ (HCO_3^-)}{Lungs\ (H_2CO_3)}$$

Arterial blood gas specimens are collected by arterial puncture and then placed in a heparinized tube (1000 units/mL). Blood should be analyzed immediately to avoid effects of cellular metabolism that might change the pH, that is, make it more acidic. If the sample is cooled on ice, the results are considered accurate for up to 1 to 2 hours after the sample is drawn.

The normal pH for *arterial* blood gases is 7.35 to 7.45 with a mean of 7.4. The normal pH for a *venous* sample is 7.32 to 7.42 with a mean of 7.37, more acidic than arterial blood because the carbon dioxide content is higher.

Beside measuring pH, ABGs also measure the partial pressures of carbon dioxide, Pa_{CO_2}, and oxygen, Pa_{O_2}. The normal Pa_{CO_2} level for arterial blood is 35 to 45 mm Hg, with a mean of 40 mmHg. Venous values are 42 to 55 mmHg, or a mean of 46 mmHg. For people under 65 years of age Pa_{O_2} is in the range of 90 to 100 mmHg. Older adults have a lower Pa_{O_2} (Table 3–2).

Other Tests

Acid-base analysis of other body fluids can also be carried out. Gastrointestinal secretions and skin are essentially acidic, although stomach secretions are much more acidic because of the presence of hydrochloric acid (HCl). The pancreas secretes bicarbonate, which, as foodstuffs enter the duodenum, lessens the acidity of the emerging stomach contents.

TABLE 3–2
ARTERIAL BLOOD GAS VALUES

Serum pH	7.35–7.45
Pa_{O_2}	90–100 mmHg (decreases with aging; 75–89 mmHg may be normal for the elderly)
Pa_{CO_2}	35–45 mmHg
HCO_3^-	22–26 mEq/L
O_2 saturation (Sa_{O_2})	95%–100%

Urine can vary from acidic (pH of 5.5 to 6.5), to neutral, to very alkaline (as much as pH of 9). Urinary pH determination is measured as part of urinalysis (UA), which also measures the amounts of sodium, chloride, and potassium that have been excreted.

Metabolic Acidosis

Metabolic acidosis is caused by a primary base deficit or an accumulation of fixed (nonvolatile) acids. In other words, acidosis may be a result of losing too much base or retaining too much acid (Fig. 3–2). It is a condition of increased hydrogen ion concentration (pH < 7.35) secondary to increased accumulation of acids formed by the metabolism of nutrients. The ratio of bicarbonate to carbonic acid is decreased from the usual 20:1 ratio to 16:1 to 5:1. The characteristic feature of metabolic acidosis is a decrease in alkaline reserve.

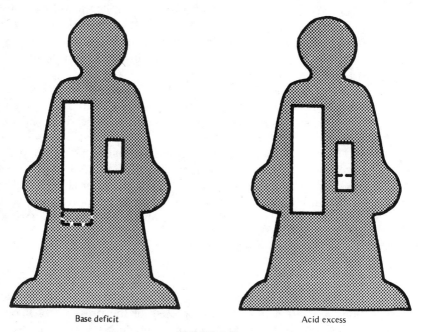

Base deficit   Acid excess

FIGURE 3–2
Acidosis.

Causative Factors

The causes of metabolic acidosis are primarily endocrine and metabolic disorders. The most typical example is diabetes mellitus. In this disorder, there is insufficient insulin to metabolize the needed amounts of glucose. As a result, the liver increases the metabolism of fatty acids, which are stronger than carbonic acid. These stronger acids consume the bicarbonate (alkaline) reserve in the sodium bicarbonate equation. The lungs compensate for the excess acid by increasing the rate of respiration, which eliminates some of the carbon dioxide, thereby decreasing the carbonic acid concentration. The kidneys respond by converting glutamic acid and other amino acids to ammonia, which is excreted with the hydrogen ions as ammonium ions. These chemical, regulatory mechanisms are all part of the body's attempt to compensate for the acidic state.

Excessive exercise, such as marathon running, can also result in acidosis. During exercise, muscle tissues can become hypoxic, in which case glucose is converted to lactic acid rather than carbon dioxide. Carbohydrates must have oxygen to be properly metabolized to carbon dioxide and water, the aerobic phase of the Krebs cycle. When sufficient oxygen is not available, the body operates in the anaerobic phase of the Krebs cycle. The result is lactic acidosis. Table 3–3 outlines the major causes of metabolic acidosis.

TABLE 3–3
CAUSES OF METABOLIC ACIDOSIS

Cause	Altered Function
Diabetes (diabetic ketoacidosis)	Insufficient insulin results in increased fat metabolism, which produces excess accumulation of ketones, acetoacetic acids, acetone, and other acids.
Renal insufficiency/failure (renal acidosis)	Increased waste products of protein metabolism (urea, creatinine, phosphoric acid, sulfuric acid) are retained; excessive acids decrease the bicarbonate available to maintain acid-base balance.
Incomplete metabolism of carbohydrates (lactic acidosis)	Insufficient oxygen for proper burning of carbohydrates to glucose and water; therefore lactic acid increases.
Salicylate intoxication	Excessive ingestion of acetylsalicylic acid.
Severe diarrhea	Intestinal and pancreatic secretions are normally alkaline, therefore excessive fluid loss results in acidosis.
Malnutrition	Improper metabolism of nutrients; fat catabolism results in excess ketones and acids.
High fat diet	Too rapid accumulation of the waste products of fat metabolism (ketones).
Addison's disease (terminal)	Insufficient amount of sodium and chloride is held by the body with buildup of potassium owing to lack of adrenocortical steroids (glucocorticoids and mineralocorticoids).

Assessment

Physical Assessment

Signs and symptoms indicative of metabolic acidosis include headache, drowsiness, nausea, vomiting, and diarrhea. If not treated, acidosis can lead to stupor, coma, twitching, and convulsions. Acidosis that results from improper fat metabolism often produces a fruity-smelling breath. Hyperpnea (increased respirations) may also be present, indicating an attempt by the body to expel extra carbon dioxide which, if retained, raises the pH of the serum. This type of deep, rapid respiration is known as Kussmaul's respiration.

A complication that may develop during metabolic acidosis is gastric dilation. Fluids ingested remain in the stomach, resulting in vomiting and possible aspiration of the vomitus. Signs of gastric dilation are a feeling of fullness and distention.

As the serum potassium level increases, physical manifestations include progressive muscle weakness and paresthesias of the face or extremities.

Diagnostic Assessment

In a patient with metabolic acidosis, the blood pH is below 7.35. Other indications of acidosis include a decreased bicarbonate level, together with a normal or lowered $PaCO_2$. If the lungs are healthy, the respiratory rate is increased in order to reduce acidosis by expelling carbon dioxide, and the $PaCO_2$ falls below 35 mmHg. (In contrast, during respiratory acidosis, the kidneys perform the compensation.) In metabolic acidosis, laboratory values may show an increase in the partial pressure of oxygen (PaO_2) as a result of hyperpnea.

Serum sodium and chloride levels may be decreased below 130 mEq/L and 98 mEq/L, respectively, especially if gastrointestinal fluids are lost. The serum potassium level is usually above 5.5 mEq/L with acidosis (hyperkalemia), indicating a false positive unless potassium has been lost from the body, for example, when vomiting and diarrhea occur (see Chap. 4). Polyuria also lowers potassium levels in patients experiencing early diabetic ketoacidosis. Urinary pH decreases as the kidneys attempt to get rid of excess hydrogen ions.

Nursing Diagnoses

The nursing diagnoses for metabolic acidosis depend on the underlying cause of the imbalance. If the serum potassium level increases, activity intolerance related to muscle weakness may be present. A complete list of nursing diagnoses associated with hyperkalemia is found in Chapter 4.

Collaborative Management

Medical Treatment

In most cases, metabolic acidosis is treated by resolving the underlying cause. However, there are some guidelines to follow regarding acidosis in general. Emergency treatment of a patient can be accomplished quickly by administering a hypertonic bicarbonate solution, usually intravenously, to add to the buffer base of the blood. Usually bicarbonate is given as sodium bicarbonate solution, 1 to 3 ampules (50 mEq $NaHCO_3$ per ampule). It is usually reserved for patients whose pH is below 7.2 or whose bicarbonate level is less than 8 mEq/L (Gahart, 1995). Caution must be exercised when administering bicarbonate because rapid conversion from acidosis to alkalosis may cause disequilibrium between the pH of the cerebrospinal fluid and plasma, and lead to neurologic symptoms.

The potassium level in the serum should be closely monitored when acidosis is being treated because potassium will shift from the blood into the cells and the blood concentration may become too low. This is especially true if potassium has been lost through the urine and gastrointestinal tract.

An intravenous solution, such as isotonic saline, 5% dextrose in 0.45% saline, sodium lactate, or bicarbonate, may be used to increase the base level. Lactate and bicarbonate are the primary treatments. The liver converts sodium lactate to bicarbonate.

The most common causes of metabolic acidosis are diabetes mellitus and renal failure, both of which require continuous medical management. In diabetes mellitus, the insulin hastens the movement of glucose out of the serum and into the cell, thereby lessening any concurrent ketosis. When glucose is being properly metabolized, the body stops converting the fats to glucose. Insulin lessens the amount of ketone bodies by decreasing the release of fatty acids from fat cells. In diabetes, it is necessary to watch for circulatory collapse because of the polyuria that may result from the hyperglycemic state. Polyuria or diuresis may lead to an extracellular volume deficit (dehydration), which is normally corrected by fluid and electrolyte replacement.

In renal failure, dialysis can be used to remove the protein waste products and thereby *lessen* the acidotic state. Although controversial, a diet low in protein and high in calories (carbohydrates and fats) will lessen the amount of protein waste products that are a result of protein catabolism. This, in turn, helps to resolve the acidosis.

Although the prognosis for the patient with lactic acidosis is poor, the administration of oxygen and fluids helps to adjust the lactic acidotic state.

Nursing Interventions

The role of the nurse is ongoing assessment, especially while the patient is being treated for acidosis. Monitoring fluid and electrolyte balance as well as acid-base balance is important.

If sodium bicarbonate is administered, special precautions regarding the rate of infusion must be followed. For instance, the usual rate of administration is 2 to 5 mEq/kg body weight. No more than 50 mEq/h should be given to adults. Rapid administration can cause hypokalemia, hypernatremia, hypocalcemia, and metabolic alkalosis (Gahart, 1995). Table 3–4 lists the major nursing interventions associated with sodium bicarbonate administration.

TABLE 3–4

NURSING INTERVENTIONS ASSOCIATED WITH
INTRAVENOUS SODIUM BICARBONATE ADMINISTRATION

- Check the solution percentage carefully because sodium bicarbonate is available in various strengths, e.g., 4.2%, 5%, 7.5%, 8.4%.
- Ensure that the solution is clear; discard discolored or cloudy solutions.
- Give diluted sodium bicarbonate solutions.
- Flush the IV tubing thoroughly before and after administration because sodium bicarbonate is incompatible with most other IV medications, including lactated Ringer's solution.
- Do not exceed a dose of 50 mEq/h; use an electronic infusion device for administration, unless used for a cardiac arrest.
- Monitor electrolyte and arterial blood gas values carefully during treatment because sodium bicarbonate, if given too rapidly or in excess, can cause hypernatremia, hypokalemia, hypocalcemia, or metabolic alkalosis.
- Use with extreme caution in patients with cardiac, liver, or renal disease.
- Monitor the IV site carefully for signs of venous irritation.

Metabolic Alkalosis

Metabolic alkalosis may have one of two causes: a malfunction of metabolism that results in an increase in the amount of available basic solutions in the blood or a reduction of available acids in the serum (Fig. 3–3). Both abnormalities cause a decrease in the number of hydrogen ions, resulting in an increase in the blood pH to a level above 7.45. The bicarbonate anion concentration is increased in relationship to carbonic acid.

Metabolic alkalosis is also known as primary base bicarbonate deficit.

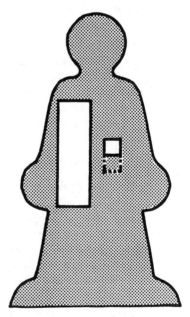

Base excess   Acid deficit

FIGURE 3–3
Alkalosis.

Shallow breathing results as the body attempts to reduce the amount of carbon dioxide expelled from the lungs and thereby allow the level of carbonic acid in the blood to increase. The kidneys react by excreting sodium bicarbonate in the urine. The mechanism for excretion of hydrogen ions in the kidney tubules restricts the formation of ammonia. Lactic acid and ketoacids are reabsorbed. As this compensation occurs, the urine becomes alkaline.

Causative Factors

Metabolic alkalosis is commonly associated with excessive loss of chloride [usually through vomiting and prolonged nasogastric (NG) suction] or excessive intake of sodium bicarbonate. Urinary potassium chloride loss results in metabolic alkalosis and is called paradoxical aciduria. Patients with paradoxical aciduria excrete acid in the urine, that is, hydrogen and potassium ions, and reabsorb sodium and bicarbonate. Patients on long-term diuretic therapy lose potassium, sodium, and chloride and consequently develop metabolic alkalosis. Table 3–5 outlines the major causes of metabolic alkalosis.

Assessment

Physical Assessment

The signs and symptoms of metabolic alkalosis are difficult to distinguish from those of other conditions occurring with the alkalosis, particularly the changes caused by potassium and chloride deficits. However, some of the major signs and symptoms include shallow breathing, nausea, vomiting, and diarrhea. Central nervous system symptoms, such as confusion, irritability, agitation leading to coma, and possible convulsions, are not uncommon. In some cases, symptoms similar to those that accompany calcium imbalance may be present. These include restlessness and twitching of the extremities.

TABLE 3–5
CAUSES OF METABOLIC ALKALOSIS

Cause	Altered Function
Ingestion of excess sodium bicarbonate	Rapid absorption of bicarbonate through gastric mucosa
Excessive vomiting	Loss of potassium and hydrochloric acid from the stomach
Gastrointestinal intubation (especially Levin tube) or gastric lavage	Loss of upper gastrointestinal secretions which are high in hydrochloric acid and secondarily in potassium
Administration of potent diuretics	Loss of hydrogen and chloride ions causes a compensatory increase of bicarbonate in the blood
Prolonged hypercalcemia	Relationship unidentified

Diagnostic Assessment

The pH level in patients with metabolic alkalosis is above 7.45. The serum bicarbonate (base) level increases and the potassium often decreases (hypokalemia). Properly functioning kidneys have the ability to secrete some of the bicarbonate.

The partial pressure of carbon dioxide ($Paco_2$) does not change unless the lungs attempt to compensate for the alkalosis. In *compensated* metabolic alkalosis, the partial pressure of carbon dioxide rises as the body attempts to increase the carbonic acid concentration to a level where it will neutralize the basic state.

Serum chloride levels are also decreased. In metabolic alkalosis, the serum chloride concentration will decrease to a greater extent than the serum sodium reading. In other imbalances, the sodium and chloride decrease in a proportionate ratio.

The electrocardiogram (ECG) reading may show a sinus tachycardia. Other changes in the ECG indicative of metabolic alkalosis are low or flat T waves, peaked P waves, and elevated U waves, a result of hypokalemia.

Nursing Diagnoses

Common nursing diagnoses for metabolic alkalosis include:

- Diarrhea, related to GI disorder or disease
- Breathing Pattern, ineffective, related to attempt to conserve carbon dioxide
- Sensory-perceptual alterations, related to effects of low potassium and sodium

Collaborative Management

Medical Treatment

The aim in the treatment of alkalosis is to correct the pH imbalance and eliminate the underlying cause of the alkalosis. In addition, prevention of alkalosis is also a major concern. For example, a nasogastric (NG) tube should be irrigated with saline (sodium chloride) rather than water because saline is isotonic to the body and does not affect acid-base balance. A person who takes baking soda for an ulcer or to alleviate an acid stomach should be reminded of the negative side effects of excessive ingestion of sodium bicarbonate and cautioned against consuming large amounts. If a patient with congestive heart failure is taking a diuretic preparation, the potassium level should be checked frequently to alert the physician to any developing potassium deficit.

If preventive measures fail and alkalosis develops, there are several common alternative methods of treatment. The method used depends on the specific case.

If there has been excessive loss of potassium and chloride, replacement of these electrolytes is necessary. This can be accomplished either intravenously or orally. Foods high in potassium are fruits (banana), dried peas, beans, baked potato, nuts, molasses, meats, and especially fish and poultry. Potassium replacement should be governed by the amounts of potassium and chloride lost and by daily intake requirements. Replacement needs can be calculated by two methods. Loss can be estimated according to a standard number of milliequivalents of electrolytes per liter of fluid. For example, the average amount of potassium lost per liter of gastric fluid is 5 to 10 mEq. The second method, in which the exact milliequivalents per liter of fluid loss are measured, requires laboratory measurement and is the more accurate of the two methods.

The diuretic acetazolamide (Diamox) may be given to promote kidney excretion of bicarbonate. Although used infrequently, acidifying solutions such as ammonium chloride and arginine chloride may be given intravenously or orally. The added hydrogen ions present in these solutions increase the available acids in the serum. Ammonium chloride is particularly helpful if the alkalosis is caused by diuretics of the potent thiazide and mercurial groups.

Sodium chloride may be administered to replace chloride orally or intravenously unless contraindicated by other conditions such as congestive heart failure.

Nursing Interventions

As for all acid-base imbalances, nursing interventions include monitoring accurate intake and output records, particularly with reference to electrolyte losses contributing to alkalosis; potassium and hydrogen ion losses are especially important. In addition, the nurse should:

- Check for changes in vital signs and ECG readings suggestive of hypokalemia
- Check for muscle weakness and bowel sound changes that may be related to hypokalemia or its underlying cause
- Take any safety precautions warranted by the confusion often present in alkalotic patients
- Implement seizure precautions for possible tetany and convulsions

Respiratory Acidosis

To understand the involvement of the lungs in the process of maintaining pH balance, it is necessary to be familiar with the movement of volatile waste products through the system during metabolism. Cells, as the metabolic units of the body, burn nutrients. Two specific nutrients, carbohydrates and fats, produce carbon dioxide and water. Carbon dioxide is a waste product of metabolism that must be removed from the body. It diffuses from the cells through the interstitial fluid space into the blood. The carbon dioxide is transported by the blood to the lung capillary bed and excreted through the lungs. The respiratory system regulates pH via two components of a negative feedback system: (1) a chemical reaction involving carbonic acid, water, and carbon dioxide ($H_2CO_3 \rightleftharpoons H_2O + CO_2$), and (2) the process of increasing or decreasing the speed and depth of respirations.

As described earlier in the chapter, respiratory control of pH is based on negative feedback mechanisms in which changes in the hydrogen ion level cause a corresponding change in the respiratory rate. The objective of the negative feedback mechanism is to maintain body functions within a normal range. Hydrogen ions affect the respiratory center in the medulla, which cause an increase in the respiratory rate in response to an increased concentration of hydrogen ions. The respiratory rate decreases when the hydrogen ion concentration decreases. The rate of respiration, in turn, affects the hydrogen ion level. For example, a rise in pH from 7.40 to 7.63 can be brought about by increasing the speed of respirations from 20 to 40 per minute. In contrast, a drop in pH from 7.40 to 7.00 can be induced by decreasing the speed of respirations from 20 to 5 per minute. As the pH returns to normal, so does the respiratory rate. This gradually fading stimulus causes respiratory correction of pH to be partial rather than total.

Although the kidneys and chemical buffers are important in maintaining the correct pH level, the lungs play the most vital role. The lungs normally excrete a quantity of CO_2 gas equal to 15,000 to 20,000 mEq of hydrogen ions per day; in comparison, the kidney's normal ion excrete is equal to 40 to 80 mEq of hydrogen ion per day. The lungs react more rapidly than the chemical buffers, responding to changes in the pH in 1 to 2 minutes. For example, if the blood pH drops from 7.40 to 7.00, the respiratory system can restore the pH to 7.20 and 7.30 in 1 minute and 2 minutes, respectively. In the normally functioning body, respiratory regulation of pH will correct 50% to 75% of the imbalance.

Carbon Dioxide

A key indicator of adequate ventilation is carbon dioxide tension, which is measured in terms of $Paco_2$. Hyperventilation decreases $Paco_2$; hypoventilation increases $Paco_2$. Chemoreceptors located in the brainstem, particularly the pons and medulla oblongata, react to changes in carbon dioxide (primary respiratory stimulus) and pH of the cerebrospinal fluid. For example, under normal conditions, when the chemoreceptors are stimulated because of increased carbon dioxide, the patient begins to breathe rapidly and deeply, enhancing carbon dioxide removal. In abnormal conditions, such as chronic airflow limitation, the primary respiratory stimulus (CO_2) is compromised and hypoxia (decreased Pao_2) becomes the stimulus for the respiratory system. Increased carbon dioxide concentration (hypercapnia) can be accompanied by hypoxia.

Oxygen

Oxygen concentration in arterial blood is also controlled by the respiratory system and the regulatory systems. Oxygen is essential for cellular metabolism. The level at which the partial pressure of oxygen (Pao_2) provides adequate stimulation and is, therefore, within an acceptable range varies with age.

Arterial Pao_2 for the newborn infant may range from 40 to 70 mmHg. A normal acceptable reading for children and adults is 90 mmHg or above. In the average adult, the Pao_2 usually must fall below 80 mmHg before a respiratory response is initiated.

Implications for Elderly Persons

The normal range for elderly patients encompasses lower values because of the development of degenerative lung changes, weakened respiratory muscles, and the resulting decrease in ventilatory efficiency. Elderly patients often have PaO_2 levels of 75 to 90 mmHg without dyspnea.

Respiratory imbalances are caused by primary defects in the function of the lungs or by changes in the normal respiratory pattern caused by secondary problems. In either case, the result of the dysfunction is an increase or decrease in the serum concentration of carbonic acid (H_2CO_3).

Respiratory acidosis occurs when carbon dioxide, which is a volatile substance, is retained because of hypoventilation (Fig. 3–4). Any disruption of the pulmonary gas exchange may result in respiratory acidosis.

Causative Factors

The most common cause of respiratory acidosis is chronic lung disease, most currently called chronic airflow limitation (CAL). It is also known as chronic obstructive pulmonary disease (COPD) or chronic obstructive lung disease (COLD). Examples of specific diseases that are included in this category are pulmonary emphysema, chronic bronchitis, and chronic asthma. In patients with these diseases, the suppression of respiration causes retention of carbon dioxide, which combines with water to form carbonic acid. This increased acid causes the pH reading to decrease, reflecting the acidotic state. Table 3–6 lists the causes and altered functions of respiratory acidosis.

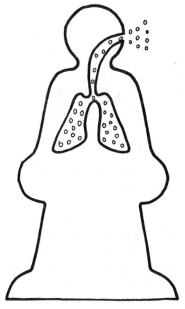

Normal ventilation

Hypoventilation

FIGURE 3–4
Hypoventilation.

TABLE 3-6
CAUSES OF RESPIRATORY ACIDOSIS

Cause	Altered Function
CHRONIC DISEASES	
Emphysema	Loss of elasticity in alveolar sacs; destruction dominant over obstruction; restricted air flow in and out, primarily out, results in elevated Pa_{CO_2}
Asthma	Spasms due to allergens, irritants, or emotions cause smooth muscles of bronchioles to constrict
Bronchitis	Inflammation obstructs airway
Pulmonary edema	ECF accumulation in acute congestive heart failure causes decreased alveolar diffusion and perfusion
Bronchiectasis	Bronchi dilated due to inflammation; destructive changes and weakness in walls of bronchi
ACUTE DISEASES	
Infection	Aeration decreased due to obstruction of airway caused by inflammation and bacterial agents
Sedatives (narcotics)	Depressed respiratory center leads to CO_2 narcosis
Pneumonia	Inadequate oxygenation due to fluid accumulation caused by infection, irritants, immobility
Atelectasis	Excessive mucus with collapse of alveolar sacs due to mucus plugs, infectious drainage, or anesthetic drugs following surgery causes decreased respiration
Brain trauma/tumor	Excessive pressure on respiratory center or medulla oblongata depresses respiration

Assessment

Physical Assessment

Signs and symptoms indicating respiratory acidosis include decreased ventilation, somnolence, and changes in sensorium and mentation. Restlessness and disorientation are the earliest signs of hypoxia. As hypoxia becomes more acute, such recognizable manifestations as diaphoresis, cyanosis, and rapid, irregular pulse may occur.

Diagnostic Assessment

In respiratory acidosis, the serum pH is below 7.35, indicating that the concentration of carbonic acid in the serum is increased. The partial pressure of carbon dioxide (Pa_{CO_2}) is well over 45 mmHg, because CO_2 is retained. The partial pressure of oxygen (Pa_{O_2}) may be normal (90 to 100 mmHg) or decreased if hypoxia results. Patients with long-standing chronic lung disease usually have a chronically low Pa_{O_2} and chronically high Pa_{CO_2}. The bicarbonate level is normal if the respiratory acidosis is uncompensated; it will be elevated with partial or complete compensation.

Healthy kidneys can compensate for respiratory acidosis; however, the compensation may take from a few hours to several days. To compensate for respiratory acidosis, the kidneys retain bicarbonate and return it to the extracellular fluid compartment. Usually, patients with chronic airflow limitation have a partial compensation with a serum bicarbonate reading above 30 mEq/L. This elevated reading reflects the action of the kidneys, which are retaining bicarbonate to buffer or neutralize the excess acids. Serum potassium will be elevated relative to the potassium level before the acidosis developed.

Nursing Diagnoses

Common nursing diagnoses related to respiratory acidosis include:

- Gas Exchange, impaired, related to alveolar or bronchial damage
- Airway Clearance, ineffective, related to copious mucus secretions
- Tissue Perfusion, altered, risk for, related to effects of hypoxia
- Fatigue, related to effects of hypoxia
- Cardiac Output, decreased, risk for, related to effect of hypoxia on myocardium
- Anxiety, related to inability to breathe without difficulty

Collaborative Management

Medical Treatment

Respiratory acidosis can develop very quickly into an emergency situation. The primary goals are to improve ventilation, increase aeration, and provide adequate tissue oxygenation by whatever means the clinical situation suggests. Nebulizer treatments, antibiotics, low-dose oxygen via nasal cannula, and chest physiotherapy are commonly used. In more severe cases, an endotracheal tube or tracheostomy with assisted or controlled ventilation may be necessary. Bronchodilators may be given intravenously until the patient is able to progress to oral theophylline or another bronchodilator. Corticosteroids and albuterol (Proventil) may also be administered on a regular basis. Inhalers and nebulizer treatments to deliver bronchodilators and steroids are used most often in patients with asthma.

Tranquilizers, opioid analgesics, and hypnotics further depress respiration and should be avoided. High amounts of oxygen must also be avoided. The patient with chronic lung disease who retains carbon dioxide depends on the hypoxic state to breathe. If the rate of oxygen delivery exceeds 2 to 3 L/min, the patient will lose the drive to breathe and respiratory arrest will ensue.

Nursing Interventions

In collaboration with respiratory specialists and therapists, the primary role of the nurse is to provide aggressive pulmonary hygiene, including deep breathing and coughing, suctioning, chest physiotherapy, and other respiratory treatments as needed. Additionally, the nurse provides adequate hydration to thin secretions, unless contraindicated (as with congestive heart failure). The oxygen flow is monitored very carefully and indicators of increased hypoxia are reported. Ongoing respiratory assessments, including vital signs, are essential.

The patient is assisted into a position of comfort for breathing. The head of the bed remains elevated at all times. Some patients are able to breathe only while sitting upright in an orthopneic position. In this position, the patient is usually supported by the customary table over the bed and a pillow.

Another vital intervention is to provide emotional support and a caring presence to decrease anxiety. Patients who experience difficulty breathing are typically very anxious and do not like to be left alone.

Respiratory Alkalosis

Respiratory alkalosis is not as common as respiratory acidosis. It occurs in patients who hyperventilate and is often neurogenic in origin. Rapid respiration causes expiration of an increased amount of carbon dioxide, resulting in a deficit of carbonic acid. Alkalosis can contribute to constriction of the coronary blood vessels. Angina, exposure to cold, or emotional stress can cause alkalosis due to involuntary hyperventilation.

Causative Factors

Table 3–7 lists causes and altered functions of respiratory alkalosis.

Implications for Pregnant Women

Mild, compensated respiratory alkalosis is associated with pregnancy as a result of hyperventilation. In early pregnancy, progesterone begins to stimulate the respiratory center in the medulla of the brain to increase the respiratory rate. In late pregnancy, the size of the uterus places pressure on the diaphragm, causing frequent, rapid respirations.

Assessment

Physical Assessment

Respiratory alkalosis can be recognized by the abnormally rapid respiration that accompanies hyperventilation. The causes listed in Table 3–7 may precede or accompany hyperventilation. The patient may experience headache, vertigo, paresthesia, tetany, carpopedal spasm, and syncope.

TABLE 3–7
CAUSES OF RESPIRATORY ALKALOSIS

Cause	Altered Function
Hysteria	Psychoneurosis causes vigorous breathing, resulting in excessive exhaling of CO_2 (hyperventilation)
Pain; brain trauma	Overstimulation of respiratory center with resultant H_2CO_3 deficit (hyperventilation)
Drugs (acetylsalicylic acid)	Salicylates initially stimulate respiratory center, causing hyperventilation; secondary development of metabolic acidosis with compensatory hyperventilation
Oxygen lack	Hypoxia may cause respiratory stimulation with resultant H_2CO_3 deficit (hyperventilation)
Ventilators	Iatrogenic; overadministration of O_2 and depletion of CO_2 occur while adjusting ventilator to the patient (hyperventilation)

Diagnostic Assessment

The blood pH of a patient with respiratory alkalosis is above 7.45, indicating a decreased concentration of carbonic acid in the serum. The blood gas carbon dioxide ($Paco_2$) is therefore below 35 mmHg. The partial pressure of oxygen (Pao_2) will probably be normal. The bicarbonate level is normal if the respiratory alkalosis is uncompensated; it will be lower with compensation. The kidneys compensate for respiratory alkalosis by decreasing both tubular acid secretion and reabsorption of bicarbonate. In other words, the kidneys retain hydrogen ions and excrete bicarbonate in proportion to the amount of carbon dioxide that has been expelled during hyperventilation.

Nursing Diagnoses

The common nursing diagnosis for patients with respiratory alkalosis include:

- Sensory-perceptual alterations, related to stress, weather exposure, hyperventilation, or hysteria
- Anxiety, related to pain or other stressor

Collaborative Management

Medical Treatment

Like all acid-base imbalances, the underlying cause needs to be identified. Having the patient breathe into a paper bag to increase carbon dioxide concentration is an inexpensive and usually successful intervention. Respiratory alkalosis can also be treated by sedation with a tranquilizer, reassurance, carbon dioxide treatments, voluntary breath-holding, or use of a rebreathing mask.

Nursing Interventions

The nurse provides emotional support and educates the patient regarding breathing patterns. The use of breathing techniques is encouraged and breathing aids are applied as indicated.

CRITICAL THINKING ACTIVITIES

A 54-year-old man was admitted to the hospital with a diagnosis of long-standing chronic obstructive pulmonary disease (COPD) and acute bronchitis. His arterial blood gases on admission were pH 7.27, $Paco_2$ 72 mmHg, and Pao_2 68 mmHg.

QUESTIONS

1. *What abnormal laboratory results were present on admission and what do they indicate?*

2. *What is the pathophysiologic basis for the laboratory results?*

3. *Should the physician's order for oxygen by nasal cannula be questioned if the rate ordered is 2 L/min?*

4. *Would you expect the serum bicarbonate level to be increased or decreased? Why might a deviation in bicarbonate occur?*

REFERENCES AND READINGS

Amundson, DE, & Diamant, J (1994). Severe metabolic alkalosis. Southern Medical Journal 87(2), 275–277.

Anderson, S (1990). ABGs: Six easy steps to interpreting blood gases. American Journal of Nursing 90(8), 42–45.

Booker, MF, & Ignatavicius, DD (1996). Infusion Therapy: Techniques and Medications. Philadelphia: WB Saunders.

Chernecky, CC, Krech, RL, & Berger, BJ (1993). Laboratory Tests and Diagnostic Procedures. Philadelphia: WB Saunders.

Gahart, BL (1995). Intravenous Medications. St. Louis: Mosby Year-Book.

Grap, MJ, Glass, C, & Constantino, S (1994). Accurate assessment of ventilation and oxygenation. MEDSURG Nursing 3(6), 435–444.

Guyton, AC (1991). Textbook of Medical Physiology. Philadelphia: WB Saunders.

Ichakawa, I (1990). Pediatric Textbook of Fluids and Electrolytes. Baltimore: Williams & Wilkins.

Janusek, L (1990). Metabolic alkalosis. Nursing90 20(6), 49–50.

Janusek, L (1990). Metabolic acidosis. Nursing90 20(7), 52–53.

Morris, L (1990). The POLK method of ABG interpretation. Critical Care Nurse 10(1), 18–20.

Russell, J (1991). Successful method for arterial blood gas interpretation. Critical Care Nurse 11(5), 14–19.

Scherer, P (1990). Get true blood gas values with less blood loss. American Journal of Nursing 90(2), 28.

Tasota, F, & Wesmiller, S (1994). Assessing A.B.G.s: Maintaining the delicate balance. Nursing94 24(5), 34–46.

Taylor, DL (1990). Respiratory alkalosis. Nursing90 20(8), 60–61.

Taylor, DL (1990). Respiratory acidosis. Nursing90 20(9), 52–53.

Yeaw, E (1992). Good lung down? How position affects oxygenation. American Journal of Nursing 92(3), 27–29.

PART ONE **QUESTIONS**

• • • • • • • • • •

1. Which of the following physiologic changes associated with aging causes elderly patients to be at high risk for dehydration?
   a. Decreased renal function
   b. Decreased thirst response
   c. Poor skin turgor
   d. Decreased cardiac output

2. A patient is admitted to the emergency department with hemorrhage from a gunshot wound. For which condition should the nurse assess the patient in this situation?
   a. Fluid overload
   b. Metabolic acidosis
   c. Hyperkalemia
   d. Fluid deficit

3. Which of the following assessment findings is commonly associated with fluid deficit?
   a. Decreased pulse
   b. Peripheral edema
   c. Moist crackles
   d. Decreased blood pressure

4. What is the most reliable indicator of fluid loss or gain?
   a. Intake and output
   b. Skin turgor
   c. Weight
   d. BUN

5. Where is the most reliable location to assess for skin turgor in an elderly patient?
   a. Sternum
   b. Abdomen
   c. Forearm
   d. Lower leg

6. How would the central venous pressure (CVP) reading be expected to change in a patient with dehydration?
   a. Stay the same
   b. Decrease
   c. Increase
   d. Drop to zero

7. Which of the following assessment findings is typically associated with isotonic fluid overload?
   a. Decreased blood pressure
   b. Bounding pulse
   c. Decreased respirations
   d. Poor skin turgor

8. In a patient with congestive heart failure, which of the following drugs would be most appropriate in the treatment of the fluid overload?
   a. Digoxin
   b. Heparin
   c. Lasix
   d. Potassium

9. A patient is admitted with these arterial gas values: pH = 7.28, $Paco_2$ = 39 mmHg, HCO_3^- = 19 mEq/L. What acid-base imbalance do these values suggest?
   a. Metabolic acidosis
   b. Metabolic alkalosis
   c. Respiratory acidosis
   d. Respiratory alkalosis

10. For severe acidosis, what drug might be used?
    a. Potassium chloride
    b. Calcium gluconate
    c. Iron sulfate
    d. Sodium bicarbonate

11. An elderly woman is admitted with shallow breathing, nausea, and agitation. Her daughter reports that she has been taking large amounts of baking soda and water for a "sour stomach." What acid-base imbalance is she probably experiencing?
    a. Respiratory acidosis
    b. Respiratory alkalosis
    c. Metabolic alkalosis
    d. Metabolic acidosis

12. A patient with chronic airflow limitation (chronic lung disease) experiences dyspnea, lethargy, and increasing confusion. The arterial blood gas values are pH = 7.25, $Paco_2$ = 73 mmHg, and HCO_3^- = 26 mEq/L. What acid-base imbalance do these assessment findings suggest?
    a. Respiratory acidosis
    b. Respiratory alkalosis
    c. Metabolic alkalosis
    d. Metabolic acidosis

13. What amount of supplemental oxygen is recommended for patients with chronic lung disease who retain carbon dioxide?
    a. 1 L/min
    b. 2–3 L/min
    c. 5–6 L/min
    d. Over 6 L/min

14. Which of the following is the most important nursing intervention for the patient with chronic lung disease?
    a. Forcing fluids
    b. Chest physiotherapy
    c. Ongoing respiratory assessment
    d. Providing rest

15. According to the arterial blood gas results, a highly anxious patient is experiencing respiratory alkalosis. What is the most likely cause of this acid-base imbalance?
    a. Chronic lung disease
    b. Intubation
    c. Hyperventilation
    d. Drug therapy

PART TWO

ELECTROLYTES AND OTHER ELEMENTS: BALANCE AND IMBALANCE

CHAPTER 4

. . . .
. . . .
. . . .
. . . .
. . . .
. . . .
. . . .
. . . .
. . . .
. . . .

Potassium

POTASSIUM BALANCE

Potassium, a cation, is the major electrolyte of the intracellular compartment. It acts as part of the body's buffer system and affects all types of neuromuscular activities.

Distribution

The total amount of potassium in the body is related to the size of the individual but is generally considered to be about 42 mEq/kg of body weight. Variation in this ratio occurs because potassium makes up a large portion of the muscular tissue, and the proportion of this varies from individual to individual.

The intracellular compartment contains most (98%) of the body's potassium, and the extracellular compartment contains the other 2%. Of the potassium found in the intracellular compartment, 70% (or about 3000 mEq) is located in the skeletal muscle and 28% (about 150 mEq) in the liver and red blood cells. The 2% present in the extracellular compartment (about 50 mEq) is reflected by the serum potassium reading. The normal serum potassium reading is between 3.5 and 5.3 mEq/L.

In contrast to sodium, almost the entire potassium content of the body is freely exchangeable. Extracellular losses (e.g., from diarrhea, vomiting, or overuse of laxatives or diuretics; or in acidosis) are always rapidly compensated from the intracellular space, so that the potassium concentration in the extracellular space often varies very little over long periods. This requires alertness on the part of healthcare personnel because a critical potassium deficiency with the threat of cardiovascular and neuromuscular disturbances can easily be overlooked in routine examinations if only the plasma potassium concentration is tested.

Because potassium is primarily located in cells, it can be used to estimate the total cell mass of the body. The relationship of potassium to body mass indicates that precipitous loss of body weight, especially lean or muscular tissue, will result in potassium loss. Thus, tissue wasting from malnutrition, stress, or chronic disease leads to potassium losses.

Approximately 5 to 10 mEq of potassium is found in each liter of gastrointestinal (GI) secretions. Therefore, excessive losses of GI contents can cause a potassium deficit.

Implications for Pregnant Women

The pregnant woman retains potassium, which is stored in the fetus, breasts, uterus, and red blood cells (Kokko & Tannen, 1990). The serum potassium does not typically change unless prolonged vomiting or the use of diuretics causes a decreased serum concentration (hypokalemia), which is described later in this chapter.

Sources

Potassium must be ingested daily because the body has no effective method of storing it. The daily turnover of potassium in the healthy adult averages 50 to 150 mEq, representing about 1.5 to 5% of the total potassium content of the body. Intake of potassium is mostly from lean meat, whole grains, green leafy vegetables, potatoes, beans, and fruits, especially bananas, melons, and oranges (Table 4–1). Raw or baked foods contain more potassium than foods prepared by boiling or frying.

Recommended Daily Allowance

The minimum daily allowance for potassium is approximately 50 mEq (2000 mg). A typical diet in the United States contains 60 to 100 mEq per day. Therefore, balance is maintained by a daily intake and output of 50 to 90 mEq/day, which is equivalent to 2000 mg to 3500 mg of potassium per day. Evidence suggests that dietary potassium has a beneficial effect on hypertension; thus, recommendations for increases in the potassium level to 3500 mg per day may be indicated (National Research Council, 1989).

Control

The normal movement of potassium between the intracellular and extracellular compartments is controlled by the sodium-potassium pump. This pump is an active transport system that moves substances from an area of lower solute concentration to an area of higher concentration. The pump causes the level of potassium to remain higher inside the cell than outside the cell. Conversely, the sodium remains higher outside the cell than inside the cell.

Potassium levels are dynamic, that is, potassium is constantly moving

TABLE 4–1
COMMON FOOD SOURCES OF POTASSIUM*

Food Source	Amount (mg)
Corn flakes (1¼ c)	26
Cooked oatmeal (¾ c)	99
Egg (1 large)	66
Codfish, raw (4 oz)	400
Salmon, pink, raw (3½ oz)	306
Tuna fish (4 oz)	375
Apple, raw with skin (1 medium)	159
Banana (1 medium)	451
Cantaloupe (1 c pieces)	494
Grapefruit (½ medium)	175
Orange (1 medium)	250
Raisins (½ c)	700
Strawberries, raw (1 c)	247
Watermelon (1 c pieces)	186
White bread (1 slice)	27
Whole-wheat bread (1 slice)	44
Beef (4 oz)	480
Beef liver (3½ oz)	281
Pork, fresh (4 oz)	525
Pork, cured (4 oz)	325
Chicken (4 oz)	225
Veal cutlet (3½ oz)	448
Whole milk (8 oz)	370
Skim milk (8 oz)	406
Avocado (1 medium)	1097
Carrot (1 large)	341
Corn (4-inch ear)	196
Cauliflower (1 c pieces)	295
Celery (1 stalk)	170
Green beans (1 c)	189
Mushrooms (10 small)	410
Onion (1 medium)	157
Peas (¾ c)	316
Potato, white (1 medium)	407
Spinach, raw (3½ oz)	470
Tomato (1 medium)	366

*U.S. Department of Agriculture recommended daily allowance for adults: 1875–5625 mg.

Sources: Pennington, J (1992). Bowe's and Church's Food Values of Portions Commonly Used (16th ed). JB Lippincott, Philadelphia.

Ignatavicius, DD, Workman, ML, and Mishler, MA: Medical-Surgical Nursing: A Nursing Process Approach, ed 2. WB Saunders, Philadelphia, 1995. Reprinted with permission.

in and out of the cells according to the body's needs. This process is identified as either pumping or pulling. Both terms are correct.

Potassium balance is regulated by the kidneys; therefore, the major loss of potassium from the body is through the urine. Eighty percent of potassium elimination occurs via the kidneys, with only 20% being excreted in the feces.

There are two basic processes by which the kidneys regulate the potassium balance. In the first process, potassium and hydrogen ions compete for exchange with sodium ions in the renal tubules as urine salt content is adjusted by the distal convoluted tubules of the kidneys. The second process involves the mineralocorticoid aldosterone. Aldosterone, released by the adrenal cortex, regulates the sodium and potassium balance by its effect on the distal convoluted tubules. It causes the kidneys to retain sodium, which in turn causes water retention. In order to retain the sodium, the body excretes potassium ions. This is particularly significant with the release of adrenal hormones in a stress situation, in which potassium is lost while sodium and water are retained to maintain adequate blood volume.

Blood levels of potassium are regulated by the kidney inversely with other chemicals such as hydrogen, bicarbonate, sodium, and magnesium. This control mechanism is primarily related to the adrenocorticotropic hormone (ACTH) response, which causes the kidney to hold sodium and excrete or eliminate potassium in response to hormonal regulation. The kidneys respond to maintain electrolyte equilibrium by excreting or eliminating potassium, but the kidneys do not conserve potassium even when a total body potassium disequilibrium state exists.

Insulin, secreted by the pancreas, also reduces the serum potassium level by increasing the rate of cellular potassium uptake. In the presence of a potassium excess, insulin may be administered to help reduce serum potassium.

Functions

Potassium plays an important function in cellular metabolism, especially in protein and glycogen synthesis, and in the enzymic processes involved in cellular energy production. It also has a key function in neuromuscular excitability, which is related to the charge potentials in the sodium-potassium pump mechanism.

Neuromuscular Activity

Potassium, in combination with sodium and calcium, also maintains normal heart rhythm. Either an excess or a deficit may affect the myocardium, resulting in dysrhythmias. Potassium is necessary for the transmission and conduction of nerve impulses and for skeletal and smooth muscle contraction. Therefore, most of the signs and symptoms of hyperkalemia and hypokalemia are related to changes in the neuromuscular system. Signs and symptoms are usually nonspecific and may be absent, even with fairly severe imbalance. The level of potassium excess or deficit at which these signs and symptoms become apparent varies from patient to patient.

Acid-Base Balance

Potassium not only affects neuromuscular activity but also acts as part of the body's buffer system. The serum potassium level rises with acidemia (decreased serum pH) and falls with alkalemia (increased serum pH). This can be explained by considering that in acidemia the body protects itself from the acidic state by moving hydrogen ions into the cells. When hydrogen ions move into the cells, potassium ions move out in order to preserve

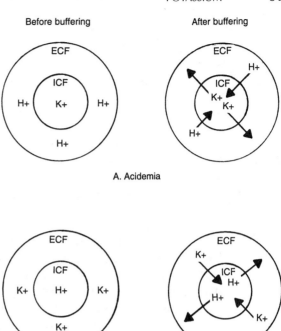

A. Acidemia

B. Alkalemia

FIGURE 4–1
Movement of potassium during buffering.

the ionic balance. In Figure 4–1A, the hydrogen content in the extracellular fluid is high, indicating acidemia. When the cells soak up hydrogen ions in an attempt to detoxify the extracellular fluid, potassium ions move out of the cells into the surrounding interstitial fluid, and eventually into the intravascular fluid. Thus, acidosis promotes movement of potassium from the intracellular fluid (ICF) to the extracellular fluid (ECF). Consequently, the serum level of potassium becomes elevated. Plasma potassium increases about 0.6 mEq/L for each fall of 0.1 unit in blood pH. This elevated potassium reading is referred to as a false positive because the *total* body potassium level is not actually elevated.

As shown in Figure 4–1B, the opposite is true in alkalemia. In this case, the plasma is low in hydrogen ions. The cells, therefore, release hydrogen ions into the plasma in an attempt to increase the acidity of the blood and combat alkalinity. As a result, potassium moves out of the plasma and into the cells, thereby lowering the serum potassium level. Plasma potassium falls about 0.6 mEq/L for each rise of 0.1 unit in blood pH. This decreased potassium is referred to as a false negative value because the total body potassium level is not actually deficient.

POTASSIUM IMBALANCES

The two major potassium imbalances are hypokalemia (potassium deficit) and hyperkalemia (potassium excess). In each case, the problem needs to be identified as an actual clinical disorder, not just a spurious or false elevation in serum levels of potassium.

Potassium Deficit

Potassium deficit, or hypokalemia, is the loss of potassium from the bloodstream that results in a serum potassium reading below 3.5 mEq/L. Potassium deficit is the most common electrolyte imbalance and is potentially life-threatening.

Causative Factors

Hypokalemia may result from a decreased intake of potassium or an excessive loss of potassium from the body. Potassium deficit may also result from a potassium shift from the ECF to the ICF.

The most common cause of potassium deficiency is increased renal loss from the inappropriate or zealous use of non–potassium-sparing diuretics such as loop (high-ceiling) and thiazide diuretics. Furosemide (Lasix) and hydrochlorothiazide (HydroDIURIL) are common examples of diuretics that increase potassium loss. Chronic use of corticosteroids, such as prednisone, also causes hypokalemia.

The second major reason for potassium loss is gastrointestinal (GI) disease, especially those diseases that cause losses of potassium-rich GI fluids through diarrhea or vomiting. Laxative abuse also increases potassium loss through diarrhea. The average potassium loss caused by diarrhea is 5 to 10 mEq/L, whereas vomiting may cause losses ranging from 30 to 50 mEq/L. Children and elderly persons are more vulnerable to GI losses of potassium because their kidneys are not able to regulate fluid balance as effectively as those of other age groups. Movement of potassium from the ECF to the ICF is typically associated with conditions that cause alkalosis and intravenous insulin administration. Chapter 3 discusses alkalosis in detail.

Another factor that contributes to a potassium deficit is insufficient potassium intake. Because the body has no effective method of storing potassium, it must be ingested daily. Elderly patients with chronic illness and patients undergoing surgery are at a high risk for hypokalemia because of insufficient intake and/or loss during surgery. Malnutrition and rapid weight loss diets also cause potassium deficit. Table 4–2 lists the common causes of hypokalemia.

Assessment

Physical Assessment

Low potassium levels manifest as failure in myoneural conduction. In hypokalemia, as well as in hyperkalemia, many of the signs and symptoms originate in the nervous and muscular systems and are usually nonspecific. The early features of hypokalemia are fatigue and lack of strength. These manifestations result from intracellular potassium levels that are insufficient to maintain cellular metabolism.

Because the sodium-potassium pump is responsible for neuroexcitability, a lowered potassium level can lead to severe muscular weakness and paralysis, ventilation problems, bradycardia, atrial dysrhythmias, and kidney dysfunction. Muscle weakness proceeding to paralysis is usually not accompanied by paresthesia, as is seen in hyperkalemia. Tetany and loss of

TABLE 4–2

COMMON CAUSES OF HYPOKALEMIA

ACTUAL POTASSIUM DEFICITS

Excessive loss of potassium

- Inappropriate or excessive use of drugs
  - Diuretics
  - Digitalis
  - Corticosteroids
- Increased secretion of aldosterone
  - Cushing's syndrome
- Diarrhea
- Vomiting
- Wound drainage (especially gastrointestinal)
- Prolonged nasogastric suction
- Heat-induced excessive diaphoresis
- Renal disease impairing reabsorption of potassium

Inadequate potassium intake

- Nothing by mouth (NPO)

RELATIVE POTASSIUM DEFICITS

Movement of potassium from extracellular fluid to intracellular fluid

- Alkalosis
- Hyperinsulinism
- Hyperalimentation
- Total parenteral nutrition

Dilution of serum potassium

- Water intoxication
- IV therapy with potassium-poor solutions

Source: Ignatavicius, DD, Workman, MI, and Mishler, MA: Medical-Surgical Nursing: A Nursing Process Approach, ed 2. WB Saunders, Philadelphia, 1995. Reprinted with permission.

deep tendon reflexes may occur later. Some patients experience a depressed mental state. Because some of these manifestations are similar to those of potassium excess, a serum potassium level is essential to determine the status of the potassium balance. Neurologic and cardiovascular assessments are essential components of the physical assessment.

In hypokalemia, death is usually caused by the anoxia, which results from paralysis of the respiratory muscles and subsequent cardiac arrest. Therefore, frequent respiratory assessments are essential when providing care for the patient with a potassium deficit.

Diagnostic Assessment

A diagnosis of hypokalemia is confirmed when the serum potassium level is below 3.5 mEq/L. The serum potassium must be assessed along with the serum pH to determine if the deficit is actual (i.e., represents the state of the entire body) or relative as a result of potassium shifting into the cell. Arterial blood gas analysis indicates whether or not the patient has an acid-base imbalance, usually metabolic alkalosis.

Urinary potassium values are also important when appraising potassium balance. The usual amount of potassium in the urine should be about 40 to 90 mEq/L. Reference values that are acceptable for urinary potassium are elimination of 24 to 125 mEq in 24 hours. The amount of potassium excreted during a 24-hour period varies with the diet, serum aldosterone, or cortisol levels. The primary reason for using a 24-hour urine collection for potassium determination is to assess hormonal functioning, because a hormonal imbalance may also be present. This test also determines whether the potassium imbalance has a renal or nonrenal etiology. A 24-hour measurement is also important because potassium amounts vary by time of day because of the effect of circadian rhythms on cortisol secretion.

Another early diagnostic sign of hypokalemia is a change in the electrocardiogram (ECG). As the serum potassium drops, peaked P waves, flat T waves, depressed ST segments, and elevated U waves appear (Fig. 4–2). The U wave is an unusual wave that is a specific sign of hypokalemia. During hypokalemia the flat T wave may invert, a sign that is also correlated with ischemic states.

Nursing Diagnoses

The common nursing diagnoses for patients with potassium deficit include:

- Injury, risk for, related to skeletal muscle weakness
- Nutrition: altered, less than body requirements, related to inadequate potassium intake or increased potassium loss
- Cardiac Output, decreased, related to dysrhythmias
- Breathing Pattern, ineffective, risk for, related to respiratory muscle weakness
- Fatigue, related to skeletal muscle weakness

Collaborative Management

Medical Treatment

The management of conditions that could cause a potassium deficit is the primary objective in preventing hypokalemia. For example, the serum potassium level should be checked frequently when potent diuretics, particularly the high-ceiling and thiazide diuretics, are being administered. If the potassium level falls, potassium supplements may be given or the physi-

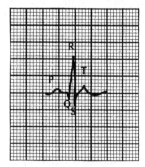

A. Normal

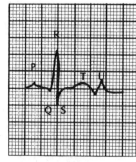

B. Hypokalemia

FIGURE 4–2
ECG tracings demonstrating changes that occur with hypokalemia.

cian may switch the patient to a potassium-sparing diuretic, such as spironolactone.

When a patient is allowed nothing by mouth (NPO) after surgery, a maintenance dose of potassium should be added to the intravenous (IV) fluid, plus additional amounts to compensate for loss by other means. At least 40 mEq in 1000 mL of IV fluid should be administered to compensate for a transient loss in which the intracellular need for potassium takes initial priority over the extracellular need.

If potassium loss cannot be prevented, then the potassium must be replaced. Potassium can be administered orally as a liquid, powder, or tablet. Most oral preparations of potassium contain potassium chloride (KCl). In many cases, the patient's chloride level is low when potassium is lost. Brands of KCl differ in dosage; therefore, one potassium preparation should not be substituted for another without a physician's directive.

Potassium chloride has an unpleasant taste and should be taken with juice or other desired liquid. Because potassium preparations can cause gastrointestinal irritation, they should not be taken on an empty stomach.

Small bowel lesions may occur because of the ingestion of potassium. Oral administration of potassium may need to be discontinued if any of the following symptoms occur: abdominal pain, distention, nausea, vomiting, diarrhea, or gastrointestinal bleeding.

Potassium can also be administered intravenously in acute conditions of deficit: for example, in NPO situations in which the vital daily intake requirement must be maintained. The usual replacement dosage is 60 to 120 mEq plus an additional quantity based on calculations for actual losses. Before IV administration, however, the patient should be assessed for sufficient renal output because the main outlet for potassium from the body is through the kidneys. Adequate kidney function must be determined before administering any potassium-containing fluids. If the kidneys are not working properly, a potassium excess could result.

Intravenous potassium is usually administered in a solution whose concentration is no greater than 40 mEq/L of fluid. A maximum concentration of 80 mEq/L can be given with caution. The recommended rate of infusion for adults is 5 to 10 mEq/h unless the patient is on a cardiac monitor. With severe serum potassium deficit (less than 2.5 mEq/L), 40 mEq/h have been given. These extremely high doses require administration through a central line to prevent phlebitis. The usual dose for children is 3 mEq/kg of body weight (Gahart, 1995). Rapid infusion can cause cardiac arrest, and therefore potassium is *never* given as an IV push medication.

A burning sensation at the site of administration indicates that the concentration is irritating to the veins and phlebitis can result. If phlebitis or infiltration occurs, the IV should be stopped immediately and restarted at another site. The physician may need to change the order for the hourly amount of potassium to be administered.

Nursing Interventions

Because the primary objective is prevention of hypokalemia, nursing interventions are directed at identifying patients at high risk for potassium deficiency, especially the elderly and surgical patients.

Elderly persons and others prone to laxative abuse need to be taught about the need to avoid excessive use of laxatives, which can deplete potassium.

Nursing care for any patient with hypokalemia, or at risk for hypokalemia, includes monitoring vital signs, especially pulse and respirations, and monitoring neuromuscular activity.

Another important nursing intervention is patient education regarding foods high in potassium and precautions that should be taken with potassium supplements. Patients should be instructed to report gastrointestinal (GI) or other side effects and to take oral potassium before meals. Table 4–3 summarizes the important points that should be stressed as part of patient education.

For patients receiving IV potassium, the nurse carefully checks the physician's order to ensure that the appropriate amount that can be safely administered is prescribed. In some acute care settings, "K runs" that deliver 40 mEq or more in one hour may be ordered. In an emergency situation, a maximum of 200 mEq of potassium is recommended per day (Gahart, 1995). Any patient receiving more than 10 mEq/h should be placed on a cardiac monitor, and the infusion should be controlled by an electronic infusion device. The nurse should always consult the pharmacist and hospital policy before administering these extremely high concentrations of potassium. A list of nursing interventions for patients receiving intravenous potassium replacement is found in Table 4–4.

Implications for Elderly Persons

Patients receiving digitalis preparations can quickly develop cardiac dysrhythmias when hypokalemia develops. Many elderly patients receive diuretics and digoxin for congestive heart failure, and therefore must be frequently monitored for changes in serum potassium levels. Elderly patients may not be able to tolerate higher doses of intravenous potassium because

TABLE 4–3
PATIENT EDUCATION TIPS FOR PATIENTS
RECEIVING ORAL POTASSIUM SUPPLEMENTS

- Do not substitute one potassium supplement for another.
- Dilute powders and liquids in juice or other desired liquid to improve taste and to prevent gastrointestinal irritation. Follow manufacturer's recommendations for the amount of fluid to use for dilution, most commonly 4 ounces per 20 mEq of potassium.
- Do not drink diluted solutions until mixed thoroughly.
- Do not crush potassium tablets, such as Slow-K or K-tab tablets. Read manufacturer's directions regarding which tablets can be crushed.
- Administer slow-release tablets with 8 ounces of water to help them dissolve.
- Do not take potassium supplements if taking potassium-sparing diuretics like spironolactone or triamterene.
- Do not use salt substitutes containing potassium unless prescribed by the physician.
- Take potassium supplements with meals.
- Report adverse effects, such as nausea, vomiting, diarrhea, and abdominal cramping to the physician.
- Have frequent laboratory testing for potassium levels as recommended by the physician.

TABLE 4–4

NURSING INTERVENTIONS FOR PATIENTS
RECEIVING INTRAVENOUS POTASSIUM REPLACEMENT

- Always dilute intravenous (IV) potassium; *never* give by the IV push method.
- Do not give more than 60–80 mEq/L in a peripheral vein. Elderly patients may not be able to tolerate over 40 mEq/L due to venous irritation or phlebitis.
- If potassium is added to an IV solution already hanging (not the preferred method), shake the IV bag vigorously and invert it to ensure that the potassium is evenly distributed throughout the solution.
- Label the bag with the correct amount of potassium added.
- Use an electronic infusion device for the administration of intravenous potassium.
- For large doses of potassium, use a central line rather than a peripheral one.
- Monitor the IV site and the surrounding tissue carefully for signs of irritation (redness, tenderness). Discontinue the solution if these signs are present and change the site.
- Place patients receiving more than 10 mEq/h on a cardiac monitor. Because authoritative sources and hospital policies vary regarding these recommendations, check the policy of the institution regarding monitoring and the rate of potassium infusions.
- Watch for signs and symptoms of hyperkalemia, especially in patients with impaired renal or adrenal insufficiency.
- Monitor laboratory values carefully.

their veins are fragile. Moreover, they typically have a reduced cardiac reserve because of the effects of aging. The nurse should, therefore, carefully observe the administration site for irritation and monitor cardiac rate and rhythm.

Potassium Excess

Hyperkalemia is a serum potassium level of over 5.3 mEq/L. It is not as common as potassium deficit.

Causative Factors

The most common health problem that causes hyperkalemia is renal disease in which potassium cannot be excreted adequately. It is almost impossible for hyperkalemia to develop when the patient has an adequate kidney function. Other common causes are excessive oral or parenteral administration of potassium to correct a deficit, excessive use of potassium-based salt substitutes, Addison's disease, and crushing injuries in which there is significant muscle damage. As discussed earlier, tissue destruction causes loss of potassium from the intracellular fluid (ICF), with an accompanying shift of potassium into the extracellular fluid (ECF). Metabolic acidosis also promotes movement of potassium from the ICF to the ECF.

Assessment

Physical Assessment

Potassium levels are related to neuromuscular irritability, and the primary signs and symptoms of potassium excess involve the nervous and muscular systems. The signs or symptoms are usually nonspecific. The most common sign is vague muscle weakness, which can proceed to paralysis. The muscle weakness may first be noted in the legs, then progress to the trunk and arms. Involvement of facial and respiratory muscles is a late symptom. The paralysis is usually flaccid and is often accompanied by paresthesia of face, tongue, or extremities, followed by anesthesia. Muscles innervated by the 12 cranial nerves are usually spared. The muscles can be tested for irritability by percussion or palpitation. The patient usually remains alert, and consciousness persists until cardiac arrest occurs.

There is no direct correlation between the degree of neuromuscular signs and the serum potassium concentration level. In many cases, clinical signs and symptoms are absent, so that recognition must often be based on combinations of nonspecific signs and symptoms and laboratory analysis.

Death in hyperkalemia usually results when the toxic state causes cardiac dysrhythmia (ventricular fibrillation or atrial standstill).

Diagnostic Assessment

A serum potassium level of over 5.3 mEq/L confirms a diagnosis of hyperkalemia. Arterial blood gas analysis determines whether an acid-base imbalance is present. In the case of hyperkalemia, this is usually metabolic acidosis.

Early recognition of hyperkalemia is possible if the patient is on a cardiac monitor, because the cardiac tracings will show very definite and specific changes. Hyperkalemia is denoted by a wide, flat P wave. In addition, the slowing of processes associated with ventricular depolarization results in a QRS complex widened beyond the standard of 0.8 to 1.2 on lead II. Electrocardiogram (ECG) changes include tented T waves, elevated T waves, widened QRS complexes, prolonged P–R intervals, flattened or absent P waves, and ST segment depression (Fig. 4–3). Heart block and dysrhythmias may occur at an extracellular level of 8 to 10 mEq/L. Elevation of T waves in an ECG may be detected at approximately 7 mEq/L.

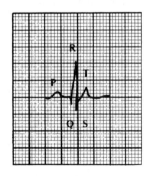

A. Normal

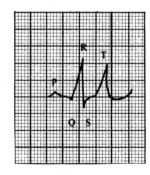

B. Hyperkalemia

FIGURE 4–3
ECG tracings demonstrating changes that occur with hyperkalemia.

Nursing Diagnoses

Clinical manifestations are very nonspecific with hyperkalemia. Common nursing diagnoses for patients include:

- Cardiac Output, decreased, risk for, related to dysrhythmias
- Nutrition: altered, more than body requirements, related to decreased renal function or excessive potassium intake
- Sensory-perceptual alterations (tactile and kinesthetic), related to muscle weakness and paresthesias
- Diarrhea, related to smooth muscle irritability
- Activity Intolerance, related to muscle weakness

Collaborative Management

Medical Treatment

The treatment of hyperkalemia varies according to the severity of the problem. Emergency therapy can include rapid intravenous administration of 250 mL of 10 to 20% dextrose with 10 to 20 units of regular insulin. This hypertonic solution should be given through a central venous catheter. The purpose of this treatment is to move the excess potassium into the cells, thus reducing the toxic state in the serum. Insulin facilitates the movement of glucose, potassium, and phosphate into the cell, and glucose is given to prevent the hypoglycemia that would result from insulin treatment.

Once the potassium is in the cell, it is no longer toxic. This is a temporary solution in an emergency situation and will control an acute condition until the underlying cause of the hyperkalemia can be corrected or its effect diminished. Infusion should not be stopped suddenly because of the stimulation of the secretion of endogenous insulin.

Another option for emergency or chronic treatment is hypertonic Kayexalate (sodium polystyrene sulfonate), which may be given orally or as an enema to be retained for a period of time. Kayexalate is an ion exchange resin that absorbs the potassium into the gastrointestinal tract. The patient needs to be observed for calcium and magnesium loss when using an exchange resin. When Kayexalate is used as an enema, the potassium is evacuated along with the solution. A bulb catheter may be used to hold the solution in the body following instillation.

Because a high potassium level depresses the action of the heart, intravenous administration of calcium may also be ordered in emergencies to stimulate the heart. Calcium should not be administered if the patient is taking a digitalis preparation because the combined treatments may result in cardiac dysrhythmias. Examples of calcium preparations are calcium gluconate and calcium gluceptate.

Peritoneal and hemodialysis may be the treatment of choice for acute conditions (see Chaps. 16 and 17). By different methods, both procedures filter the potassium and waste products from the extracellular fluid.

Because a sodium deficiency may also cause hyperkalemia, increasing the sodium increases the loss of potassium through the kidneys, thereby eliminating the excess.

If necrotic tissue is present, debridement may be necessary to remove injured and dead cells that are releasing potassium into the vascular system. Intramuscular injection of testosterone propionate may help prevent excess protein breakdown that accompanies tissue destruction.

Excessive or too-rapid administration of potassium may also cause hyperkalemia; in such a case, administration must be discontinued at once. Potassium should never be given in a direct undiluted injection. The average dose is 40 mEq/L of intravenous fluid, not to be administered faster than 10 mEq/h unless specifically ordered. The patient should be placed on a cardiac monitor if the dosage or rate of administration exceeds these averages.

Kidney function should always be checked before potassium is ingested or administered. Potassium should not be given to any patient with oliguria or anuria and should not be administered to patients with acidosis who show a potassium deficit. Once the severity of the acidosis is decreased and urinary output is adequate, then potassium can be administered.

When blood transfusions are ordered for a hyperkalemic patient, the patient should receive fresh blood if possible. Transfusions of stored blood may elevate the potassium level because the breakdown of older blood cells releases potassium.

When hyperkalemia and acidosis exist concurrently, the lactate or bicarbonate used to correct the acidosis will also drive the potassium back into the cells.

Nursing Interventions

Nursing care is directed to both preventive and episodic care. The nurse should:

- Be alert to conditions that can lead to hyperkalemia
- Take preventive measures whenever possible
- Observe for pulse changes and/or ECG changes, especially bradycardia
- Observe for signs and symptoms of potassium excess
- Check the serum potassium reports; notify the physician when potassium is above 5.3 mEq/L
- Give fresh blood transfusions if possible
- Avoid giving calcium if the patient is also taking a digitalis preparation
- Teach the patient to avoid foods high in potassium, including organ meats, fish, fruits, red meat, pork, and poultry
- Teach the patient to avoid the use of salt substitutes or other potassium-containing substances

CRITICAL THINKING ACTIVITIES

A 20-year-old woman was admitted to the hospital following a car-train collision. She appeared to have internal injuries and was thought to be hemorrhaging. She remained hypotensive throughout emergency surgery for removal of a ruptured spleen. Following surgery and blood transfusions, her condition remained stable for 4 days. Urine output then began decreasing, and BUN, creatinine, and serum potassium levels showed elevation.

QUESTIONS

1. *Discuss the relationship between decreased urinary output and elevation of BUN and serum potassium levels.*

2. *The urinary output continued to decrease, and the patient became oliguric. It was determined that she had developed an acute renal failure and should be placed on hemodialysis (artificial kidney) for 6-hour periods. This procedure was tolerated well, but the elevation of the serum potassium level continued to be a major problem between dialyses. What other options are available for the treatment of hyperkalemia?*

3. *The physician ordered 300 mL sodium lactate and 50 mL of 5%/D/W with 10 units of insulin to be given intravenously immediately. What is the action of this medication?*

4. *What would you expect to find in this patient's ECG pattern as a result of hyperkalemia?*

REFERENCES AND READINGS

Booker, MF, & Ignatavicius, DD (1996). Infusion Therapy: Techniques and Medications. Philadelphia: WB Saunders.

DeAngelis, R, & Lessig, ML (1991). Hypokalemia. Critical Care Nurse 11 (7), 71–72, 74–75.

Dennison, R, & Blevins, B (1992). Myths and facts about electrolyte imbalance. Nursing92 22(2), 26.

Gahart, BL (1995). Intravenous Medications (11th ed). St. Louis: Mosby-Year Book.

Guyton, AC (1991). Textbook of Medical Physiology (8th ed). Philadelphia: WB Saunders.

Hawthorne, J, Schneider, S, & Workman, ML (1992). Common electrolyte imbalances associated with malignancy. AACN Clinical Issues in Critical Care Nursing 3(3), 714–723.

Hutchinson, R, Barksdale, B, & Watson, R (1992). The effects of exercise on serum potassium levels. Chest 101(2), 398–400.

Ignatavicius, DD, Workman, ML, & Mishler, MA (1995). Medical-Surgical Nursing: A Nursing Process Approach ed 2. Philadelphia: WB Saunders.

Innerarity, S (1992). Hyperkalemic emergencies. Critical Care Nursing Quarterly 14(4), 32–39.

Kokko, JP, & Tannen, RL (1990). Fluid and Electrolytes (2nd ed). Philadelphia: WB Saunders.

National Research Council (1989). Diet and Health: Implications for Reducing Chronic Disease Risk. Washington, DC: National Academy Press.

National Research Council (1989). Recommended Dietary Allowances. Washington, DC: National Academy Press.

Terry, J (1994). The major electrolytes: Sodium, potassium, and chloride. Journal of Intravenous Nursing 17(5), 240–247.

Vallerand, A, & Deglin, J (1991). Nurse's Guide for IV Medications. Philadelphia: FA Davis.

CHAPTER 5

· · · · · · · · ·
· · · · · · · · ·
· · · · · · · · ·
· · · · · · · · ·
· · · · · · · · ·
· · · · · · · · ·
· · · · · · · · ·
· · · · · · · · ·
· · · · · · · · ·

Sodium

SODIUM BALANCE

Sodium exists in both the extracellular fluid (ECF) and the intracellular fluid (ICF), with about 99% in the ECF and the remaining 1% in the ICF. Thus, sodium is the major cation and mineral of the extracellular fluid compartment. The combination of sodium and chloride in water constitutes a saline solution.

Sodium and chloride levels in the body may deviate from normal independently, or in combination with each other. A sodium imbalance can result from a loss or gain of sodium or a change in water volume. A change in the combination of the electrolytes (sodium and chloride) also causes a change in the volume of the extracellular fluid because both tend to hold water.

Three factors can create a sodium imbalance. The first factor is a change in the sodium content of the extracellular fluid, such as a deficit caused by excessive vomiting or an excess caused by a failure to excrete sodium. The second factor is a change in the chloride content, which can affect both the sodium concentration and the amount of water in the extracellular fluid. When the ratio of chloride to sodium deviates from normal, it is reflected as an acid-base imbalance. The third factor is a change in the quantity of water in the extracellular fluid. Because only one-third of the total body fluid is extracellular, a serum sodium change indicates the major problem is a total body water imbalance.

Distribution

Sodium comprises about 90% of the total cation content of the body. The normal serum concentration is approximately 135 to 145 mEq/L, or 135 to 145 mmol/L in extracellular fluid. The average man has 52 to 60 mEq/kg of sodium in comparison to the average woman, whose body contains 48 to 55 mEq/kg. Each milliequivalent of sodium is equal to 23 mg. About 35 to 40% of the total body sodium is in the skeleton, and this skeletal sodium is not freely exchanged with serum sodium. Thus skeletal sodium, unlike sodium in the serum, does not function as an electrolyte. In this it differs from skeletal calcium, which can be mobilized at any time and thus functions as an electrolyte.

Implications for Pregnant Women

Pregnant women retain both sodium and water. About 300 to 400 mEq of the retained sodium are stored in the extracellular fluid compartment. The remaining 600 to 700 mEq are stored in the fetus, placenta, and amniotic fluid (Kokko & Tannen, 1995). Sodium retention and the resulting water retention during pregnancy account for the majority of the mother's weight gain.

Sources

The average adult consumes approximately 5 to 6 g of sodium per day even though the minimum daily requirement is approximately 2 g. The most common source of sodium is sodium chloride (salt). One teaspoon of salt per day meets the minimum daily dietary requirement of sodium.

Sodium is found naturally in some foods and in local water supplies, and salt is usually added during the cooking process for food prepared at home; however, the National Research Council indicates that most of the average sodium consumption results from salt added during the manufacturing of processed foods (Table 5-1).

Other common food sources of sodium include bacon, corned beef, bouillon cubes, bread stuffing mixes, pasteurized processed American cheese and cheese spread, corn flakes, canned crab, crackers, processed oat cereals, green olives, pickles, pretzels, most processed salad dressings,

TABLE 5-1
COMPARISON OF SODIUM CONTENT
IN FRESH AND PROCESSED FOODS

Fresh Food	Sodium (mg)	Processed Food	Sodium (mg)
Natural Swiss cheese, 1 ounce	74	Pasteurized, processed Swiss cheese, 1 ounce	388
Lean roast pork, 3 ounces	65	Lean ham, 3 ounces	930
Whole raw carrot, 1	25	Canned carrots, ½ cup	176
Tomato juice, canned without salt, 1 cup	24	Tomato juice, canned with salt, 1 cup	881

Source: Lutz, CA, & Przytulski, KR: Nutritional and Diet Therapy. FA Davis, Philadelphia, 1994.

frankfurters, lunch meat, soy sauce, and catsup. All of these items have over 1000 mg of sodium per 100 g of edible food. Some items on this list have over 7000 mg of sodium per 100 g of edible food.

Recommended Daily Allowance

Estimates for the recommended allowance are based on the amount considered to be needed for growth and replacement of losses. A minimum average requirement is considered to be 115 mg of sodium or approximately 300 mg of sodium chloride per day. This would be the average for healthy adults without active sweating. Recognizing that climate and physical activity could increase this minimum, a safe minimum is considered to be 500 mg of sodium chloride per day (Lutz & Przytulski, 1994).

Control

Sodium is absorbed actively by the intestines and excreted through the kidneys and skin. The regulation of salt intake and output is primarily controlled by the need for extracellular fluid (ECF) volume rather than by osmolality or osmolarity. Thus, craving for salt is not a reliable guide to sodium levels. Thirst is indicative of osmolarity increase and water loss.

The amount of sodium in the ECF is controlled by the release of aldosterone through a feedback mechanism that functions in the following way: aldosterone is secreted by the adrenal cortex and causes the renal distal tubules to reabsorb sodium. When sodium is reabsorbed, the sodium content in the extracellular fluid is increased. The increased concentration of sodium in the extracellular fluid, in turn, causes a decrease in the amount of aldosterone secreted. Because the feedback mechanism is a two-way relationship, a decrease in the amount of aldosterone secretion results in an increase in the amount of sodium lost via the urine (Fig. 5–1).

A low serum sodium level is just one of the factors that triggers the renal tubules to absorb sodium. As the serum sodium concentration rises, the secretion of aldosterone decreases. Aldosterone secretion is influenced not only by sodium levels but also by stress mechanisms, blood volume baroreceptors, and potassium levels. Thus, even though the serum sodium level may not be reduced, stress and other factors can trigger the retention of sodium and the loss of potassium.

Functions

The major functions of sodium are to work in combination with chloride to maintain blood volume, to regulate charging potential in neuromuscular function, and to influence acid-base balance.

Because sodium does not easily penetrate the cell membrane, it is a primary regulator in the relationship of body fluids. Both sodium and chloride are important factors in the body's ability to retain water. When sodium is reabsorbed by the kidneys, chloride is usually reabsorbed with it. The effect of their combined reabsorption is to increase the amount of water retained by the body. Changes in the body fluid level cause corresponding changes in the extracellular fluid volume. Thus, the amount of extracellular fluid, both

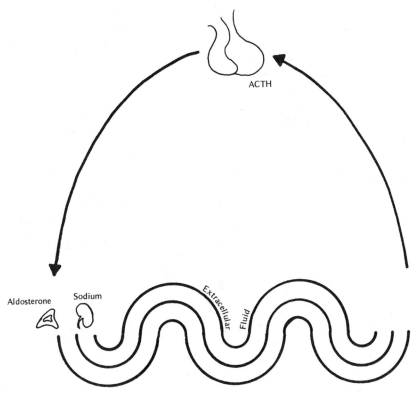

FIGURE 5–1
Negative feedback mechanism. Decreased sodium in the ECF results in the release of ACTH from the anterior pituitary gland, which causes an increase in aldosterone levels. Aldosterone in turn causes the retention of sodium.

blood volume and interstitial fluid volume, increases as the amount of water increases. Based on the process of its reabsorption in the kidneys, sodium is able to control blood volume and interstitial fluid volume, regulate the shift of water between compartments, and influence the excretion of water.

Sodium also helps regulate acid-base balance, because it readily combines with chloride and bicarbonate and can, therefore, help maintain the balance between cations and anions. This process is very important in determination of the anion gaps in pH imbalance.

Sodium, in conjunction with potassium, is responsible for maintaining the normal balance of electrolytes in intracellular and extracellular fluids by means of an active transport mechanism called the sodium-potassium pump. The sodium-potassium pump also influences the irritability of nerves and muscles. Sodium is, therefore, an important factor in nerve conduction. It aids in establishing the potential of nerve membranes to excite charges.

A nerve membrane in a relaxed or resting stage has a positive charge outside and a negative charge inside (Fig. 5–2). When the membrane becomes more permeable, a small number of potassium ions move out of the cell, and simultaneously, a large number of sodium ions move into the cell. The membrane is now depolarized; the outside of that portion of the membrane is now negatively charged. The depolarization is propagated along

Nerve cell

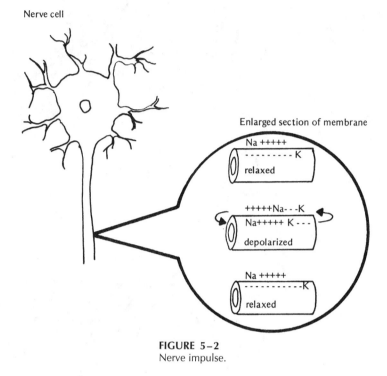

Enlarged section of membrane

FIGURE 5-2
Nerve impulse.

the membrane of the nerve cell. This depolarization wave is called the nerve impulse. Repolarization immediately follows depolarization, thus reestablishing the resting stage. This electrolyte change forces the nerve to conduct electrical impulses to the muscle and causes the muscle to contract. Sodium, therefore, influences the irritability of nerves, muscles, and the heart. Any disturbance in the sodium balance can disturb neuromuscular function.

Even though less than 1% of the body's sodium exists inside the cell, it is an important factor in numerous vital cellular chemical reactions.

The osmolality of the blood is also significant when discussing sodium imbalances. Osmolality measures the tendency of water to move across a membrane that separates solutions with different concentrations of the same solute. A solution with a higher concentration has a higher osmolality, that is, water will tend to move into the compartment with higher concentration. Osmolality is measured both in the serum and urine. Deviations from the normal osmolality of the blood, known as hyperosmolality and hyposmolality, also represent increases or decreases in the sodium concentration in the blood.

SODIUM IMBALANCES

There are two types of sodium electrolyte imbalances: hyponatremia and hypernatremia. Both of these imbalances reflect changes in fluid balance. Although sodium disorders and fluid disorders are two separate entities, they must be assessed together insofar as they relate to sodium imbalance.

Sodium Deficit

Hyponatremia, sodium deficit, is a condition in which there is too little sodium in the serum, less than 135 mEq/L. Although hyponatremia can occur in a normovolemic, hypovolemic, or hypervolemic state, it is generally synonymous with hypervolemia in which there is an excess amount of water. This situation is sometimes referred to as dilutional hyponatremia. The actual amount of sodium in the extracellular fluid compartment has not changed, but the fluid volume has increased. The reduced serum sodium reading may be accompanied by a reduced chloride reading. Chapter 2 describes fluid excess in detail.

Causative Factors

Sodium losses may be actual or relative. *Actual* losses can be caused by gastrointestinal, renal, and third space losses. Skin losses are most commonly due to profuse perspiration, draining skin lesions, and fibrocystic diseases of the pancreas (cystic fibrosis), which affect primarily infants, children, adolescents, and young adults. Prolonged use of thiazide and loop (high-ceiling) diuretics causes sodium loss as well as potassium loss.

Sodium deficit may also be related to a lack of sodium intake, low-salt diets, and excessive loss with trauma. Loss of sodium is usually associated with decreased circulating plasma proteins (hypoproteinemia) and hyperlipidemia.

Bile is very rich in sodium. Therefore, bile loss caused by drainage, fistulas, excessive vomiting as in bulimia, or gastrointestinal (GI) surgery can cause sodium deficiencies. Prolonged nasogastric suction can also cause sodium loss.

Relative sodium loss occurs in the presence of a fluid overload. Iatrogenically induced water excess is usually related to excessive oral intake of water without adequate excretion. Another iatrogenic cause is administration of electrolyte-free solutions, especially when these are administered in conjunction with documentable electrolyte losses such as those that occur with sweating, GI losses, renal losses, and whole blood losses. Irrigation of body cavities with hypotonic solutions such as sterile water can also lead to hyponatremia.

The syndrome of inappropriate antidiuretic hormone secretion (SIADH) is an endocrine disorder in which the antidiuretic hormone (ADH) normally secreted by the posterior pituitary gland is released unexpectedly. This disorder is often caused by malignancies, such as oat cell carcinoma of the lung and leukemia; central nervous system problems, such as brain tumors or head injury; and drugs, such as cyclophosphamide (Cytoxan), vincristine (Oncovin), and first-generation sulfonylureas, such as chlorpropamide (Diabinese) and tolbutamide (Orinase). Chapter 20 discusses SIADH in more detail.

Hyponatremia is also a common electrolyte imbalance in postoperative patients. During surgery, the individual loses blood and other fluids that are rich in sodium. A complete list of causative factors is found in Table 5–2.

TABLE 5–2
CAUSATIVE FACTORS IN HYPONATREMIA

Thiazide and loop diuretics (especially when used in the elderly)
Profuse sweating
Draining skin lesions and fistulas
Cystic fibrosis
Trauma
Excessive bile loss
Vomiting
Gastrointestinal suction
Hepatic failure (late)
Water intoxication
Renal disease
Syndrome of inappropriate antidiuretic hormone secretion (SIADH)
Surgery
Inadequate oral intake
Use of hypotonic or electrolyte-free oral or IV solutions
Low birth weight (infants)
Use of oxytocin for labor and delivery

Implications for Pregnant Women

Severe hyponatremia can occur in pregnant women who are treated with oxytocin during labor and delivery. Oxytocin stimulates the uterus to contract, but can cause severe water retention with resulting dilutional hyponatremia. This complication of oxytocin therapy can be prevented by administering the drug in an intravenous electrolyte solution to compensate for the water retention. A continuous infusion pump should be used to control the rate of drug administration.

Assessment

Physical Assessment

Patients with hyponatremia may not manifest any signs or symptoms if fluid volume is unchanged and the serum sodium is between 125 and 134 mEq/L. If both sodium and water are lost, however, the signs and symptoms are the same as those for an extracellular fluid deficit. These assessment findings include postural blood pressure changes, poor skin turgor, flat hand and neck veins, nausea, vomiting, and anorexia. As the deficit increases, oliguria can occur. A more pronounced or severe deficit can cause vasomotor collapse with symptoms such as hypotension, rapid thready pulse, cold clammy skin, and cyanosis. In the early stages of severe deficit, the patient's skin is flushed, but as the patient reaches the shock stage, it becomes cold and clammy.

In a dilutional hyponatremic state, the signs and symptoms are similar to those of fluid excess. Early signs are headache, faintness, and giddiness, followed by muscle cramps. Further signs and symptoms involve the central nervous system; these include mental confusion, muscle twitching, coma, and convulsions. Blurred vision may also be present. Body weight and blood pressure may stay the same or they may increase. Urinary output may decrease, leading to oliguria. The client must be assessed for symptoms related to combined disorders of sodium deficit and water imbalances.

Diagnostic Assessment

Hyponatremia is confirmed when the serum sodium level is below 135 mEq/L. Urinary sodium can also be measured. A urinary sodium level of less than 15 mEq/L suggests that the kidneys are conserving sodium and that loss of the electrolyte is from a nonrenal source. Patients with SIADH typically have a urinary sodium concentration greater than 20 mEq/L.

When the specific gravity of the urine is decreased (below 1.010), the actual sodium concentration in the extracellular fluid is usually decreased. This test is significant because when sodium is decreased in the body, the kidneys conserve sodium and, therefore, less sodium is present in the urine. In both hypernatremia and hyponatremia with a fluid deficit, the red blood cell count is usually elevated because of hemoconcentration.

Nursing Diagnoses

The common nursing diagnoses associated with hyponatremia include:
- Fluid Volume deficit, related to sodium imbalance
- Fluid Volume excess, related to sodium imbalance
- Injury, risk for, related to seizures or confusion
- Skin Integrity, impaired, potential, related to dry skin and poor skin turgor
- Cardiac Output, decreased, risk for, related to hypovolemia

Collaborative Management

Medical Treatment

Preventing hyponatremia in high-risk patients is a priority because death can result from severe hyponatremia, especially in elderly persons. The goal of treatment for an actual or relative sodium deficit is to treat the cause and replace the lost sodium.

In determining individualized treatment, it is necessary to establish whether the sodium change is an electrolyte problem only or a combined electrolyte and fluid problem. A situation in which both the fluid and sodium state decrease together is called isotonic dehydration. Isotonic dehydration may be treated by administering either saline (at the isotonic level, 0.9%) or lactated Ringer's solution.

If the fluid volume is normal or excessive, it may be necessary to administer a hypertonic solution such as 3% saline. In SIADH, in which water is retained, the treatment is fluid restriction rather than sodium replacement. Diuretics that promote water excretion rather than sodium loss, such as osmotic diuretics (e.g., mannitol), may also be used.

Nursing Interventions

The nurse assesses patients for the risk of hyponatremia. Unless otherwise contraindicated, foods high in sodium should be encouraged for patients with mild hyponatremia. For patients on a liquid diet, broth and tomato juice are nutrients high in sodium. Laboratory values must be monitored for indications of further fluid and electrolyte imbalance.

Assessment of fluid balance by measuring intake and output, monitoring weights, and checking skin turgor and mucous membranes is critical. If patients lose fluids that are rich in sodium, such as gastrointestinal (GI) juices and sweat, intravenous sodium replacement will most likely be required. Irrigations of the GI and urinary tracts should be done with normal (isotonic) saline rather than sterile water. Isotonic ice chips are recommended for patients on nasogastric suction.

In patients taking lithium, the nurse monitors drug levels very carefully. Hyponatremia can cause diminished lithium excretion and result in toxicity. To avoid this problem, patients on lithium should not take diuretics and should be carefully monitored if they experience GI losses (vomiting and diarrhea) or have profuse sweating, such as occurs with fever.

Patients with severe sodium deficit may need an IV infusion of hypertonic 3 or 5% saline. These solutions must be infused slowly using an electronic infusion device. Table 5–3 summarizes the nursing interventions associated with the administration of IV hypertonic saline solutions.

Sodium Excess

The condition in which there is too much sodium in the serum is called hypernatremia. It is generally synonymous with a hyperosmolar state in which there is a fluid deficit. Hypernatremia is defined as the condition in which serum sodium concentration greater than 145 mEq/L. Hypernatremia is actually a reflection of the intracellular fluid (ICF) state in the body because increased serum sodium draws water from the cells and causes cellular dehydration.

Causative Factors

The usual cause of water deficit is inadequate intake or excessive loss of water. Inadequate fluid intake may be related to inability to obtain adequate intake or diminished thirst response, especially in aging adults or infants not responsive to minor physiologic changes. Also, total enteral nutrition (tube feedings), total parenteral nutrition (TPN), and other high solute loads may deplete the cells of water.

TABLE 5–3
NURSING INTERVENTIONS ASSOCIATED
WITH HYPERTONIC SALINE INFUSIONS

- Recognize that hypertonic solutions are usually 3% saline (30 g NaCl or 513 mEq each of Na and Cl; 1030 mOsm/L) or 5% saline (50 g NaCl or 855 mEq each of Na and Cl; 1710 mOsm/L).
- Infuse solution until the serum sodium level reaches 130 mEq/L.
- Give one half of the calculated dose over at least 8 hours.
- Do not exceed a rate of 100 mL/h.
- Assess for local pain and venous irritation; slow rate if needed for comfort.
- Use with extreme caution in elderly patients with renal or cardiac insufficiency.
- Give solution only when sodium is severely decreased.
- Hypertonic saline is incompatible with mannitol, amphotericin B (Fungizone), and norepinephrine (Levophed).
- Assess carefully for signs and symptoms of hypernatremia.
- Monitor laboratory test values for sodium and chloride carefully.

Actual retention of sodium may be caused by Cushing's syndrome or hyperaldosteronism, in which excessive secretion of steroids results in the retention of sodium and the loss of potassium. Individuals taking corticosteroids for a prolonged period may experience the same sodium retention effect. Patients with severe renal failure also experience hypernatremia because the kidneys cannot excrete sodium.

Iatrogenic causes are generally related to excessive administration of substances such as sodium bicarbonate during cardiac arrest, or the use of commercial feeding formulas in infants. A hypernatremic state may also be caused by administering sodium chloride intravenously without replacement of water.

Assessment

Physical Assessment

Signs and symptoms of sodium excess result from changes in excitable cell membrane activity, especially the neuromuscular and cardiovascular tissues, and from the decrease in fluid volume. Changes in personality, agitation, and confusion are early assessment findings. Later, seizures, a decreased level of consciousness, or death can occur. Early muscular changes include twitching and irregular muscle contractions. As hypernatremia progresses, skeletal muscle weakness may be present. Decreased myocardial contractility with resulting diminished cardiac output may lead to heart failure. Increased serum sodium concentration prevents movement of calcium into the myocardium.

In patients with accompanying hypovolemia, signs and symptoms include dry, sticky mucous membranes, flushed skin, intense thirst, and oliguria. Temperature may become elevated and tachycardia may occur. Death is usually caused by an excessive rise of osmotic pressure and respiratory arrest.

Diagnostic Assessment

A serum sodium level above 145 mEq/L confirms hypernatremia. The increased sodium concentration means that osmolality is also increased, and serum osmolality is usually over 300 mOsm/kg. The specific gravity of the urine is above 1.030 because the kidneys attempt to maintain homeostasis by conserving water.

The osmolality of the urine can also be checked and is a more accurate measurement of kidney function than specific gravity. The control mechanisms of the body respond to changes in the serum, not to changes in characteristics of urine, such as specific gravity. Specific gravity does not accurately reflect the concentration of urine because the weight of substances varies. For example, heavier substances, such as glucose and albumin, cause a rise in specific gravity out of proportion to their concentration in the urine.

Urine-serum osmolality ratios can also be computed. A normal ratio is greater than 1, usually 3 or higher. In chronic renal disease, this falls to 1 or below. In obsessive-compulsive water drinking (a form of water intoxication), the ratio rises to 2 or 3. In hypernatremia the urinary sodium level is decreased; the red blood count and the hematocrit as well as the serum protein count are elevated.

Nursing Diagnoses

The common nursing diagnoses associated with hypernatremia include:

- Fluid Volume deficit, related to actual or relative sodium excess
- Injury, risk for, related to neuromuscular effects of excess sodium, confusion, and seizures
- Cardiac Output, decreased, related to decreased contractility

Collaborative Management

Medical Treatment

The usual treatment for hypernatremia is to increase water intake either orally or intravenously with salt-free solutions. The serum sodium is monitored until it returns to normal.

Nursing Interventions

Nursing interventions are directed towards ongoing nursing assessment of patients during the treatment phase of the hypernatremia. Laboratory test results are monitored to ensure that overly aggressive treatment of the condition does not result in a sodium deficit.

To prevent hypernatremia in patients receiving tube feedings, additional water must be given. Monitoring of intake and output is especially important for the elderly, for infants, and for children.

CRITICAL THINKING ACTIVITIES

A 21-year-old woman went to the physician's office complaining of urinary frequency and burning on urination. Examination revealed a healthy woman with no other apparent health problems. Vital signs were as follows: temperature 98.4°F; pulse 78; respirations 20; blood pressure 120/64. Laboratory tests yielded the following results: CBC, normal; urinalysis, cloudy, packed RBCs and WBCs.

Her physician instructed her to take the prescribed medication and drink "lots of fluids" for her cystitis. Approximately 12 hours later, she was admitted by ambulance to the emergency room with convulsions. Her husband gave the following account: "She came home from the doctor's office, took the prescribed medication, and started drinking lots of water. She continued drinking water throughout the afternoon and going to the bathroom frequently." In retrospect, he remembered that "by early evening she was still drinking lots of water but not going to the bathroom any more. Around 10 PM she appeared tired and somewhat confused. About 10:30 PM she began convulsing and an ambulance was called."

QUESTIONS

1. *What electrolyte imbalance does the patient probably have as a result of drinking excessive amounts of water?*

2. *What would the laboratory values for sodium, chloride, and potassium probably be for a patient with this imbalance?*

3. *Why did the patient convulse? (Specify the pathophysiology.)*

4. *Because the patient has a severe electrolyte imbalance, what treatment will she likely receive and why?*

REFERENCES AND READINGS

Ayus, JC, & Arieff, AI (1993). Pathogenesis and prevention of hyponatremic encephalopathy. Endocrinology and Metabolism Clinics of North America, 22(2), 425–446.

Bryce, J (1994). S.I.A.D.H. Nursing94 24(4), 33.

Chernecky, CC, Krech, RL, & Berger, BJ (1993). Laboratory Tests and Diagnostic Procedures. Philadelphia: WB Saunders.

Gahart, BL (1995). Intravenous Medications. St. Louis: Mosby–Year Book.

Kelso, LA (1992). Fluid and electrolyte disturbances in hepatic failure. AACN Clinical Issues in Critical Care Nursing 3(3), 681–685.

Kokko, JP, & Tannen, RL (1995). Fluids and Electrolytes. Philadelphia: WB Saunders.

Lee, CA (1992). Organic mental disorders. Lippincott's Review Series: Mental Health and Psychiatric Nursing. Philadelphia: JB Lippincott.

Lutz, CA, & Przytulski, KR (1994). Nutrition and Diet Therapy. Philadelphia: FA Davis.

Mendyka, BE (1992). Fluid and electrolyte disorders caused by diuretic therapy. AACN Clinical Issues in Critical Care Nursing 3(3), 672–680.

Norris, MK (1992). Evaluating sodium levels. Nursing92 22(7), 20.

Statter, MB (1992). Fluids and electrolytes in infants and children. Seminars in Pediatric Surgery 1(3), 208–211.

Theunissen, IM, & Parer, JT (1994). Fluids and electrolytes in pregnancy. Clinical Obstetrics and Gynecology 37(1), 3–15.

Vieweg, VR (1994). Treatment strategies in the polydipsia-hyponatremia syndrome. Journal of Clinical Psychiatry 55(4), 154–160.

CHAPTER 6

. . . .
. . . .
. . . .
. . . .
. . . .
. . . .
. . . .
. . . .
. . . .
. . . .

Chloride

CHLORIDE BALANCE

Chloride is the major inorganic, negative ion (anion) of the extracellular fluid (ECF). As was mentioned in Chapter 5 when discussing sodium, chloride levels may vary independently of or in relation to sodium and water balances. The sodium and chloride in water make up the saline solution of the ECF compartment and are, therefore, key components determining osmotic pressure. Both sodium and chloride tend to hold water.

Like the serum sodium reading, the serum chloride reading generally reflects a change in dilution or concentration of the ECF (Fig. 6–1). There is also an important interaction between sodium, chloride, and bicarbonate. The normal relationship between the three electrolytes is given by the equation

$$[Na^+] = [HCO_3^-] + [Cl^-] + 12 \text{ mEq/L}$$

where $[Na^+]$, $[HCO_3^-]$, and $[Cl^-]$ are the concentrations in milliequivalents per liter for sodium, bicarbonate, and chloride ions, respectively. For example, if the concentrations for sodium and chloride are 140 and 101 mEq/L, respectively, the concentration for bicarbonate will be 27 mEq/L.

$$[Na^+] = [HCO_3^-] + [Cl^-] + 12 \text{ mEq/L}$$

$$140 \text{ mEq/L} = 27 \text{ mEq/L} + 101 \text{ mEq/L} + 12 \text{ mEq/L}$$

In general, then, a disturbance in any one of the three may affect the other two.

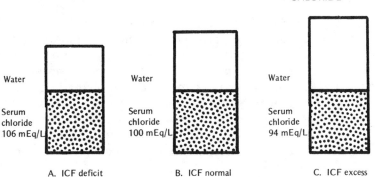

FIGURE 6-1.
Chloride excess-deficit. Actual chloride content remains the same, only the chloride reading varies as the ICF level varies.

Distribution

The normal amount of chloride in the blood ranges from 97 to 109 mEq (or 97 to 109 mmol) per liter. Chloride is found in greater proportion in the interstitial and lymph fluid compartments than in blood. Transcellular fluids such as gastric juice, sweat, and pancreatic juice also contain chloride ions (Terry, 1994). Although the intracellular chloride is very small in quantity, it is significant because it is found in certain specialized cells, such as nerve cells.

Sources

Chloride comes almost entirely from sodium chloride (salt) and, to a lesser extent, potassium chloride. The dietary sources of chloride are the same as those for sodium (see Chap. 5).

Recommended Daily Allowance

According to the National Research Council (1989), the exact daily minimum requirement for chloride has not been established. However, it is estimated that the average adult over 18 years of age should consume approximately 500 mg per day.

A diet restricted in sodium often leads to reduction in chloride. However, this reduction in chloride rarely leads to a state of inadequacy because the kidneys *selectively* secrete chloride *or* bicarbonate depending on the acid-base balance.

Control

Chloride concentration is regulated secondary to the regulation of sodium concentration. With each sodium ion reabsorbed in the renal tubules, a chloride *or* bicarbonate ion is also reabsorbed. The amount of chloride or bicarbonate reabsorbed depends on the acid-base balance of the extracellular fluid. (Acid-base balance is discussed in detail in Chap. 3.)

Aldosterone, a mineralocorticoid, causes reabsorption of sodium; because of the physiologic relationship between sodium and chloride, chloride ions are also reabsorbed. The effect of aldosterone is to increase reabsorption of both sodium and chloride (salt), and thus water, in the extracellular fluid compartment. When aldosterone causes reabsorption of sodium and chloride, it allows some potassium or hydrogen to be released. The overall result, in which chloride is only one factor, is to increase or decrease the osmotic pressure of the blood, the total blood volume, and arterial pressure. When both sodium and chloride levels are elevated or reduced but retain their normal ratio relative to one another, a water imbalance is usually present.

Abnormal changes in the serum chloride level are *most commonly* the result of changes in dilution or concentration because of water excess or water deficit. The normal ratio of sodium to chloride is 3 to 2. Deviations from this ratio may be caused by gastrointestinal and renal losses or retention due to renal disease.

Functions

Chloride has five major functions in the body. First, chloride along with sodium is responsible for maintaining the osmolality of the extracellular fluid. When serum osmolality increases (greater than 295 mOsm/L), the concentration of sodium and chloride ions is higher than normal. With decreased osmolality, fewer sodium and chloride ions are present.

Second, chloride helps to maintain fluid balance. When sodium and chloride are retained by the body, water is also retained. When these ions are lost via the kidneys, water is also lost.

Third, chloride is important in maintaining acid-base balance in the body. As mentioned earlier, when positive sodium ions are reabsorbed by active transport, negative ions (chloride or bicarbonate) are also reabsorbed to maintain ionic balance. Whether the negative ions reabsorbed are chloride or bicarbonate depends on the needs of the body. If bicarbonate ions are reabsorbed the pH of the blood increases. If chloride is reabsorbed the pH remains the same.

Chloride also plays a part in the exchange of oxygen and carbon dioxide in red blood cells. Carbon dioxide diffuses into red blood cells, where the enzyme carbonic anhydrase catalyzes its conversion into carbonic acid. The carbonic acid ionizes to form hydrogen and

$$CO_2 + H_2O \longrightarrow H_2CO_3$$

$$H_2CO_3 \longrightarrow H^+ + HCO_3^-$$

bicarbonate ions. The diffusion of bicarbonate back into the plasma is facilitated by the movement of chloride from the plasma into the red blood cell. This movement of chloride, known as the *chloride shift*, maintains ionic balance; that is, it ensures that there is no net in electric charge in the process. The hydrogen ions generated in this process facilitate the release of oxygen from oxygenated hemoglobin.

Finally, chloride is a major component of gastric juice, which is primarily hydrochloric acid. Loss of gastric secretions leads to a deficit of chloride.

CHLORIDE IMBALANCES

When chloride and sodium concentrations change, the ratio of chloride to sodium usually remains constant. Increased serum chloride levels mirror increased serum sodium levels and commonly reflect a water deficit. Decreased serum chloride levels mirror decreased serum sodium levels and commonly reflect a water excess. An exception to this proportional variation occurs in metabolic alkalosis, in which chloride decreases independently of sodium.

Chloride imbalances are known as hypochloremia (chloride deficit) and hyperchloremia (chloride excess). Hypochloremia may coexist with metabolic alkalosis. Hyperchloremia may occur concomitantly with metabolic acidosis.

Chloride Deficit

When chloride is decreased, the bicarbonate increases in compensation because the total number of anions of the extracellular fluid must always be equal to the total number of cations in order to preserve electrical neutrality. Thus when chloride decreases, extra bicarbonate ions are retained by the kidneys to balance the sodium ions. When the chloride is disproportionately reduced in comparison to sodium and the bicarbonate is increased (either from excess bicarbonate administration or loss of gastrointestinal chloride), a hypochloremic metabolic alkalosis occurs.

Causative Factors

The most common causes of chloride deficit are gastrointestinal (GI) loss, renal loss, and loss of chloride through excessive sweating. Vomiting and gastric suction are the usual ways in which chloride is lost from the GI tract. Chloride levels may also decrease as a result of using loop (high-ceiling) or thiazide diuretics, as well as from the zealous use of bicarbonate.

Chronic lung disease can also result in a hypochloremia. In clients with chronic lung disease (respiratory acidosis), renal compensation results in reabsorption of bicarbonate to reduce the acidotic state. The chronic elevation of bicarbonate resulting from renal reabsorption and the chloride shift leads to a decrease in the serum chloride level. Excess bicarbonate from overuse of sodium bicarbonate as an antacid results in the same process. Table 6–1 summarizes the common causes of hypochloremia.

Assessment

Physical Assessment

Clinical manifestations of hypochloremia include hyperexcitability of muscles, tetany, and tremors. Respirations are depressed to conserve carbon dioxide (carbonic acid) and decrease the serum pH. In severe cases of both fluid and chloride loss, the blood pressure may be decreased as well.

TABLE 6–1
NURSING ASSESSMENT OF RISK
FACTORS FOR CHLORIDE IMBALANCES

Chloride Deficit	Chloride Excess
Excessive vomiting	Excessive diarrhea
Gastric suction	Increased chloride intake
Diuretics	Acute renal tubular necrosis
Excessive sweating	Head injury
Chronic lung disease (respiratory acidosis)	Cortisone preparations/corticosteroids
Fluid overload (dilutional increase)	Dehydration (relative effect)
Excessive intake of sodium bicarbonate as antacid	

Diagnostic Assessment

Hypochloremia occurs when the serum chloride level falls below 97 mEq/L. In addition to measuring serum chloride, levels of sodium and potassium are also assessed because they are often lost with chloride. Arterial blood gas determination identifies acid-base imbalance, usually metabolic alkalosis. In this condition, the serum pH and bicarbonate level are typically elevated.

A urinary chloride level may also be assessed. The normal range is 110 to 250 mEq/L in 24 hours for adults, 15 to 40 mEq for children, and 2 to 10 mEq for infants (Pagana & Pagana, 1992). In the patient with a chloride deficit, the urinary chloride level is usually decreased.

Nursing Diagnoses

The most common nursing diagnosis associated with hypochloremia is Altered Nutrition, less than body requirements related to chloride loss or inadequate intake.

If concomitant metabolic alkalosis develops, the following nursing diagnoses are pertinent:

- Injury, risk for, related to mental status changes
- Cardiac Output, decreased, risk for, related to dysrhythmias

Collaborative Management

Medical Treatment

The best way to treat any electrolyte balance is to prevent it. To prevent chloride deficit, intravenous solutions should contain normal saline for dilution rather than water. Patients having surgery or those experiencing trauma are at a particularly high risk for electrolyte deficits. These patients usually receive additional chloride in the form of potassium chloride in the intravenous solution. Chapter 4 discusses potassium supplementation and its associated nursing interventions in detail.

The primary goal of treatment for the patient who has hypochloremia is to alleviate the cause of the deficit. For example, if the patient is vomiting, antiemetics are administered to resolve this problem. If the patient is receiving loop or thiazide diuretics, the physician may discontinue them. Chloride is replaced by administering an intravenous normal saline (NS; 0.9% solution) or ½ NS (0.45% solution). In severe cases, hypertonic saline (3% or 5% NaCl) may be administered slowly in small doses. If excessive chloride is replaced, the patient may rebound to hyperchloremia with metabolic acidosis. Hypertonic saline administration is discussed in more detail in Chapter 5.

Nursing Interventions

Patients should be instructed to eat foods high in chloride, including salt. They should also be taught to use antacids sparingly, especially sodium bicarbonate. Accurate intake and output records are important so that the physician can calculate the intravenous or oral replacement of chloride, with consideration of the possibility of creating a water imbalance. Ongoing nursing assessment for early signs and symptoms of chloride imbalance is crucial during treatment.

Chloride Excess

When caring for patients with hyperchloremia, it is necessary to keep in mind that hyperchloremia may manifest as a metabolic acidosis. Metabolic acidosis is generally a result of one of two situations: a decrease in bicarbonate, which results in an excess of chloride (in contrast to alkalosis in which bicarbonate is increased and chloride decreased), or a marked increase in hydrogen ions (accumulation of acids).

Causative Factors

An excess of chloride ions may result from chloride being retained or ingested. Excessive intake of nutritional or parenteral chloride may be in the form of such common chloride-containing salts as sodium chloride, calcium chloride, and ammonium chloride. The additional chloride causes bicarbonate ions to be released in the kidney tubules, thus lowering the base bicarbonate circulating in the serum. This leads to a hyperchloremic metabolic acidosis as a result of the primary deficit in the base bicarbonate concentration in the extracellular fluid.

Cortisone preparations can cause hyperchloremia because they cause sodium retention and subsequently chloride retention. Chloride excess can also result from severe diarrhea (loss of bicarbonate), head injury (sodium retention), and renal tubular necrosis or interstitial nephrosis (acute renal failure). If fluid loss (dehydration) occurs, serum electrolytes, including chloride, become concentrated and the serum chloride value increases. (See also Table 6–1.)

Assessment

Physical Assessment

The major signs and symptoms of hyperchloremia are essentially those of metabolic acidosis and include weakness, lethargy, and deep, rapid breathing. In severe acidosis, stupor and unconsciousness develop, and there is a risk for dysrhythmias.

Diagnostic Assessment

A serum chloride value of more than 109 mEq/L confirms hyperchloremia. An adult with chloride excess will excrete more than 250 mEq of urinary chloride in 24 hours; for children, the value is over 40 mEq; and for infants, the value is more than 10 mEq.

In patients with metabolic acidosis, two types of metabolic disorders can be distinguished on the basis of the anion gap: normochloremic acidosis disorders (anion gap acidosis) and hyperchloremic acidosis (non–anion gap acidosis). The anion gap is the concentration of plasma anions not routinely measured by laboratory screening. It accounts for the difference between the concentrations of the routinely measured anions and cations. The anion gap (AG) is computed using the following formula:

$$AG = ([Na^+] + [K^+]) - ([Cl^-] + [HCO_3^-])$$

where $[Na^+]$ is the sodium ion concentration in milliequivalents per liter (mEq/L), $[K^+]$ is the potassium ion concentration, and so on.

Nursing Diagnoses

The key nursing diagnoses for the patient with hyperchloremia are related to metabolic acidosis and/or fluid deficit, depending upon the cause of the problem. (See Chapter 3 on acid-base imbalances and Chapter 2 on fluid imbalances.) In addition, the most common diagnosis for all patients with hyperchloremia is Nutrition, altered, more than body requirements, related to chloride retention or excessive chloride ingestion.

Collaborative Management

Medical Treatment

The treatment of hyperchloremia is essentially that for metabolic acidosis and is very complex. In many cases, treatment depends on the underlying cause of the acidosis. Therefore, the primary goal is to eliminate or resolve the cause of hyperchloremia. For example, when the change in chloride level is greater than the change in sodium level, and this imbalance is accompanied by a change in the volume of extracellular fluid, renal failure is usually involved, and the kidneys are not able to excrete chloride properly.

In an emergency, when acidity must be reduced rapidly, an intravenous injection of a bicarbonate solution quickly increases the buffer base of the blood, thereby relieving the acidosis. A rebound metabolic alkalosis must be avoided by careful monitoring of arterial blood gases. Other drugs should not be mixed with bicarbonate, as many drugs will precipitate when the pH is alkaline. Certain drugs, such as adrenalin, are effective only when the pH is normal.

In less urgent cases, an intravenous lactated Ringer's solution can be used. This solution increases the base level of the blood because a healthy liver can convert the sodium lactate to bicarbonate.

Nursing Interventions

Patients with chloride excess should be instructed to limit salt intake. Nursing assessment for early signs and symptoms of chloride excess is essential, especially when patients are receiving intravenous solutions containing chloride. Monitoring for clinical manifestations of metabolic acidosis is also important.

CRITICAL THINKING ACTIVITIES

An 82-year-old woman was hospitalized with a diagnosis of syncope, hypertension, and supraventricular tachycardia. She was alert and oriented and did not know why she "fainted." Her home medications include Cardizem (diltiazem) 120 mg BID (discontinued shortly after admission), Hytrin (terazosin) 2 mg QD, Ogen (estropipate) 1.25, ASA (aspirin) 2.5 g QD, Calan (verapamil) SR 24 mg QD, and Clinoril (sulindac) 200 mg BID. Her heart rate is currently between 80 and 90. Admission laboratory values are:

Carbon dioxide	25.9 mEq/L
Chloride	107 mEq/L
Sodium	139 mEq/L
Potassium	4.36 mEq/L
Serum calcium	9.3 mg/dL
Anion gap (AG)	9.96 mEq/L

Three days later, the patient became confused and lethargic, and was placed on telemetry. She exhibited a sinus rhythm heart rate of 160 and atrial fibrillation. The laboratory report at this time was:

Carbon dioxide	26.3 mEq/L
Chloride	80.1 mEq/L
Sodium	116.1 mEq/L
Potassium	3.22 mEq/L
Serum calcium	9.6 mg/dL
Anion gap (AG)	12.9 mEq/L

The patient was taken to the ICU immediately. At 2300 patient experienced a generalized seizure lasting 20 sec that resolved by itself. The laboratory report after the seizure showed:

Carbon dioxide	17.8 mEq/L
Chloride	81.2 mEq/L
Sodium	116.1 mEq/L
Potassium	2.46 mEq/L
Anion gap (AG)	19.6 mEq/L

The patient was given intravenous 3% NaCl in very small doses, then placed on NS at a rate of 100 mL/h.

QUESTIONS

1. *What was the progression of the anion gap and what were the contributory factor(s) in its change?*

2. *What electrolyte imbalances did the patient have?*

3. *What was the purpose of the hypertonic saline infusion (3% NaCl)?*

REFERENCES AND READINGS

Chernecky, CC, Krech, RL, & Berger, BJ (1993). Laboratory Tests and Diagnostic Procedures. Philadelphia: WB Saunders.

Lutz, CA, & Przytulski, KR (1994). Nutrition and Diet Therapy. Philadelphia: FA Davis.

National Research Council (1989). Recommended Dietary Allowances. Washington, DC: National Academy Press.

Pagana, KD, & Pagana, TJ (1992). Mosby's Diagnostic and Laboratory Test Reference. St. Louis: Mosby–Year Book.

Terry, J (1994). The major electrolytes: Sodium, potassium, and chloride. Journal of Intravenous Nursing 17(5), 240–247.

Zeman, FJ (1991). Clinical Nutrition and Dietetics (2nd ed). New York: Macmillan.

CHAPTER 7

. . . .
. . . .
. . . .
. . . .
. . . .
. . . .
. . . .
. . . .
. . . .
. . . .

Calcium

CALCIUM BALANCE

There is a growing concern among health care practitioners about the role of calcium in relation to the homeostasis of body fluids and electrolytes. Calcium, a cation found in many minerals, is the most abundant electrolyte in the human body.

Calcium and phosphorus compose the bulk of the mineral content of the skeleton. However, both minerals are also present in the serum. Calcium circulates both as a free ion and bound to the plasma proteins.

Distribution

Over 99% of the total body calcium is located in the bones, teeth, and nails in the form of calcium phosphate and calcium carbonate. In the average adult, this represents approximately 1.2 kg of calcium, so that calcium, on average, composes 20 to 25 g/kg of fat-free tissue. The remaining 1% circulates in the blood and some soft body tissues.

Serum calcium exists in both a free and a bound state. The amount of free calcium is a function of the amount of serum albumin, that is, the amount of free Ca^{2+} is equal to the total serum calcium concentration minus the amount that is bound to albumin. The normal total serum calcium level is 9 to 11 mg/dL, 4.5 to 5.5 mEq/L, or 2.05 to 2.54 mmol/L, depending on the laboratory technique used for measurement, and the units used for reporting the laboratory value. The values for children are slightly higher than those given above, and those for the elderly are slightly lower. Because of the binding of Ca^{2+} to protein, the value obtained for total serum Ca^{2+} is clinically useful only if the protein concentration is also known. This is because the clinical symptoms of calcium deficit or excess are correlated with the free Ca^{2+} serum concentration.

Sources

Calcium must be ingested daily through normal dietary intake because without daily intake, calcium-containing tissues in the body are vulnerable to depletion. Dairy products represent 55% of the average dietary source of calcium in the United States. Table 7–1 summarizes a list of dairy products whose calcium content equals that of 1 cup of milk.

Other foods high in calcium include oysters, sardines, molasses, macaroni, almonds, filberts, collards, white mustard cabbage, mustard greens, spinach, turnip greens, some creamed soups, kale, and rhubarb. These foods are rated as high in calcium based on an approximate serving size for each food rather than on identical weights.

Less traditional sources of calcium include antacids containing calcium salts, calcium-fortified foods, calcium-precipitated tofu, and fish and poultry bones.

Some foods that are high in calcium are not recommended as ideal sources because the calcium is not easily absorbed by the body. Spinach and kale are two foods in this group.

TABLE 7–1
QUANTITIES OF FOOD CONTAINING
CALCIUM EQUAL TO ONE CUP OF MILK*

Food	Amount	Kilocalories	Fat (Grams)
CHEESE			
Blue	2 ounces	200	16
Cheddar	1.5 ounces	171	14
Cottage			
Creamed, large curd	2.25 cups	529	22
2% low-fat	2 cups	410	9
Grated Parmesan	4.3 tablespoons	99	6
Processed American	1.7 ounces	180	15
Swiss	1.1 ounces	118	8
FROZEN DESSERTS			
Hard ice cream, vanilla	1.7 cups	459	24
Soft ice cream	1.3 cups	479	29
Sherbet	2.9 cups	786	11
Milkshake, vanilla	0.9 cups	273	9
MILK			
Skim	1 cup	86	0.4
2%	1 cup	121	5
Whole	1 cup	150	8
YOGURT, LOW-FAT			
Plain	0.7 cups	101	2
With fruit	0.9 cups	199	2

*One cup of milk contains approximately 300 milligrams of calcium.
Source: Lutz, CA, & Przytulski, KR: Nutrition and Diet Therapy. FA Davis, Philadelphia, 1994.

Recommended Daily Allowance

The mean adult intake of calcium in Western populations is estimated at between 400 and 1300 mg daily, with men of all ages ingesting more than women. Some individuals ingest 2 to 4 times the recommended daily allowances. The National Health Promotion and Disease Prevention Objectives for the year 2000 are to increase calcium intake so that at least 50% of young people aged 12 through 24 and 50% of pregnant and lactating women consume three or more servings daily of foods rich in calcium (U.S. Department of Health and Human Services Public Health Service, 1990). At least 50% of people aged 25 or older should consume two or more servings daily.

The recommended dietary allowance as set by the National Research Council is 1,200 mg per day for males and females aged 11 to 24 years, for pregnant women, and for lactating women. Healthy adults over the age of 24 years and children up to 10 years should consume 800 mg/day. Infants need 400 to 600 mg daily. Many experts recommend higher doses for premenopausal women (at least 1000 mg), and 1500 mg or more for postmenopausal women.

In addition to knowing the appropriate levels of calcium intake, it is important to understand the relationship between calcium and vitamin D. Without vitamin D, calcium cannot be absorbed properly from the intestines. Moreover, vitamin D plays a vital role in the movement of calcium from bone and in the control of calcium ion concentration in the plasma. The most readily available source of vitamin D is sunlight, since the ultraviolet rays of sunlight enable vitamin D to be synthesized in the skin. In the United States, most milk is fortified with vitamin D. Other dietary sources include eggs, butter, and fortified margarine. Vitamin D supplements are also available for use. Although vitamin D is necessary to form and maintain healthy teeth and bones, and to maintain appropriate levels of calcium in the plasma, excessive vitamin D can be toxic. Symptoms of vitamin D overdose are fatigue, heart and kidney dysfunction caused by deposition of calcium in these tissues, nausea, and high blood pressure.

Implications for Elderly Persons

Aging individuals need to consume more calcium because of the declining efficiency of their intestines, negative calcium balance associated with osteoporosis, increased renal excretion, and lactose intolerance from intestinal lactose deficiency (malabsorption syndrome). When elderly persons have a calcium deficiency, parathyroid hormone (PTH) secretion is increased to stimulate release of calcium from the bone. The result is irreversible bone loss, or osteoporosis, which leads to fractures of hips, wrists, and the vertebral column (compression fractures). Because they impose prolonged immobility, these fractures are debilitating and are associated with a high mortality rate.

Elderly patients and others who are homebound or institutionalized may also have a vitamin D deficiency, since the primary source of vitamin D is sunlight. Bone softening caused by lack of vitamin D is called osteomalacia. Many elderly patients have both osteomalacia and osteoporosis because they lack both vitamin D and calcium.

Control

Absorption of calcium is affected by intestinal integrity, endocrine balance, aging, renal efficiency, and other factors. The average adult absorbs 30 to 40% of the calcium ingested through the small intestines. About 15% (130 mg) of daily ingested calcium is considered nonabsorbable. Active children absorb up to 75% of ingested calcium.

Absorption of calcium by the intestines is affected by many variables. A person's sex and age help determine how much ingested calcium the body assimilates. Adequate levels of ultraviolet light and vitamin D must be present to maintain normal absorption.

Calcium absorption from the intestine can be *increased* by adding amino acids (lysine and arginine) to the diet and by increasing the consumption of vitamin D, fat, and lactose. Most antibiotics and estrogens enhance the absorption of calcium. Calcium absorption is also increased in an acid environment (Lutz & Przytulski, 1994).

Calcium absorption can be *decreased* by rapid peristalsis, stress, immobilization, increases in thyroid and steroid hormone concentrations, and aging. Food items that decrease absorption are cocoa, soybeans, and spinach (oxalic acid), and foods high in phosphates such as rice, bran, and wheat meal (dietary fiber). Ingestion of alkalis also inhibits calcium absorption.

Calcium is excreted through the urine and feces. Urinary excretion of calcium is about 100 to 130 mg per day, depending on daily dietary intake and efficiency of intestinal absorption. High-protein diets increase the excretion of calcium through the urine. Urinary excretion shows diurnal variation, and most excretion occurs during the day. Fecal excretion of calcium averages 150 mg per day. Calcium can also be excreted through the skin (20 to 350 mg per day), particularly with overt hyperthyroidism. Individuals working in high temperatures may lose up to 1 g per day via the skin.

The kidneys can selectively control excretion of calcium. The glomeruli filter about 10,000 mg (10 g) of calcium every 24 hours. Ninety-nine percent of this load is reabsorbed in the renal proximal convoluted tubules. Calcium excretion is dependent on a concurrent excretion of sodium and is influenced by the reciprocal relationship with phosphorus. Urine studies can measure the amount of calcium excreted. Because of diurnal variation in excretion 24-hour excretion studies are needed to provide an accurate daily average.

Factors that increase calcium excretion include saline diuresis, increased carbohydrates, decreased phosphates, metabolic acidosis, specific hormone levels (steroids, thyroid, growth), and diets with excessive protein and magnesium.

Lifestyle also has an impact on calcium levels in the body. Lack of exercise, abuse of alcohol, and smoking can all lead to calcium depletion. Heredity and body build are other factors. Petite women with thin bone structures are most likely to experience osteoporosis.

Calcium balance is controlled primarily by the two key hormones, calcitonin and PTH. Calcitonin is secreted by the thyroid, whereas PTH is secreted by the parathyroid gland. Calcitonin's main function is to protect against an excess of calcium in the serum by inhibiting withdrawal of calcium from the bone.

The role of PTH is to regulate the concentration of extracellular calcium by a feedback mechanism controlling urinary excretion of calcium. When calcium concentrations increase, PTH secretion is inhibited; when calcium concentrations are low, PTH secretions increase and stimulate the kidneys to reabsorb Ca^{2+}. At the same time that PTH stimulates reabsorption of Ca^{2+}, it facilitates the excretion of phosphate and decreases the concentration of phosphate in the serum.

PTH also affects osteoclastic cells, which are located in the bone cavities. PTH increases both the size and number of osteoclastic cells, whose activity increases the release of calcium and phosphate into the extracellular fluid. Osteoblastic cells perform just the opposite function. They draw calcium into the bone matrix, thus increasing the calcium deposit in the bones. Normally, osteoclastic activity and osteoblastic activity are properly balanced.

When osteoclastic activity is more vigorous than osteoblastic activity, bone loss occurs. Estrogen has long been credited with the ability to decrease osteoclastic activity. Researchers have found that lack of estrogen stimulates production of interleukin-6, a specific chemical in the immune system. Interleukin-6 stimulates the growth of osteoclasts. Estrogen apparently has the effect of suppressing interleukin-6 production, thereby maintaining the balance in the activity of osteoblasts and osteoclasts. This explains why postmenopausal women, who have a reduction in estrogen levels, are predisposed to bone loss.

The parathyroid glands enable vitamin D to play a major role in the regulation of calcium balance. When the serum calcium levels fall, the parathyroid glands signal the kidneys to convert vitamin D from its inactive form to the active hormonal form called calcitrol. Calcitrol, which is secreted by the parathyroid, should not be confused with calcitonin, which is secreted by the thyroid. Calcitrol facilitates the absorption of calcium from the small intestines. Simultaneously, the parathyroid hormone alerts the kidneys to reabsorb more calcium. If insufficient calcium can be made available by reabsorption, the body begins to break down bone; that is, the body will "leach" calcium from bones and teeth to keep the serum calcium level balanced.

Functions

Calcium is important for the proper development of bones and teeth, neuromuscular function, blood clotting ability, acid-base balance, and the activation of certain enzymes.

In conjunction with phosphorus, the most common function associated with calcium is, of course, its role in the formation of bones and teeth. The bone matrix is composed of protein, which provides tensile strength for flexing. The matrix is filled primarily by the calcium phosphate compound hydroxyapatite, which provides hardness and bulk strength for supporting weight.

Intracellular calcium has an important neurologic function in the initiation of muscular contraction. When calcium enters the muscle fiber, it combines with special molecules to activate adenosine triphosphate (ATP), the energy source of all cells. The muscle stays contracted until the energy source is depleted or until the free calcium is transported to the sarcoplasmic reticulum by the calcium pump. Because the heart is a muscle, it is also affected by the role of calcium in muscular contractions. Changes in blood pressure have been associated with changes in levels of serum calcium and the resulting changes in heart polarization.

Calcium also interacts neurologically with the cell membranes of the nervous system so that membranes are not overly charged.

Although calcium functions are usually associated with the neurologic and skeletal functions described above, calcium also has a functional role in many other processes throughout the body. For example, calcium performs the function of a clotting agent in the process of blood coagulation. A group of 13 reactants, including calcium, participates in the reaction, which converts prothrombin, a blood protein, into thrombin. Thrombin activates fibrinogen to be converted to fibrin. The amount of prothrombin activator is related to the amount of calcium available in the serum.

The free calcium concentration in the blood is dependent on the pH. In an acidotic state (pH less than 7.35), the amount of calcium bound to protein is less, so that free serum calcium levels are elevated. In an alkalotic state (pH greater than 7.45), more calcium is bound to protein, so that the free serum calcium level is low, even if the total body calcium is normal.

In addition, calcium strengthens the capillary membrane, participates in the activation of enzyme reactions and the stimulation of hormone secretion, and facilitates the metabolism of vitamin B_{12}.

CALCIUM IMBALANCES

Calcium Deficit

A serum calcium deficit is called hypocalcemia. It can result from a variety of etiologies, some of which are interrelated.

Causative Factors

Inadequate dietary intake can cause a calcium deficit. Repke (1991) reports substantial evidence that the average American diet does not contain sufficient calcium. This may be particularly true in children, whose recommended dietary allowance is 800 mg per day (National Research Council, 1989).

Approximately 46% of the serum calcium is bound to protein, primarily albumin, and a *decrease* in the albumin level can lead to hypocalcemia. Elderly patients in hospitals and nursing homes often have hypoalbuminemia. Hypocalcemia associated with a decrease in serum albumin reflects a decrease in protein-bound calcium but not necessarily in ionized calcium. Only a decrease in *ionized* calcium blood levels leads to signs and symptoms of hypocalcemia.

An *increase* in dietary protein (protein greater than 0.9 to 1.0 mg/kg of body weight) can have an impact on calcium levels, since the increased protein causes ionized calcium to move out of the bones, thereby allowing it to be excreted through the urine. Western diets rich in animal protein and phosphorus increase excretion of calcium in the urine and enhance release of parathyroid hormone, which stimulates reduction of bone mass. The low calcium-to-phosphorus ratios of these high-protein diets increase the loss of calcium because of the reciprocal relationship of phosphorus and calcium levels.

Prolonged immobility also promotes bone demineralization (loss). Initially, bone demineralization creates a state of hypercalcemia because, as the calcium moves into the serum, the serum calcium level reflects this increase. However, the long-term effect of immobilization and bone demineralization is hypocalcemia. In situations of both physical and psychological stress, the serum calcium level may decrease, because calcium is mobilized from storage in the bones and teeth and is carried to the site of the injury or excreted. Any deficiency in bone calcium can ultimately result in hypocalcemia, because bone calcium is no longer a ready source for stabilizing the serum calcium.

Deficits also result from actual calcium losses. Excessive gastrointestinal losses can occur as a result of diarrhea and wound exudation. Other major causes of calcium loss include extraction of calcium from the extracellular fluid, such as occurs in acute pancreatitis, generalized peritonitis, or massive infections of the subcutaneous tissues. During these types of infections, calcium is withdrawn from the extracellular fluid but is not absorbed by the intestines and is therefore lost with the feces. The calcium level is also decreased during the diuretic phase of renal failure.

Other factors which contribute to hypocalcemia include:

- Excessive administration of citrated blood or too rapid correction or overcorrection of alkalosis
- Thyroidectomy in which all four parathyroids have been removed
- Inadequate vitamin D consumption, or insufficient exposure to the sun resulting in an inability of the body to produce vitamin D, especially in the very young and the aged.
- Hyperphosphatemia associated with renal failure
- Calcium-excreting drugs, such as loop (high-ceiling) diuretics, caffeine, anticonvulsants, heparin, laxatives, and nicotine

Implications for Infants and Children

Children with insufficient calcium intake may experience a higher than average frequency of bone fractures. Rickets is a sign of severely insufficient intake or absorption; presently it is not frequently seen in the United States because vitamin D fortified milk is readily available.

Implications for Pregnant Women

The pregnant woman and her fetus are at risk for hypocalcemia. Proper nutritional intake of calcium, including calcium supplementation, is recommended based on the noted propensity of calcium to reduce prematurity and the risk of pre-eclampsia (Repke, 1991). Preterm infants may be hypocalcemic at birth as a result of low calcium intake, poor intestinal absorption of calcium and vitamin D, and a diet of unsupplemented human milk (Shiao, 1992).

Implications for Elderly Persons

Calcium deficit leading to osteomalacia and osteoporosis is also of concern among menopausal women. Dawson-Hughes (1991) reports that responsiveness of postmenopausal women to calcium supplementation is dependent upon their menopausal age. He reports maximal effect on the patient with supplemental doses of about 1000 mg per day of elemental calcium. Estrogen in combination with calcium is the most effective measure to minimize bone loss. Supplementation within the first 5 years of menopause attenuates bone loss but does not arrest it. In early postmenopausal women, the spine is unresponsive to supplementation; however, in late menopausal stages, spinal bone loss can be minimized by increased calcium intake. Studies regarding the effect of supplementation on the hip are incomplete at this time.

Men have adequate secretions of testosterone until the age of 75 to 80. Testosterone is a bone builder and protects men from osteoporosis until around 80 years of age.

Transcultural Considerations

Thin, petite Caucasian women are at the highest risk for hypocalcemia and subsequent bone loss. African-American women have more bone reserve than Caucasian women and therefore do not experience this problem as often.

Assessment

Physical Assessment

The major assessment findings associated with hypocalcemia are attributed to increased neuromuscular excitability. Symptoms of calcium deficit include muscle cramps, memory impairment, insomnia, irritability, anxiety, depression, tingling sensations, tetany, and convulsions caused by changes in neuromuscular irritability (Table 7–2). Lack of calcium causes the mem-

---

TABLE 7–2
ASSESSMENT FINDINGS ASSOCIATED
WITH HYPOCALCEMIA

NEUROMUSCULAR ASSESSMENT

Skeletal muscle irritability (twitching, cramping, tetany)
Seizures
Hyperactive deep tendon reflexes
Trousseau's phenomenon and positive Chvostek's sign
Paresthesias
Anxiety, psychosis

RESPIRATORY ASSESSMENT

Shallow respirations
Respiratory failure resulting from seizures or tetany

CARDIOVASCULAR ASSESSMENT

Increased heart rate and dysrhythmias
Hypotension
Diminished pulses

GASTROINTESTINAL ASSESSMENT

Hyperactive bowel sounds
Abdominal cramping
Diarrhea

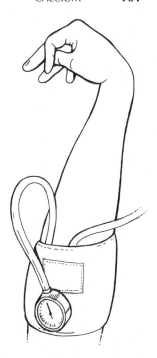

FIGURE 7–1
Palmar flexion, demonstrating a positive Trousseau's sign in hypocalcemia. (From Ignatavicius, DD, Workman, ML, and Mishler, MA: Medical-Surgical Nursing: A Nursing Process Approach, ed 2. WB Saunders, Philadelphia, 1995, p 307. © 1992 M. Linda Workman. Reprinted with permission.)

branes of nerve fibers to become partially charged; this results in the transmission of repetitive and uncontrolled impulses and leads to spasms of several muscle groups.

Long-term calcium deficit leads to bone loss, osteomalacia, and osteoporosis, as well as development of cataracts, sparse hair, and scaly, dry skin. In children, stunted growth may indicate a calcium deficiency.

Renal or gastrointestinal bleeding may be observed, and if the blood calcium level is extremely low, changes in clotting time may be observed.

A calcium deficit is also suspected if Trousseau's phenomenon is observed or if Chvostek's sign is positive. The carpopedal spasm (hand folding in) of Trousseau's phenomenon may be observed while the arm is constricted with a blood pressure cuff (Fig. 7–1). To observe Chvostek's sign, the nurse taps the face over the facial nerve, which is in front of the temple. If the face on that side twitches, the results are positive (Fig. 7–2).

Diagnostic Assessment

The diagnosis of hypocalcemia is confirmed when the serum calcium level is below 9 mg/dL or 4.5 mEq/L. If calcium excretion by the kidneys is increased (above 150 mg per day), the patient also has hypercalciuria. Excessive excretion is usually associated with destructive bone disease and a resultant hypocalcemia.

A decreased calcium level causes the contraction of the heart muscle to be weak. Prolongation of the isoelectric phase of the Q–T interval in the ECG is usually indicative of low calcium.

Radiographic examinations of bone do not show bone changes, especially bone loss, until as much as 25% of the bone is demineralized. More sensitive bone testing, such as with a bone densitometer or photon absorp-

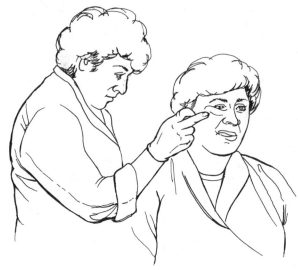

FIGURE 7–2
Facial muscle response, demonstrating a positive Chvostek's sign in hypocalcemia. (From Ignatavicius, DD, Workman, ML, and Mishler, MA: Medical-Surgical Nursing: A Nursing Process Approach, ed 2. WB Saunders, Philadelphia, 1995, p 308. © 1992 M. Linda Workman. Reprinted with permission.)

tiometer, can reveal bone density changes much earlier. Bone density measurements, especially in screening clinics, are being promoted in women's health programs.

Nursing Diagnoses

Common nursing diagnoses for a patient with hypocalcemia include:

- Nutrition, altered, less than body requirements, related to inadequate calcium intake or increased renal excretion
- Injury, high risk for, related to tetany and convulsions
- Cardiac Output, decreased, risk for, related to delayed ventricular repolarization

Collaborative Management

Medical Treatment

The physician prescribes specific treatment directed to the underlying and complex etiologies of hypocalcemia. The treatment of hypocalcemia usually includes administration of calcium orally or intravenously, the route depending upon the severity of the condition. For patients with adequate renal function, replacement therapy recommendations are 1 to 2 g of elemental calcium per day, in the form of citrate, gluconate, carbonate, or lactate. Vitamin D may need to be administered as well because it aids in the absorption of calcium from the intestinal tract. As discussed earlier, estrogen replacement therapy (ERT) for osteoporosis increases the intestinal absorption of calcium. To decrease the risk of reproductive cancers, progesterone may also be prescribed (hormone replacement therapy, or HRT). In addition to oral medication, topically administered estrogen is now available as a skin patch.

Other facets of treatment for osteoporosis include physical therapy to keep the bones physiologically active and supplements of vitamin D. Use of fluorides and testosterone to replace bone is reserved for the most severe cases because adverse effects of such treatment are common.

In an *acute* calcium deficit, calcium should be given intravenously using a 10% solution of calcium gluconate or gluceptate. A rate of infusion of 30 to 60 mL of 10% calcium gluconate mixed in 1000 mL of IV fluid over 6 to 12 hours may be indicated. This treatment is especially necessary if the patient has had tetany or a convulsion. In an emergency, treatment consists of 10 to 20 mL of 10% calcium gluconate carefully administered over a 4-minute period. Ten mL of 10% calcium gluconate solution contain 4.65 mEq (93 mg) of calcium (Gahart, 1994). Some hospitals require the first dose to be administered by a physician. Table 7–3 summarizes the nursing interventions associated with intravenous calcium replacement.

If hyperphosphatemia is present, dietary restriction of phosphorus may be indicated. Calcium replacement is more crucial than phosphate restriction, however. Oral phosphorus-binding agents, such as aluminum hydroxide, may be utilized. The aluminum antacid is more efficient than the magnesium antacid in binding phosphates. As the phosphorus levels return to normal, the calcium level will increase.

Since hypocalcemia often accompanies hypomagnesemia, the patient may also require treatment for magnesium deficiency.

Nursing Interventions

For mild hypocalcemia or to prevent hypocalcemia, nursing interventions include teaching the patient to consume foods high in calcium. Exposure to

TABLE 7–3

NURSING INTERVENTIONS FOR PATIENTS RECEIVING INTRAVENOUS CALCIUM REPLACEMENT

- Check the calcium solution for clarity; it should be clear and free of crystals.
- Warm the solution to room temperature before administration.
- When giving calcium gluconate, do not exceed a dosage of more than 200 mg/min.
- Administer the form of calcium ordered; do not substitute. For example, calcium gluconate is one third as potent as calcium chloride.
- Use the smallest IV access device possible since IV calcium can cause vein irritation.
- Keep the patient in bed after administering calcium because it can cause postural hypotension.
- Monitor serum calcium levels carefully.
- Monitor vital signs, being especially alert for bradycardia and other dysrhythmias, and hypotension.
- Instruct the patient that a metallic or chalky taste may be experienced when receiving IV calcium.
- Be aware that IV calcium administration increases the risk of digitalis toxicity; in a patient on digitalis, give small amounts very slowly and carefully assess the patient for signs of toxicity.
- Recognize that IV calcium may reduce plasma levels of atenolol (Tenormin) and reverse the clinical effects of verapamil.

the sun, in both summer sun and winter, is important for homebound or institutionalized patients to ensure vitamin D intake. Because dairy products other than milk are usually *not* fortified with vitamin D, vitamin D supplementation may be necessary as well. Calcium supplements should be taken 1 to 2 hours after meals to maximize intestinal absorption. The nurse should encourage the use of mineral waters containing calcium, such as Evian, San Pellegrino, Apollinaris, US Mountain Valley, or Mendocino.

The nurse also needs to teach the importance of properly managing or avoiding calcium-excreting drugs. When diuretics are used, for instance, the serum calcium levels should be periodically evaluated.

When intravenous calcium is ordered, the nurse must monitor the patient's response, including assessing for signs and symptoms of hypercalcemia. The calcium replacement should be controlled by an electronic infusion device and, when possible, the patient should be on a cardiac monitor. The nurse assesses the patient frequently for vital signs and the presence of Chvostek's sign and Trousseau's phenomenon.

Because the patient with hypocalcemia experiences neuromuscular excitability, the nurse needs to provide a quiet environment and avoid overstimulation. Comfort measures that relax the patient are very important. Seizure precautions should be implemented and emergency equipment should be readily available.

Calcium Excess

Excess calcium in the serum is called hypercalcemia. A serum calcium level is considered in excess when it exceeds 11 mg/dL, or 5.5 mEq/L.

Causative Factors

Calcium excesses are not common. When hypercalcemia does occur, it is usually associated with metabolic changes such as increased bone resorption or destruction related to neoplasms or multiple myeloma, Paget's disease, hyperparathyroidism, and thyrotoxicosis. Multiple fractures can also elevate serum calcium.

In patients with cancer, neoplastic production of prostaglandins may lead to elevated serum calcium. Hypercalcemia in cancer patients can also result from direct invasion of the malignancy into tissues where calcium is stored.

Spurious serum calcium elevations may occur with hemoconcentration, use of lithium, adrenal insufficiency, and albumin changes.

Other pharmacologic etiologies are related to the use of thiazide diuretics, excessive administration of vitamin D (over 50,000 units daily) leading to vitamin D intoxication, or excessive intake of calcium supplements. Thiazide diuretics enhance the reabsorption of calcium in the proximal convoluted tubules of the kidneys; therefore, if they are used for a long period of time, thiazides can cause hypercalcemia. Acquired or inherited sensitive hyperabsorption of vitamin D can also cause a calcium excess.

Calcium excess may also result from renal disease in which the calcium cannot be properly excreted because of decreased renal function. Familial hypocalciuric hypercalcemia is a genetic disorder in which improper excretion results in a calcium excess.

During prolonged immobilization, calcium moves from the bones, teeth, and intestines into the bloodstream to make up for hypocalcemia. Although osteoporosis and osteomalacia are long-term manifestations of immobility, symptoms of hypercalcemia may occur during the early stages when calcium is moving out of the bones into the serum.

Excess ingestion of milk and antacid products, primarily calcium carbonate, may produce a milk-alkali syndrome which results in hypercalcemia. Excessive ingestion of calcium may also lead to iron, zinc, and magnesium deficiencies. (See Table 7–4.)

Assessment

Physical Assessment

Classic manifestations of hypercalcemia are reflected in the neuromuscular and skeletal systems, renal function, and the gastrointestinal (GI) system.

Neuromuscular changes include hypotonicity of the muscles, especially relaxed skeletal muscles; deep bone pain associated with possible pathologic fractures; and flank pain indicative of developing renal stones. Excessively high levels of ionized calcium in the extracellular fluid act have a sedative effect on the neuromuscular system and can produce skeletal and muscle weakness, lethargy, confusion, somnolence, stupor, and coma. Emotional disturbances and psychosis may also be present.

TABLE 7–4
COMMON CAUSES OF HYPERCALCEMIA*

Concentrational	In water deficit state, may exhibit a false increase in serum calcium
Immobilization	Without appropriate weight-bearing ambulation and minimal exercise, bone demineralization occurs, leading to possible pathologic fracture and renal calculi
Hyperparathyroidism with hypophosphatemia	Pathology in the parathyroid leads to decreased serum phosphorus level, which induces reciprocal elevation in serum calcium, urinary hyperphosphaturia, and renal stones
Vitamin D intoxication	Megadoses of vitamin D lead to increased intestinal absorption of calcium
Malignancies/neoplastic disorders	Stress response in human body activates the transportation of calcium to sites of injury or pathology; some therapeutic drugs may increase calcium levels
Thiazide diuretics	Exacerbate hypercalcemia
Addison's disease	Hormonal imbalance in which adrenal cortex insufficiency leads to saline deficit, hyperkalemia, and hypercalcemia
Milk-alkali syndrome	Excessive ingestion of milk and absorbance antacids, also known as Burnett's syndrome

*Overall, the primary cause is excessive amounts of calcium entering the extracellular fluid from bones and, less commonly, from the intestines.

An increase in ionized calcium also results in depressed nerve and muscle activity and alteration in the action of the cardiac muscle. The heart responds to elevated levels of calcium with increased contractility and ventricular dysrhythmias (Porth, 1990).

The skeletal system may manifest hypercalcemia through bone loss. The first sign of a calcium imbalance may be a pathologic fracture, which can be related to serum calcium excess and phosphorus loss. Radiologic examination may reveal thinning of the bones just under the periosteum.

Renal stones may develop because of the excretion of a high concentration of calcium through the parenchyma of the kidneys. Polyuria accompanied by polydipsia may be present. Late signs of long-term calcium excess include impaired renal function (azotemia), precipitation of calcium salts leading to renal calculi, calcific conjunctivitis and keratopathy of the eyes, and calcium deposits under the skin. Straining the urine can help the examiner to detect renal stones.

Gastrointestinal symptoms may also appear in patients with hypercalcemia. These assessment findings include anorexia, nausea, vomiting, thirst, ileus, abdominal pain, and constipation. Constipation is related to the hypotonicity of muscles, since the smooth muscles of the gastrointestinal tract lack the necessary tension for adequate peristalsis.

Diagnostic Assessment

A serum calcium level above 11 mg/dL or 5.5 mEq/L confirms a diagnosis of hypercalcemia. Ionized calcium levels vary somewhat in relationship to the levels of the protein albumin because some calcium is bound to albumin. Thus it is important to compare the ionized calcium reading with the serum albumin level. The normal albumin range is 3.5 to 5.0 g/dL. The results of serum calcium readings may be misleading if calcium and albumin levels are not compared.

The calcium level also varies in relationship to the pH balance. Calcium readings may need to be taken more than once, as prolonged venous stasis or prolonged delay in testing of the specimen (without being separated from RBCs because they absorb calcium) may indicate a false decrease.

Renal tests, such as blood urea nitrogen (BUN) and creatinine, may be ordered to determine stone development. Urinary levels of calcium excretion vary with the pathology and the treatment. Two methods are used to evaluate urinary calcium measurement: quantitative (24-hour urine sample) and qualitative (the Sulkowitch test).

In the quantitative method, the 24-hour urine is collected in a container with 10 mL of hydrochloric acid to keep the pH of the urine low (a pH of at least 2 to 3) because calcium precipitates in alkaline urine. For testing, the client remains on the customary diet. If the patient is on a high-protein diet, this should be noted because a high-protein diet will cause an increase in the excretion of calcium. For an adult the normal level of urinary calcium is 50 to 300 mg/dL, depending on nutritional intake. For children, the value is calculated on the basis of 5 mg/kg of body weight.

The Sulkowitch urine test is a rough qualitative indication of the amount of calcium in the urine. The findings are reported as normal, increased, or decreased. The test is performed by adding a few drops of calcium oxalate to a specimen of urine to observe for the level of precipitation.

Results are assessed as follows:

Observation	*Indication*
Heavy white precipitate	Increased (hypercalciuria)
Fine white cloud	Normal
Clear specimen	Decreased (hypocalciuria)

In urinary calcium assessment, it is important to obtain fasting specimens in order for the test to be a strong indicator of persistently high calcium levels. If the specimen remains clear as a postprandial check, then hypocalcemia may be present. Assessment over a period of time, such as several weeks, is important because of nutritional fluctuations. Because phosphorus affects calcium, diurnal variations also need to be taken into account.

An electrocardiogram (ECG) can also be useful in diagnosing a calcium excess. Because the heart overcontracts when an excess of calcium is present, the ECG will indicate neuroelectrical changes in the heart. In hypercalcemia, cardiac dysrhythmias occur and are recognizable by a shortening of the Q–T interval. An increased serum calcium level is thought to enhance early ventricular repolarization.

Nursing Diagnoses

Common nursing diagnoses associated with hypercalcemia include:

- Nutrition, altered, more than body requirements, related to impaired renal excretion, metabolic disease, excessive calcium intake, prolonged immobility, or prolonged use of glucocorticoids
- Constipation, related to intestinal hypotonicity
- Cardiac Output, decreased, risk for, related to early ventricular repolarization (especially with concomitant digoxin therapy)
- Physical Mobility, impaired, related to pathologic fractures
- Thought Processes, altered, related to the effects of excess calcium on the central nervous system
- Fatigue, related to muscle weakness

Collaborative Management

Medical Treatment

Medical treatment is directed at increasing the excretion of calcium and alleviating the underlying cause of the calcium excess. Dietary restriction of calcium and diuresis are key therapies.

Pharmacologic treatment can include administration of saline and sodium sulfate to cause diuresis and thereby encourage renal excretion of calcium. This treatment reduces the serum calcium level until the underlying cause can be diagnosed. Intravenous or oral phosphate may also be administered to induce calcium excretion, since the body responds to high phosphate levels by increasing the excretion of calcium via the kidneys. Treatment with phosphates is a slow process and normal renal function is important to the success of this therapy choice.

For transitory hypercalcemia, a synthetic preparation of calcitonin called Calcimar may be used. Calcimar is used primarily to increase bone formation in patients with Paget's disease. Allergic reactions may occur with parenteral administration. Calcitonin, a thyroid hormone, normally works to increase incorporation of calcium into the bones, thus keeping it out of the serum.

For emergency treatment of hypercalcemia, a forced diuresis with normal saline is recommended. Furosemide may be given concurrently to enhance excretion of calcium. Thiazide diuretics should *never* be used with calcium excess as they can worsen the condition.

When the rate of calcium excretion is increased, there is the possibility of developing renal calculi. During diuretic treatment, forcing fluids is typically recommended using a rate of 2500 mL every 24 hours to maintain proper urine volume. During diuretic treatment when the urine is alkaline, calcium may precipitate. Usually the urine is acidic ranging from 5.5 to 6.5 pH, but if a urinary infection occurs, the urine may be alkaline.

Because elevated calcium enhances digitalis intoxication, dosages of digitalis preparations may need to be decreased until the hypercalcemic state is improved.

In clients with malignancies, plicamycin (mithramycin), an antineoplastic drug, may be prescribed for the reduction of the serum calcium level. A single dose may reduce the calcium level in 3 to 15 days, but rebound hypercalcemia may occur when the drug is discontinued.

If a parathyroid tumor is responsible for hypercalcemia, its removal will correct the problem, but calcium replacement therapy may be necessary after surgery.

Nursing Interventions

Nursing interventions focus on preventing hypercalcemia in the immobolized patient, monitoring drug and intravenous fluid therapy, and providing safety measures. The nurse should:

- Encourage mobility and weight-bearing ambulation when possible through planned ambulation programs.
- Assist the patient with passive exercises when ambulation is not possible.
- Assist the patient with mobility skills and activities of daily living, providing a safe environment.
- Record levels of consciousness and be alert for confusion or other neurologic change.
- Move patients carefully and watch for the possible development of pathologic fractures, usually manifested by bone pain.
- Unless otherwise contraindicated, force fluids to hydrate clients, dilute the serum calcium, decrease the chances of renal calculi, and decrease constipation.
- Avoid the use of calcium-containing IV fluids such as lactated Ringer's solution.
- Avoid giving drugs that may contribute to hypercalcemia.
- Strain all urine to check for urinary stones.
- Observe for symptoms that are associated with the passage of a urinary stone, especially severe flank or abdominal pain.
- Take vital signs frequently and be alert for dysrhythmias (especially common in patients on digitalis preparations).

- Temporarily decrease calcium in the diet; teach the patient what foods to avoid.
- Teach the patient about drug therapy, including therapeutic and adverse effects.

CRITICAL THINKING ACTIVITIES

A 53-year-old woman is admitted to a major hospital trauma center from a local nursing home with a broken left hip resulting from a fall. She complains of severe pain in the pelvic region, has a shortened left extremity, and abduction. She also complains of nausea and vomiting, which she has had for several days but which have become worse. She has been on large doses of supplemental vitamin D and calcium for osteoporosis, and has had limited ambulation for some time. Laboratory test results show the following:

Sodium	133 mEq/L
Chloride	100 mEq/L
Potassium	3.5 mEq/L
Serum calcium	18 mg/dL
Serum phosphate	3 mg/dL
Alkaline phosphatase	25 units/L

QUESTIONS

1. *What are the contributing factors that caused the elevated serum calcium level?*
2. *What factors contributed to this patient's fractured hip?*
3. *Which lab values are abnormal and what is the significance of each finding?*
4. *What is the important medical and nursing care needed for this patient with hypercalcemia?*

REFERENCES AND READINGS

An update on calcium application for the '90s. Proceedings of a symposium. (1991). American Journal of Clinical Nutrition, 54(1), 177S–290S.

Arthurs, RS, et al (1990). Effect of low-dose calcitriol and calcium therapy on bone histomorphometry and urinary calcium excretion in osteogenic women. Mineral and Electrolyte Metabolism 16(6), 385–390.

Baker, W (1985). Hypophosphatemia. American Journal of Nursing 85(9), 999–1003.

Booker, MF, & Ignatavicius, DD (1996). Infusion Therapy: Techniques and Medications. Philadelphia: WB Saunders.

Chernecky, CC, Krech, RL, & Berger, BJ (1993). Laboratory Tests and Diagnostic Procedures. Philadelphia: WB Saunders.

Cook, JD, et al (1991). Calcium supplementation: Effect on iron absorption. American Journal of Clinical Nutrition 53(1), 106–111.

Davis, JR, & Sherer, K (1994). Applied Nutrition and Diet Therapy for Nurses. Philadelphia: WB Saunders.

Dawson-Hughes, B (1991). Calcium supplementation and bone loss: A review of controlled clinical trials. American Journal of Clinical Nutrition 54(suppl 1), 274S–280S.

Dawson-Hughes, B, et al (1990). A controlled trial of the effect of calcium supplementation on bone density in post-menopausal women. New England Journal of Medicine 323(13), 878–883.

Erdmann, E (1991). Calcium for resuscitation. British Journal of Anesthesia 77(2), 178–184.

Food and Nutrition Board Commission on Life Sciences. (1989). Recommended Dietary Allowances (10th ed). Washington, DC: National Academy Press.

Gahart, BL (1994). Intravenous Medications. St. Louis: Mosby–Year Book.

Guyton, AC (1991). Textbook of Medical Physiology (8th ed). Philadelphia: WB Saunders.

Hawthorne, JL, Schneider, SM, & Workman, ML (1992). Common electrolyte imbalances associated with malignancy. AACN Clinical Issues in Critical Care Nursing 3(3), 714–723.

Hay, EK (1991). That old hip. Nursing Clinics of North America 26(1), 43–51.

Ignatavicius, DD, Workman, ML, & Mishler, MA (1995). Medical-Surgical Nursing: A Nursing Process Approach (2nd ed). Philadelphia: WB Saunders.

Lutz, CA, & Przytulski, KR (1994). Nutrition and Diet Therapy. Philadelphia: FA Davis.

Mendyka, B (1992). Fluid and electrolyte disorders caused by diuretic therapy. AACN Clinical Issues in Critical Care Nursing 3(3), 672–680.

Miller, P (1990). New developments in prevention and treatment. Physician's Assistant 14(6), 17–28.

Nussbaum, SR (1993). Pathophysiology and management of severe hypercalcemia. Endocrinology and Metabolism Clinics of North America 22(2), 343–362.

Porth, C (1990). Pathophysiology: Concepts of Altered Health Status (3rd ed). Philadelphia: JB Lippincott.

Repke, JT (1991). Calcium, magnesium and zinc supplementation and perinatal outcome. Clinical Obstetrics and Gynecology 34(2), 262–267.

Shiao, SP (1992). Fluid and electrolyte problems of infants of very low birth weight. AACN Clinical Issues in Critical Care Nursing 3(3), 698–704.

Spencer, H, et al (1988). Do protein and phosphorus cause calcium loss? Journal of Nutrition 118(6), 657–660.

Strong, P, et al (1991). Thiazide therapy and severe hypercalcemia in a patient with hyperparathyroidism. Western Journal of Medicine 154(3), 338–340.

Terry, J (1991). The other electrolytes: Magnesium, calcium and phosphorus. Journal of Intravenous Nursing 14(3), 167–176.

Tohme, JF, & Bilezikian, JP (1993). Hypocalcemic emergencies. Endocrinology and Metabolism Clinics of North America 22(2), 363–376.

Vallerand, AH, & Deglin, JH (1991). Nurse's Guide for IV Medications. Philadelphia: FA Davis.

Walpert, N (1990). An orderly look at calcium metabolism disorders. Nursing90 20(7), 60–64.

Zeman, FJ (1991). Clinical Nutrition and Dietetics (2nd ed). New York: Macmillan.

CHAPTER 8

. . . .
. . . .
. . . .
. . . .
. . . .
. . . .
. . . .
. . . .
. . . .
. . . .
. . . .

Phosphorus

PHOSPHORUS BALANCE

In the body, phosphorus is found in the form of phosphates, both organic and inorganic. Organic phosphates [adenosine mono-, di-, and triphosphate (AMP, ADP, and ATP) and creatine phosphate] are the major anions of the intracellular fluid. Inorganic phosphates are found in the serum and in bone. Serum phosphate levels ($H_2PO_4^{2-}$) are reported in milligrams of phosphorus (P) per deciliter (mg/dL) or milliequivalents of phosphate per liter (mEq/L). Because serum calcium and phosphate levels have a reciprocal relationship—that is, when calcium levels increase, phosphate levels decrease, and vice versa—it is important to assess calcium levels when assessing phosphate levels (Fig. 8–1).

Distribution

The distribution of phosphorus in the body is somewhat different from that of calcium. The largest amount of phosphorus is in the bones (80%); the rest is found in skeletal muscle (10%), the serum (9%), and in nerve tissue (1%). Phosphorus in the bone tissue consists of organic phosphates (intracellular phosphate) and inorganic phosphates (the calcified bone matrix), with the latter predominating. In the average adult, the total body phosphorus represents approximately 11 to 14 g/kg of fat-free tissue.

Phosphate is one of the key electrolytes and minerals whose levels differ markedly between children and adults. This difference is caused by the greater amount of growth hormone present in children through puberty.

A normal serum phosphorus reading for adults below 65 years of age is

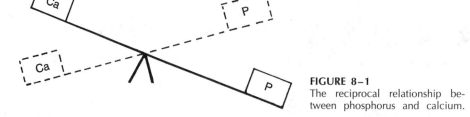

FIGURE 8–1
The reciprocal relationship between phosphorus and calcium.

3.0 to 4.5 mg/dL, 1.8 to 2.6 mEq/L, or 0.97 to 1.45 mmol/L, depending on the laboratory technique used and the units used for reporting. Phosphorus values are slightly higher in women and somewhat lower during pregnancy (Table 8–1).

Implications for Elderly Persons

Phosphate levels tend to decrease with age, but as in earlier years, the levels are lower in men than women.

Sources

Approximately 0.8 to 1.5 g of phosphorus must be ingested daily to maintain a normal balance. Approximately 70% of ingested phosphorus is absorbed, compared to an absorption rate for calcium of approximately 20% to 30%. Foods high in phosphorus include almonds, dried beans, barley, bran, pumpkin, squash, cheese, cocoa, chocolate, eggs, lentils, meats, poultry, fish, sardines, liver, milk, oatmeal, peanuts, matured dried peas, walnuts, wheat, and rye. Soft drinks are an abundant source of phosphoric acid. Only milk, natural cheeses, green leafy vegetables, and bone contain more calcium than phosphorus (Food and Nutrition Board Commission on Life Sciences, 1989).

TABLE 8–1

NORMAL SERUM PHOSPHORUS
LABORATORY VALUES BY AGE

Age	Serum Values
Adults <65 years old	3.0–4.5 mg/dL
Women >65 years old	2.8–4.1 mg/dL
Men >65 years old	2.3–3.7 mg/dL
Children	
Premature infant	5.4–10.9 mg/dL
Newborn	3.5–8.6 mg/dL
Infant	4.5–6.7 mg/dL
Child	4.5–5.5 mg/dL

Source: Chernecky, CC, et al: Laboratory Tests and Diagnostic Procedures. WB Saunders, Philadelphia, 1993.

Recommended Daily Allowance

Although the exact daily requirement for phosphorus is unknown, phosphorus intake equal to calcium intake is usually recommended. Thus, the recommended dietary allowance of phosphorus for the average healthy adult aged 25 and up is 800 mg per day, the same as the recommended dietary allowance for calcium. The exception to this 1:1 ratio is the recommended allowance for infants and lactating women, whose calcium needs exceed their phosphorus needs. Neither excessive nor insufficient intake appears to be harmful, especially if the calcium to phosphorus ratio is kept in balance.

Western diets can create problems because of the high consumption of phosphorus, which upsets the ratio of calcium to phosphorus. In most Western diets, the actual intake of phosphorus is estimated to be double that of calcium. The higher intake is attributed in part to natural foods; however, processed foods, such as processed meats, cheeses, and soft drinks contain food additives that contribute substantial amounts of phosphorus to the American diet. Such an imbalance can lead to possible secondary hyperparathyroidism. Infants on formulas may develop idiopathic hypocalcemia (tetany), because cow's milk contains 1.3 parts calcium to 1 part phosphorus, compared with a ratio of 2.3 to 1 for human milk. The recommended dietary allowance for formula-fed infants is 240 mg of phosphorus per day for the first 6 months and 360 mg per day for the latter half of the first year.

Control

Phosphorus, like calcium, is also regulated by parathyroid hormone (PTH). As the level of PTH increases, phosphate is excreted by the kidneys and the level of serum phosphate decreases. Vitamin D is needed for the absorption of both calcium and phosphorus.

In a healthy state, the renal tubules reabsorb 85% to 90% of the phosphorus ingested. Reabsorption increases with short-term cortisol therapy and use of growth hormone. Chronic renal failure may also increase phosphorus retention. Absorption may decrease with the use of digoxin, estrogen, thyroid and parathyroid hormones, long-term cortisol therapy, and elevations in the level of circulating calcium.

The kidneys excrete 600 to 900 mg of phosphorus daily averaging 0.4 to 1.3 g every 24 hours in the urine. Some phosphorus is excreted in the feces. Fecal phosphorus is composed of unabsorbed phosphorus and the phosphorus secreted into the gastrointestinal tract, averaging about 3 mg/kg of body weight per day. Some substances, such as aluminum hydroxide, can significantly increase the amount of phosphorus excreted in the feces. Dietary phosphorus is absorbed more readily than calcium; thus, the renal excretion of phosphorus is greater than that of calcium in order to maintain a constant level.

A diurnal pattern for phosphorus excretion is associated with physical exertion. The excretion of phosphorus peaks a few hours after the end of sleep. It is interesting to note that diurnal variations are lost in adrenal insufficiency states, suggesting that the diurnal rhythm is regulated by the adrenal gland. Thus 24-hour urine testing is important.

Functions

Phosphate is essential for the metabolism of nutrients such as carbohydrates, lipids, and protein. It functions as a cofactor in the numerous enzyme systems of cellular metabolism, one of whose functions is the production of high-energy compounds such as adenosine triphosphate (ATP), which facilitate normal neuromuscular activity. Phosphorus is a crucial component of DNA, which forms the genetic code, and in cell division. Other functions include the formation of bone, regulation of acid-base balance, and reciprocal action with the regulation of calcium. Phosphorus also has a modifying role in the renal excretion of hydrogen ions and B vitamins.

PHOSPHORUS IMBALANCES

Phosphorus Deficit

Hypophosphatemia, or phosphorus/phosphate deficit, caused by inadequate intake is unusual because the food sources for phosphorus are so abundant, but a phosphorus deficit is a serious concomitant condition of many medical disorders. For adults, hypophosphatemia is defined as a serum level below 3.0 mg/dL.

Causative Factors

A deficit in serum phosphorus is usually related to low or decreased nutritional intake, poor absorption from the bowel (usually due to lack of vitamin D), steatorrhea (fatty stools), or malabsorption syndrome.

Actual losses may occur through increased phosphate excretion. Increased renal excretion (usually caused by hyperparathyroidism or renal insufficiency) or disturbed carbohydrate metabolism (usually diabetic ketoacidosis [DKA]) may result in excessive phosphorus loss. When glucose metabolism is increased in patients with DKA, phosphorus binds with the excess sugar.

A poor nutritional state, such as that associated with alcoholism, can also lead to hypophosphatemia and may be a key feature in delirium tremens. Intake of nonabsorbable carbonate antacids, such as Amphojel, Gelusil, or Maalox, may also increase the loss of phosphorus. Other conditions related to phosphorus deficit include cellular shifts, such as those occurring in rapid cellular growth or hypermetabolic states (e.g., fever and pregnancy).

Phosphorus deficit may be related to specific disorders such as hyperparathyroidism. In this disorder, the serum calcium level increases, which causes a reciprocal *decrease* in serum phosphate levels.

Long-term total parenteral nutrition (TPN) may cause a deficit in phosphorus, especially if renal reabsorption is impaired. In fact, hypophosphatemia is one of the most frequent and disturbing electrolyte imbalances occurring with TPN administration.

A drop of 1 mg/dL is considered a profound deficit and is usually related to other very acute or major conditions. These include alcoholism, severe burns in the "healing" phase, diabetic ketoacidosis, bicarbonate excess, carbonic acid deficit, and hepatic disease.

There is some concern that an increase in dietary fiber will result in phosphorus deficits. However, after reviewing high-fiber diets in developing countries, Walker (1985) determined there was inadequate evidence to support the theory that high-fiber diets alone induce mineral deficiencies in people who otherwise consume a balanced diet.

Assessment

Physical Assessment

People with mild phosphorus deficits are usually asymptomatic. Recognizing the stages and abruptness of the shift to a deficient state is crucial for monitoring hypophosphatemia. Symptoms are vague or absent but can include anorexia, weakness, malaise, and skeletal pain and aches. The signs of phosphorus depletion syndrome are insidious and include neuropsychiatric, constitutional (e.g., malaise, lethargy), and gastrointestinal (e.g., anorexia, dysphagia) manifestations.

Other specific signs include cardiopulmonary manifestations including tachypnea, shallow respirations, decreased cardiac contractility, and vasodilation. *Acute* manifestations are delirium, seizures, pulmonary insufficiency, and systemic decompensation caused by tissue hypoxia and ATP (cellular energy) deficiency. It is thought that some of the symptoms attributed to hypercalcemia may actually be caused by concomitant hypophosphatemia.

Chronic phosphorus decreases are also evident in vague symptoms such as weakness, anorexia, bone pain, and pathologic fractures, all of which may be related to hypercalcemia. More specific manifestations are primarily neurologic and include hyporeflexia, malaise, irritability/anxiety, muscle weakness, and myalgia. Paresthesia may be exhibited. Patients may progress to confusion, stupor, seizures, and coma.

Muscular weakness is particularly related to a deficit in adenosine triphosphate (ATP), a crucial energy source in cellular metabolism. Another condition reflected by muscular weakness is phosphate reduction, which depletes the red blood cell chemical 2,3-diphosphoglycerate. The resulting anemia manifests in fatigue, malaise, and weakness.

Profound deficiency, a serum phosphate level below 1 mg/dL, is indicative of very serious conditions. Acute cases may be manifested by hematologic and neurologic changes, including possible respiratory failure. Extreme cases are recognized by development of hemolytic anemia, granulocyte abnormality (defective functioning of leukocytes and platelets), frank myopathy (also rhabdomyolysis [skeletal muscle destruction]), and encephalopathy. Hypercalciuria with consequent renal calculi may also occur.

Nursing assessment of key etiologic factors is imperative, since a period of vulnerability is characteristic of various conditions. For example, in patients experiencing diabetic ketoacidosis, cellular phosphates are catabolized, and the resulting phosphates, after being transported from the cells to the serum, are excreted by the kidneys. In such cases, insulin is administered to treat ketoacidosis and to decrease ketone production. As a result, serum phosphate levels may drop "abruptly" in the first 24 to 36 hours after initiation of treatment for diabetic ketoacidosis unless phosphates are simultaneously administered. In alcoholic withdrawal, severe symptomatic hypophosphatemia usually occurs in 2 to 4 days. With total parenteral nutrition and carbohydrate loading phenomenon, serum phosphates are depleted in about 3 days.

Diagnostic Assessment

A serum phosphorus level below 3.0 mg/dL in adults confirms a diagnosis of hypophosphatemia. In addition, low urinary phosphate concentration or increased urinary calcium excretion is typical. Normal urinary excretion for phosphorus is 0.4 to 1.3 g every 24 hours.

More specific assessments include hematologic changes or dysfunctions of platelets, red and white blood cells, and kidneys (e.g., glycosuria, hypercalciuria, hypermagnesuria).

Nursing Diagnoses

Common nursing diagnoses associated with hypophosphatemia depend on the cause. The key nursing diagnoses include:

- Injury, high risk for, related to weakness, paresthesias, seizures
- Sensory-perceptual alterations, related to paresthesias
- Nutrition, altered, less than body requirements, related to increased renal excretion of phosphorus, use of phosphate binders, or decreased phosphorus intake
- Breathing Pattern, ineffective, related to respiratory muscle weakness

Collaborative Management

Medical Treatment

Prevention is paramount in caring for clients on prolonged high-dose antacid therapy because many of these antacid are phosphorus binding. Patients on TPN are particularly susceptible to hypophosphatemia. Administration of 10 to 15 mL of potassium phosphate in a liter of IV fluid can assist in preventing the disorder. Careful attention must be given to catabolic states, in which there may be physiologic elevation of phosphorus due to cellular destruction. Assessing renal function before administering phosphorus is important because the renal system is the primary means of excretion of phosphorus.

Treatment of an actual phosphorus deficit relies primarily on replacement of phosphorus, usually orally, and correction of the underlying cause. Protein is a primary food source when using diet to replenish phosphorus levels. Pork, beef, legumes (dried mature peas and dried beans), eggs, and milk are good sources. Processed foods and cola beverages are also high in phosphorus but are considered less healthful replacement choices.

High phosphorus diets may also be high in calcium, so that in dietary treatment, it is important to know the serum calcium levels. It is also important to bear in mind that phosphorus is more easily absorbed than calcium, with approximately 70% of dietary phosphate being absorbed in contrast to about 20% to 30% of calcium. The recommended dietary allowance of phosphorus varies with age but is between 800 and 1200 mg per day, the same as

TABLE 8–2
NURSING INTERVENTIONS FOR PATIENTS RECEIVING
INTRAVENOUS PHOSPHATE REPLACEMENT

- Dilute potassium or sodium phosphate in a large volume of suitable IV solution before administering.
- Do not use potassium phosphate for patients with hyperkalemia or sodium phosphate for patients with hypernatremia.
- Use potassium phosphate with caution in patients with cardiac or renal disease. Avoid its use in patients taking digitalis.
- Use sodium phosphate with caution in patients with renal disease, cardiac failure, or cirrhosis.
- Infuse slowly to prevent phosphate or potassium toxicity, and hypocalcemic tetany.
- Monitor serum calcium, phosphorus, sodium, and chloride levels.

the recommended dietary allowance for calcium. The exception to the 1:1 ratio of phosphorus to calcium is for infants, whose calcium requirement is 400 mg per day between birth and the age of 6 months, and 500 mg per day between the ages of 6 months and 1 year. The corresponding amounts for phosphorus are 300 mg per day and 500 mg per day respectively (Food and Nutrition Board Commission on Life Sciences, 1989).

If the patient has no oral intake or is on a restricted intake, IV replacement of phosphorus may be used to correct hypophosphatemia. Table 8–2 summarizes the nursing interventions associated with administering IV phosphate.

Vitamins D and B also play a role in reestablishing the appropriate phosphorus level. Adequate vitamin D must be present for the proper absorption of both phosphorus and calcium. A vitamin B deficiency may be evident in hypophosphemic states as the B vitamins are effective only when combined with phosphorus.

Restorative care such as mechanical ventilation, safety measures, and seizure precautions may be necessary in severe depletion states. Careful monitoring of clients on calcium- and phosphorus-excreting diuretics is important, as well as evaluation of renal output.

Nursing Interventions

- Observe for decreased neuromuscular activity and the accompanying potential for injury.
- Observe for calcium excess and concomitant disorders especially development of kidney stones.
- Note abnormal laboratory reports with phosphorus levels below 3 mg/dL. Take immediate action if profound deficits of below 1 mg/dL occur.
- Assess for other concomitant disorders of hypomagnesemia and hypokalemia.
- Administer phosphate IV solutions, as ordered, taking great care not to exceed a rate of more than 10 mEq/h. A reciprocal hypocalcemia may occur with the administration of phosphorus. (See calcium deficit discussion in Chap. 7.)
- Monitor patients receiving total parenteral nutrition (TPN) closely for

the emergence of hypophosphatemia, hypomagnesemia, and hypokalemia related to the carbohydrate loading effect.

- Monitor phosphorus states in treatment with TPN, diabetic ketoacidosis, and alcohol withdrawal care.
- Collaborate with the dietitian in planning nutritional care, including special approaches in caring for clients with malabsorption syndrome.
- Instruct the patient about careful antacid use because of the phosphate-binding effect.

Phosphorus Excess

Excess phosphorus is called hyperphosphatemia. A serum phosphorus level over 4.5 mg/dL in the adult is indicative of hyperphosphatemia.

Causative Factors

Properly functioning kidneys excrete any excess of phosphorus; thus the incidence of hyperphosphatemia is rare and is usually associated with a chronic renal insufficiency in which the glomerular filtration rate falls below 20 mL/min. Any catabolic stress or cell breakdown worsens the elevated state. Serum levels of phosphate and calcium maintain an inverse relationship; thus, it is important to evaluate the serum calcium level and the serum phosphorus level concurrently. The calcium pathology should be assessed first.

There are three main scenarios where hyperphosphatemia is found. One instance is hypoparathyroidism, in which the serum phosphate level rises and the calcium level drops. Without parathyroid hormone (PTH), which is secreted by the parathyroid, phosphate is reabsorbed in the kidneys instead of being excreted, and serum phosphate levels rise. Lack of PTH has the opposite effect on calcium; the reabsorption of calcium in the kidneys is reduced and calcium levels fall. A second situation in which hyperphosphatemia occurs is vitamin D intoxication.

Serum calcium is increased because of increased absorption of calcium with a resulting increase in the phosphorus level. Vitamin D intoxication is an example of a nonreciprocal reaction that accounts for the fact that both calcium and phosphorus increase. Thirdly, hyperphosphatemia occurs in renal failure, where the serum calcium is decreased because the kidneys cannot reabsorb calcium and phosphorus cannot be excreted.

Although serum phosphate and calcium levels are generally coupled inversely to each other, there are situations in which the phosphate level will be increased without an accompanying calcium decrease. For example, when levels of growth hormone are increased in certain childhood diseases, the phosphorus level is elevated but the calcium level is not changed. During malignancies in which calcium is mobilized and serum levels increase the serum phosphate level may either remain normal or be slightly elevated.

Hyperphosphatemia may also occur when there is excessive bone growth in infants and children, and in acromegaly (overproduction of growth hormone) in adults. In cases of abnormal bone growth, levels of the enzyme alkaline phosphatase are elevated.

Other complex etiologies of phosphorus excess are related to dysfunctions of hormonal and metabolic balances, such as those with insulin and glucagon, thyroid and parathyroid hormones, steroids, and lysis states

TABLE 8–3

COMMON CAUSES OF HYPERPHOSPHATEMIA

Cause	Altered Function
Renal failure	Glomerular filtration reduced or tubular excretion impaired
	Associated with acute tubular acidosis or chronic renal failure with oliguria
	Reciprocal hypocalcemia; decreased urinary excretion (below 1 g/24 h)
Hypoparathyroidism	Insufficient PTH
	Associated with hypocalcemia
Adrenal insufficiency	Decreased adrenal hormones leading to decreased phosphate excretion
Increased growth hormone (children); acromegaly (adults)	Increased production of somatropin associated with increased alkaline phosphate levels
	Serum calcium usually normal
Nutritional: Vitamin D intoxication; excessive dietary intake; iatrogenic phosphate therapy	Stimulates intestinal transport of and absorption of phosphates
	Stimulates synthesis of alkaline phosphates
	Mobilizes calcium from bones and teeth
	Associated with hypercalcemia
Tissue damage or malignancies: Lactic acidosis; chemotherapy; radiation therapy; crushing injuries; rhabdomyolysis	Catabolic stress with cellular release of phosphates
	Depending on renal functioning, may lead to hypophastemia, especially in rhabdomyolysis
Overuse of phosphate-containing cathartics or enemas	Intestinal absorption of sodium phosphate
Phosphate-binding drugs	When dialysis cannot efficiently reduce phosphate level in renal failure, phosphate binders, such as aluminum hydroxide gels used to reduce phosphate levels, may also lead to hypophosphatemia

(rhabdomyolysis and autoimmune states). Further, changes in the extracellular status and use of diuretics may affect phosphorus levels. Chemotherapy may also elevate phosphates (see Table 8–3).

Assessment

Physical Assessment

The specific symptoms of hyperphosphatemia are vague and usually recognized as symptoms associated with hypocalcemia, such as neuroexcitability, tetany, and seizures. If, however, hyperphosphatemia is not associated with hypocalcemia, then phosphates may precipitate into body tissues as phosphate salts. This process is evident in eye changes such as conjunctivitis and keratopathy band, pruritus, renal deposits leading to renal failure, and arthritis.

Diagnostic Assessment

Hyperphosphatemia is indicated by a serum phosphate level greater than 4.5 mg/dL or 1.8 mEq/L. Urinary assessments can also help establish hyperphosphatemia. The urinary analysis requires a 24-hour collection of urine. The specimen need not have a preservative or be iced as is required for calcium assessments. Levels in the normal range are 0.4 to 1.3 g every 24 hours, but the levels vary with dietary intake. The average is 1 g in 24 hours.

Hypoparathyroidism results in hyperphosphatemia because it reduces the amount of phosphorus excreted; thus, the urinary phosphorus levels are reduced below the normal range. If renal disorders exist concomitantly with other disorders, then urinary analysis may not reflect the hormonal or mineral balance or the overall malnutrition problem.

When obtaining specimens for phosphorus the nurse should take care that intravenous (IV) fluids with dextrose are not infusing, because carbohydrates lower the serum phosphorus levels. If the patient is receiving dextrose IV, it must be noted on the laboratory request.

The serum alkaline phosphatase level may be elevated, especially when abnormal bone growth or breakdown has occurred.

Nursing Diagnoses

Common nursing diagnoses associated with phosphorus excess include:

- Urinary Elimination, altered patterns, related to calcium precipitates
- Nutrition, altered, more than body requirements, related to vitamin D intoxication, decreased calcium intake, or excessive phosphorus intake
- Sensory-perceptual alterations, related to concomitant calcium imbalances
- Injury, high risk for, related to effects of hypocalcemia, tetany, and seizures

Collaborative Management

Medical Treatment

The treatment for hyperphosphatemia is related to its etiology. If the cause is chronic renal failure, hypoparathyroidism, or is tumor related, and renal excretion is not possible, elimination of extra phosphorus is achieved by the cathartic effect of phosphate binders. Certain medications may be given to increase fecal excretion of phosphorus by binding phosphorus from the food in the GI tract. For example, aluminum hydroxide gels can be given by mouth. Aluminum hydroxide binds with the phosphorus to form insoluble aluminum phosphate. This insoluble phosphate compound is then evacuated via the feces. Aluminum hydroxide may be given even when dialysis has been initiated to decrease the phosphorus levels. Magnesium hydroxide may also be used to bind with phosphates to facilitate excretion, but the patient's magnesium level must be determined first. Magnesium supplementation is contraindicated in renal failure. Enemas containing sodium phosphates, such as Fleet enemas, should not be administered to patients with phosphorus excess.

If the phosphate elevation is caused by a massive influx of phosphorus not related to renal insufficiency, enhanced renal excretion may be accomplished with sodium bicarbonate solutions. In critical states, dialysis may be the therapy of choice, either peritoneal dialysis or hemodialysis. Dialysis reduces the phosphate overload, as well as the catabolic waste products (see Chaps. 16 and 17).

If the homeostatic mechanisms for maintaining the balance between phosphorus and calcium are intact, that is, if elevation of phosphate is accompanied by a decrease in calcium and vice versa, treatment focuses on correcting the calcium level. If the phosphorus level is increased, but the calcium level is normal or decreased, then calcium supplements are given, for example, in patients with hypoparathyroidism. If both calcium and phosphorous are elevated, the treatment may include a diet reducing phosphorus and calcium in the food. Although it is relatively easy to control calcium intake by regulating the diet, it is more difficult to decrease phosphorus by diet because of the abundance of phosphorus in most foods. It is especially difficult to reduce phosphorus in a diet without at the same time reducing dietary calcium. Calcium-restricted diets focus primarily on restriction of dairy products.

Growth hormone pathologies in children may result in elevations in the phosphorus level. In this situation, the calcium level would not normally be decreased. Treatment is directed toward establishing the correct hormonal balance.

Nursing Interventions

When elevations of phosphorus are not related to renal causes nursing interventions aim to prevent extracellular volume depletion and maintain normal blood volume. A thorough understanding of balanced diets that incorporate the proper ratio of calcium and phosphorus is essential. Teaching patients about foods to avoid and how to administer phosphate binders is especially important.

The nurse should be observant for neuromuscular irritability reflecting an accompanying calcium deficit. The nurse should especially assess for hyperreflexia, tetany, and seizures. (See hypocalcemia discussion in Chap. 7.) Other manifestations of phosphorus excess include evidence of acute or chronic renal disease accompanied by catabolic stress in which destruction of protein tissue mass releases more phosphorus. The nurse should be alert to the possibility of phosphorus excess and calcium deficit occurring with chemotherapy treatment of neoplasms. Table 8–4 summarizes the major nursing interventions for patients with hyperphosphatemia.

TABLE 8–4

NURSING INTERVENTIONS FOR PATIENTS
WITH HYPERPHOSPHATEMIA

- Assess patients at high risk for hyperphosphatemia, such as those with renal disease and hypoparathyroidism.
- Administer oral and IV phosphate preparations carefully to prevent hyperphosphatemia.
- Teach patients to avoid phosphate-containing medications, including laxatives and commercially prepared enemas.
- Teach patients to decrease consumption of foods high in phosphates.
- Teach the patient how to take phosphate-binding drugs, emphasizing that they should be taken with meals or immediately after meals.
- Monitor for signs and symptoms of hypocalcemia, including tetany and seizures.
- Assess for Trousseau's phenomenon and Chvostek's sign.

CRITICAL THINKING ACTIVITIES

A 45-year-old woman has renal disease from long-standing hypertension. She has been on hemodialysis for several years. Her husband found her on the floor having a seizure. On admission to the hospital, her laboratory values were:

Calcium	7.9 mg/dL
Phosphorus	5.5 mg/dL
BUN	77 mg/dL
Creatinine	2.9 mg/dL
Sodium	153 mEq/L
Potassium	5.8 mEq/L

QUESTIONS

1. *What was the cause of this patient's hyperphosphatemia? Why are her sodium and potassium levels elevated?*

2. *What assessments should you perform and what would you expect your examination to reveal?*

3. *What treatment options would be appropriate for this patient?*

4. *What teaching should you plan regarding her diet?*

REFERENCES AND READINGS

An update on calcium application for the '90s. Proceedings of a symposium. (1991). American Journal of Clinical Nutrition 54(1), 177S–290S.

Angeli, P, et al (1991). Hypophosphatemia and renal tubular dysfunction in alcoholics. Gastroenterology 100, 502–512.

Arthurs, RS, et al (1990). Effect of low-dose calcitriol and calcium therapy on bone histomorphometry and urinary calcium excretion in osteogenic women. Mineral and Electrolyte Metabolism 16(6), 385–390.

Baker, W (1985). Hypophosphatemia. American Journal of Nursing. 85(9), 999–1003.

Daily, W, Tonnesen, A, & Allen, S (1990). Hypophosphatemia—incidence, etiology, and prevention in the trauma patient. Critical Care Medicine 18(11), 1210–1214.

Food and Nutrition Board Commission on Life Sciences. (1989). Recommended Dietary Allowances (10th ed). Washington, DC: National Academy Press.

Gahart, BL (1994). Intravenous Medications. St. Louis: Mosby–Year Book.

Hodgson, SF, & Hurley, DL (1993). Acquired hypophosphatemia. Endocrinology and Metabolism Clinics of North America 22(2), 397–410.

Peppers, M, Geheb, M, & Desai, T (1991). Hypophosphatemia and hyperphosphatemia. Critical Care Clinics of North America 7(1), 201–214.

Porth, C (1990). Pathophysiology: concepts of Altered Health Status (3rd ed). Philadelphia: JB Lippincott.

Repke, JT (1991). Calcium, magnesium and zinc supplementation and perinatal outcome. Clinical Obstetrics and Gynecology 34(2), 262–267.

Terry, J (1991). The other electrolytes: Magnesium, calcium, and phosphorus. Journal of Intravenous Nursing 14(3), 167–176.

Vallerand, AH, & Deglin, JH (1991). Nurse's Guide for IV Medications. Philadelphia: FA Davis.

Workman, ML (1992). Magnesium and phosphorus: The neglected electrolytes. AACN Clinical issues in Critical Care Nursing 3(3), 655–663.

Zeman, FJ (1991). Clinical Nutrition and Dietetics (2nd ed). New York: Macmillan.

CHAPTER 9

. . . .
. . . .
. . . .
. . . .
. . . .
. . . .
. . . .
. . . .
. . . .
. . . . . .

Magnesium

MAGNESIUM BALANCE

Magnesium (MG^{2+}) is a common electrolyte that is rapidly being recognized for its important role in acute and chronic illness. It is second only to potassium in intracellular concentration.

Distribution

The adult human body contains about 20 to 28 g of magnesium. Ninety-eight percent of all magnesium is located in the intracellular compartment. Of the total magnesium, 31% is in the cells and 67% is in the bones and teeth. Most of the magnesium in the bones is not exchangeable. The remaining 2% of body magnesium is located in the extracellular compartment, making magnesium the fourth most plentiful electrolyte in the serum.

Magnesium occurs in two forms, free and bound. Over one half (55%) of the serum magnesium exists as free ions (can be measured in the blood) and about one third (33%) is bound to albumin. The remainder is bound to phosphate, citrate, or other compounds (12%). The normal serum magnesium level is 1.5 to 2.3 mEq/L, 1.8 to 2.6 mg/dL, or 0.75 to 1.0 mmol/L, depending on laboratory measurement technique.

Sources

Magnesium is supplied by nutrients in the diet. The best sources of magnesium are unprocessed cereal grains (bran and oats), nuts, and legumes. Because magnesium is part of the chlorophyll molecule, green leafy vegetables are also good sources of magnesium. Other food sources include dairy products, dried fruit, meat, and fish.

Drinking water that has not been processed through a water softener provides another source of magnesium. Water with a high mineral content (called "hard" water) has been linked to a decreased incidence of cardiovascular disease and sudden death.

Recommended Daily Allowance

In 1989, the National Research Council lowered the recommended daily allowances of magnesium for men to 350 mg and for women to 280 mg. Even with lowered requirements, up to three quarters of the United States population may be consuming less than the recommended amount of dietary magnesium. Groups at the greatest risk for magnesium deficiency are elderly persons, pregnant women (especially adolescents), African Americans, athletes, alcoholics, and persons with diabetes.

Control

Not much is known about the control of magnesium in the body. Magnesium is actively transported from the extracellular fluid compartment to the intracellular compartment when needed.

Ingested magnesium is absorbed in the small intestine with an efficiency of about 30% to 40%, but the efficiency of absorption may rise to as high as 70% in a deficiency state. Like calcium absorption, magnesium absorption may be affected by the level of vitamin D. The rate of intestinal absorption is also related to the amount of calcium present. As the amount of dietary calcium is lowered, magnesium absorption is increased. Because the absorption of magnesium is primarily in the jejunum and ileum, malabsorption syndrome or ileitis can substantially affect serum magnesium levels.

The remaining magnesium that is not absorbed in the small intestine is excreted in the feces. A small amount is excreted in perspiration.

Renal reabsorption of filtered magnesium differs from the reabsorption of other major electrolytes. More than one half is reabsorbed in the thick ascending limb of the loop of Henle in the nephron. Diuretics that act on this portion of the loop of Henle are associated with increased loss of magnesium from the body.

Like the calcium level, the magnesium level is regulated by the parathyroid glands. Calcium, magnesium, and parathyroid hormone share a complex and incompletely understood relationship. It is known, for example, that an excessive serum magnesium level inhibits bone calcification and minimizes the effects of calcium on muscle and bone tissue.

Potassium concentration is also directly affected by magnesium balance. In a state of magnesium deficiency, the kidneys excrete more potassium, so that hypokalemia develops as well as hypomagnesemia. The kidneys conserve magnesium, especially in a state of hypomagnesemia. This action helps control magnesium balance. This is in contrast to the situation with hypokalemia, where the kidneys do not conserve potassium.

Functions

Magnesium is extremely important for enzyme systems as well as for neuromuscular activity. It is a cofactor in over 300 enzymatic processes involving energy metabolism and protein synthesis.

Magnesium is essential for the proper metabolism of adenosinetriphosphate (ATP), the energy source for the body's cellular reactions. Magnesium thus is essential for the operation of the sodium-potassium pump that transports ions through cell membranes and maintains membrane potentials that activate nerves and muscles. ATP also has a pivotal role in the metabolism of carbohydrates, fats, nucleic acids, and proteins.

Two other related functions of magnesium are to facilitate neuromuscular integration and stimulate the secretion parathyroid hormone. Magnesium influences the utilization of calcium by decreasing and depressing acetylcholine release at synaptic junctions in the neuromuscular system. In the process of muscular contractility, calcium serves as a stimulator and magnesium acts as a relaxer.

Additionally, magnesium is noted for having a major influence on cardiac functioning, in that it corrects dysrhythmias and counteracts the toxic effects of selected cardiac drugs. It may also reduce postoperative episodes of uncontrolled hypertension and coronary and cerebral vasospasms.

MAGNESIUM IMBALANCES

Magnesium imbalances include hypomagnesemia, a decreased serum magnesium level of less than 1.5 mEq/L, and hypermagnesemia, an increased serum level of 2 mEq/L or greater. Hypomagnesemia is the most common magnesium imbalance.

Magnesium Deficit

Refined grain products and food grown with magnesium-poor fertilizers have contributed to the reduction of magnesium intake in the American diet. Because the usual avenues for excretion are both urinary and fecal, hypomagnesemia most commonly occurs with conditions of impaired absorption of magnesium from the intestine, as in malabsorption disorders, or in situations of excessive excretion of urine or feces (polyuria, diarrhea, nasogastric suctioning), including diuresis with drugs. Hypomagnesemia is also typically present with alcoholism and withdrawal from alcohol.

Causative Factors

Chronic alcoholism is the most common cause of hypomagnesemia in the United States. Decreased intake, gastrointestinal (GI) losses, intestinal malabsorption, and increased renal excretion in the patient with alcoholism contribute to magnesium deficit.

Another common cause for decreased serum magnesium is loss from the GI tract, including vomiting, diarrhea, and nasogastric (NG) suction. The fluid from the lower GI system is richer in magnesium than that of the upper GI tract. Therefore, diarrhea and intestinal fistulas are more likely to induce hypomagnesemia than vomiting or NG suction.

Other common causes of magnesium deficit include decreased intestinal absorption, drug-related losses, and excessive loss of other related electrolytes, such as calcium and potassium. The loop (high-ceiling) and, to a lesser extent, thiazide diuretics increase magnesium excretion via the kidneys. Other magnesium-excreting drugs include the aminoglycosides (such as gentamycin and tobramycin), amphotericin B, cisplatin, and cyclosporine. Rapid administration of citrated blood (faster than 1.5 mL/kg per minute) can cause a temporary drop in serum magnesium because citrate binds with the magnesium.

Magnesium deficiency is a common electrolyte imbalance in critical care patients, but may be misinterpreted as a potassium deficit. A decreased potassium level can cause a decreased magnesium level and vice versa.

Magnesium deficit may also be seen in patients with diabetic ketoacidosis. This is thought to be the result of increased renal excretion of magnesium caused by osmotic diuresis (high glucose) and the movement of magnesium and potassium into cells with insulin therapy. Magnesium may also be excreted in excess in patients with early renal disease.

Additional causes of hypomagnesemia include burns (because of debridement), sepsis, hypothermia, pancreatitis, and hypercalcemia. Magnesium and calcium have a mutually suppressive effect such that if calcium intake is high, magnesium absorption is reduced, and vice versa.

The relationship between magnesium deficiency and chronic fatigue syndrome is being studied, although no direct cause-and-effect relationship has yet been found.

Implications for Pregnant Women

Pregnancy causes a slight decrease in serum magnesium, probably caused by a dilutional effect. No signs and symptoms of the decrease are expected in women with adequate dietary intake.

Table 9–1 summarizes the most common etiologies for magnesium deficiency.

Assessment

Physical Assessment

The most common clinical manifestations of hypomagnesemia result from neurologic irritability. Assessment findings typically include tremors, seizures, acute confusion, weakness, muscular excitability, ataxia, and cardiac dysrhythmias. Severe hypomagnesemia can lead to laryngeal stridor, coma, or sudden death.

Patients with magnesium deficit may also develop tetany and exhibit Chvostek's sign (spasm of the facial muscles) and Trousseau's phenomenon (muscle spasm of the wrist and arm). Tetany usually reflects an associated hypocalcemia (see Chap. 7 for information on calcium imbalances).

Diagnostic Assessment

Serum testing is helpful in identifying *acute* magnesium deficiency. A patient with a serum level below 1.5 mEq/L is considered hypomagnesemic.

TABLE 9–1
COMMON ETIOLOGIES FOR MAGNESIUM DEFICIT

DECREASED INTAKE OR ABSORPTION
- Malabsorption syndrome
- Chronic diarrhea
- Bowel resection (postoperative complications)
- Alcoholism (decreased intake)
- Prolonged diuretic therapy
- Nasogastric suctioning (loss of magnesium, potassium, calcium, and hydrogen ion) and excessive vomiting
- Administration of intravenous fluids
- Magnesium-free salts

Poor intake is primarily related to gross malnutrition or starvation. It can also be caused by anorexia nervosa, hyperalimentation (TPN), and, as mentioned earlier, alcoholism.

DRUG-RELATED LOSSES
- Diuretics and digitalis therapy (loop diuretics)
- Long-term insulin-treated diabetes
- Other: aminoglycosides, amphotericin B, cisplatin, and cyclosporine

EXCESSIVE LOSS WITH CONCOMITANT ELECTROLYTE CONDITIONS
- Severe saline and water deficit following severe diuresis or diarrhea
- Concomitant occurrence with diabetic acidosis
- A "bout" of severe diarrhea related to ketoacidosis
- Hypokalemia and hypocalcemia, occurring with hyperaldosteronism (Cushing's disease) and hypoparathyroidism (medical or postthyroidectomy)
- Hyperthyroidism and hypercalcemia, which may trigger release of magnesium, causing a false low magnesium level
- Excessive release of adrenocortical hormones or treatment of steroids
- Calcium loading
- Liver disease and/or failure

Serum levels of potassium and calcium are also evaluated because deficiencies of these electrolytes often coexist with hypomagnesemia.

Unlike serum testing, the renal excretion test is useful in assessing long-term or *chronic* wasting of magnesium related to drugs or other physiologic factors. A 24-hour urine specimen is collected and analyzed. A loss of more than 2 mEq per day is considered excessive renal excretion of magnesium.

Another test, the magnesium challenge test, also involves a 24-hour urine collection. The collection is followed by parenteral administration of magnesium. After the infusion, another 24-hour specimen is collected. If less than 50% to 70% of the administered magnesium is excreted, the patient may have a chronic magnesium deficiency.

As realization of the clinical importance of magnesium is growing, newer diagnostic techniques are being developed. Two new techniques for measuring levels of ionized magnesium are the ion-selective electrode and phosphorus nuclear magnetic resonance (NMR) spectroscopy. The ion-selective electrode provides a rapid and sensitive measurement of serum magnesium and may become a routine diagnostic test in several years. NMR spectroscopy allows direct measurement of magnesium in the blood and soft tissues, based on analysis on the ATP molecule.

In addition to laboratory tests, an electrocardiogram (ECG) of the patient with hypomagnesemia usually reveals atrial or ventricular dysrhythmias, especially in patients on digitalis preparations. ECG changes associated with a moderate deficiency include flattened T waves, shortening of the ST segment, and prolonged P–R and QRS intervals. In patients with a severe deficiency, inverted T waves and prominent U waves may be seen.

Nursing Diagnoses

Common nursing diagnoses for patients with magnesium deficit include:
- Nutrition, altered, less than body requirements, related to magnesium deficit
- Acute confusion, related to central nervous system effects of hypomagnesemia
- Injury, high risk for, related to seizures
- Cardiac Output, decreased, related to dysrhythmias

Collaborative Management

Medical Treatment

The objective in the treatment of hypomagnesemia is to identify and alleviate underlying causes and conditions, such as malabsorption problems, excessive loss from diuretic therapy, concurrent hypokalemia and hypocalcemia, and hyperaldosteronism. Mild hypomagnesemia that cannot be reversed by diet is treated with oral magnesium supplements. Usual nonemergency deficits are treated with 204 mg elemental magnesium (17 to 33 mEq/L) given daily in divided doses (usually twice daily) in the form of magnesium sulfate ($MgSO_4$). Pediatric doses are calculated by weight. One gram of magnesium sulfate is equal to 8.12 mEq. Common oral preparations are magnesium oxide (Mag-Ox) and Uro-Mag. The typical side effect of magnesium preparations is diarrhea, but an emergency hypotensive crisis or changes in reflexes can occur.

Hypomagnesemia may be treated with magnesium sulfate intravenously (10 to 40 mEq/L *diluted* in IV fluid). If the condition is very severe, then 2 to 4 mL of 50% $MgSO_4$ (4 mEq/mL) is given intravenously over 15 minutes, followed by the above infusion. In extremely severe deficiency, the infusions may need to be given over a period of 3 to 7 days in order to replace magnesium stores. Calcium and potassium are also administered if these electrolyte values are low. If acute magnesium intoxication occurs as a result of treatment of the hypomagnesemia, then IV calcium gluconate is given.

Serum magnesium levels must be monitored every 12 to 24 hours, and the infusion rate must be adjusted to keep the serum magnesium levels under 2.5 mEq/L. Close observation of tendon reflexes is advisable since reduced tendon reflexes suggest hypermagnesemia. Magnesium sulfate can be given intramuscularly by the Z-tract method, but it is very painful.

Nursing Interventions

The goal of nursing care is to anticipate, monitor, and report the complications associated with a decreased magnesium level. For example, because the patient is prone to seizures, seizure precautions should be implemented. Cardiac monitoring is also essential because the patient may experience a life-threatening dysrhythmia. Monitoring of vital signs is crucial in detecting cardiac or respiratory problems. Monitoring for tetany, assessing neuromuscular irritability, and monitoring laboratory values of magnesium, potassium, and calcium are also crucial interventions.

Accurate intake and output documentation is particularly important because the kidneys increase urinary output if the patient receives too much magnesium. Emergency drugs, such as calcium gluconate, should be readily available.

Another important role for the nurse is monitoring the patient during magnesium administration. Table 9–2 describes the nursing implications for patients receiving intravenous magnesium.

As with any electrolyte imbalance, patient education is vital. The nurse teaches the patient which foods are high in magnesium, calcium, and potassium. If the patient continues on electrolyte supplements at home, instruction about the administration, side effects, and/or toxic effects of these drugs is essential. The nurse should review with the patient signs and symptoms that should be reported to the physician or other health care provider.

Patient education should also include information about how to prevent hypomagnesemia, especially the need to treat alcoholism. The signs and symptoms of hypomagnesemia should be reviewed for patients at high risk for the imbalance.

Magnesium Excess

Causative Factors

Magnesium excess, or hypermagnesemia, occurs far less frequently than magnesium deficit. It most commonly occurs in patients with advanced renal failure, or those patients with a glomerular filtration rate of less than 30

TABLE 9–2
NURSING INTERVENTIONS ASSOCIATED WITH
INTRAVENOUS MAGNESIUM ADMINISTRATION

- Always give diluted magnesium.
- Carefully check the concentration of the solution prescribed.
- Check deep tendon reflexes (such as the knee jerk) and respiratory rate before drug administration and during drug infusion; if change is noted, discontinue the drug and notify the physician.
- Monitor vital signs and laboratory assessments for magnesium.
- Teach the patient that a flushed sensation and sweating may occur during infusion.
- Monitor urine output carefully; the patient's output should be at least 30 mL/h to receive magnesium.
- Do not give high doses of opioid analgesics with magnesium because both are central nervous system depressants.
- Do not give with calcium preparations, dobutamine, aminoglycosides, hydrocortisone, alkalies, procaine, and phytonadione (Aquamephyton).
- Use the drug with caution in patients with impaired renal function; monitor BUN and creatinine levels carefully.

mL/min. With normally functioning kidneys, increased ingestion or parenteral administration of magnesium is not usually associated with an elevated magnesium level. The common causes of increased serum magnesium are:

- Chronic renal disease, including uremia, especially in elderly persons
- Excessive use of laxatives containing magnesium, especially in elderly persons, such as milk of magnesia and magnesium citrate solutions
- Overuse of antacids, such as Di-Gel, Gelusil, Mylanta, Riopan, and Maalox
- Extracellular fluid volume deficit, resulting in oliguria
- Iatrogenic reasons, such as overcorrection of hypomagnesemia
- Treatment of toxemia with magnesium
- Untreated acute diabetic ketoacidosis
- Hemodialysis with hard water or dialysate too high in magnesium
- Adrenal insufficiency
- Pheochromocytoma

Pregnant women are of special concern with regard to magnesium excess since magnesium sulfate may be utilized in the treatment of toxemia. Excessive magnesium administration can cause seriously elevated magnesium levels, both maternal and fetal, especially if the infusion rate and urinary output are not carefully monitored. To improve perinatal outcome, calcium, magnesium, and zinc supplementation may be administered instead of magnesium sulfate alone.

Assessment

Physical Assessment

Hypermagnesemia may be classified as mild (3 to 5 mEq/L), moderate (6 to 7 mEq/L), severe (10 to 11 mEq/L), and emergency (12 to 15 mEq/L) levels. A patient with a mild degree of hypermagnesemia is usually asymptomatic. Neurologic manifestations begin to occur at levels of 6 to 7 mEq/L, and are observable as symptoms of neurologic depression, such as drowsiness, sedation, lethargy, respiratory depression, muscle weakness, and areflexia. Severe hypotension is also a classic sign, usually concurrent with nausea and vomiting and occurring as early as an increased elevation of 6 to 7 mEq/L.

Severe hypermagnesemia (over 10 mEq/L) results in a paralysis of voluntary muscles, leading to a flaccid quadriplegia. When an emergency level of 12 to 15 mEq/L is reached, coma occurs and paralysis of the respiratory center leads to respiratory arrest. In addition to hypotension, other cardiac findings include bradycardia and dysrhythmias. Table 9–3 lists manifestations associated with increased levels of serum magnesium.

Diagnostic Assessment

The diagnosis of hypermagnesemia is confirmed by a serum magnesium level of 2.3 mEq/L or greater. As previously mentioned, it is also important to note elevation of potassium and calcium levels. These conditions are often

TABLE 9–3
CLINICAL MANIFESTATIONS OF HYPERMAGNESEMIA

Clinical Manifestation	Serum Mg (mEq/L)
Vague restlessness; fatigue	2–3
Nausea, vomiting, and hypotension	3–5
Sedation	4–7
Muscle weakness	4–7
Hyporeflexia; decreased deep tendon reflexes (DTRs)	4–7
Lethargy	5–7
Bradycardia; ECG changes	5–10
Voluntary muscle paralysis, absent DTRs, respiratory paralysis	10–15
Coma and respiratory arrest	12–15
Cardiac arrest	>15

associated with hypermagnesemia caused by renal failure (also Chaps. 4 and 7). If hypermagnesemia does not present with renal failure, then hypocalcemia may occur; that is, as the serum magnesium level increases and the serum calcium level decreases in the presence of intact renal function.

A patient with magnesium excess often experiences cardiac dysrhythmias, which can be detected by ECG. Typical ECG changes include peaked T waves, increased P–R intervals, and widened QRS complexes. With extreme magnesium elevations, prolongation of the Q–T interval occurs. Atrioventricular (AV) block, premature ventricular contractions (PVCs), and cardiac arrest may also occur.

Nursing Diagnoses

The major nursing diagnoses associated with hypermagnesemia are:

- Injury, high risk for, related to muscle weakness and/or paralysis
- Knowledge Deficit, related to magnesium-containing drugs
- Cardiac Output, decreased, related to dysrhythmias

Collaborative Management

Medical Treatment

Prevention is the best treatment for magnesium excess. For mild elevations, management is directed toward removing the source of the excess magnesium. For example, the patients who overuse antacids or laxatives need to be informed that such practices pose a danger of magnesium excess.

The goal of treatment for moderate or severe hypermagnesemia is to identify and correct underlying conditions, primarily by increasing renal excretion in order to eliminate excess magnesium. Thus, forcing fluids or using high-ceiling, or loop, diuretics, such as furosemide (Lasix), is usually indicated. Increasing fluid intake should always be done with caution and with careful monitoring for renal insufficiency. In the elderly, forcing fluids may contribute to congestive heart failure.

If a crisis occurs (i.e., serum levels are over 7 to 10 mEq/L, then a magnesium antagonist is utilized for temporary relief of symptoms. A typical magnesium antagonist used in emergency situations is 10% calcium gluconate, 10 to 20 mL intravenously over 10 minutes. Magnesium excretion by the kidneys, in a patient with severe renal failure, can be promoted by administration of 150 to 200 mL/h of calcium gluconate [20 mL of 10% solution added to a liter of 0.9% saline (normal saline)].

Prompt, supportive mechanical ventilation is crucial for patients with severe magnesium excess. A temporary pacemaker may be needed for bradycardia that does not respond to medication. If the excess magnesium cannot be naturally eliminated through the kidneys, or if renal function is failing, then dialysis may be necessary (see Chaps. 16 and 17). Depending on the severity of the hypermagnesemia, peritoneal dialysis or hemodialysis may be used independently or in combination. The most important aspect of this treatment is that magnesium salts should not be used in the dialysis of patients with acute or chronic renal failure. If the patient is on either peritoneal dialysis or hemodialysis, the dialysing solution should be magnesium-free.

Nursing Interventions

Hypermagnesemia may be prevented through patient education concerning the excessive use of laxatives and antacids containing magnesium. This instruction is particularly important for elderly patients, or patients with compromised renal systems.

Patients should be guided to avoid over-the-counter antacids that contain magnesium hydroxide. They can be instructed to select aluminum hydroxide gels instead. Avoiding laxatives, which contain high levels of magnesium, may mean that increased dietary fiber and fluids need to be considered to maintain comfortable bowel function. Dietary changes to restrict magnesium are rarely recommended.

Nursing care also includes monitoring of vital signs, level of consciousness, and neurologic and muscular activity. Frequent blood pressure and pulse monitoring are essential to note hypotension, bradycardia, and dysrhythmias. Neurologic assessment includes determining the level of consciousness and reflex ability. Any such neurologic changes require protection of the patient from injury.

CRITICAL THINKING ACTIVITIES

A 36-year-old trauma patient is admitted to the critical care unit as a result of a motor vehicle accident. According to the family, he has a history of drug and alcohol abuse. In addition to a number of fractures, the patient has a ruptured spleen and concussion. On the second day of hospitalization, laboratory values are as follows:

Hemoglobin (Hgb)	10.6 g/dL
Blood urea nitrogen (BUN)	48 mg/dL
Sodium	140 mEq/L
Potassium	3.2 mEq/L
Magnesium	1.1 mg/dL

QUESTIONS

1. *What electrolyte imbalances does this patient have?*

2. *What factors possibly contributed to the electrolyte imbalances?*

3. *What interventions are needed at this time, including assessments?*

4. *What teaching should the nurse plan considering the patient's health status?*

REFERENCES AND READINGS

Kelso, LA (1992). Fluid and electrolyte disturbances in hepatic failure. AACN Clinical Issues in Critical Care Nursing 3(3), 681–685.

Ricci, J, et al (1990). Influence of magnesium sulfate-induced hypermagnesemia on the anion gap: Role of hypersulfatemia. American Journal of Nephrology 10, 409–411.

Rude, RK (1993). Magnesium metabolism and deficiency. Endocrinology and Metabolism Clinics of North America 22(2), 377–396.

Salem, M, Munoz, R, & Chernow, B (1991). Hypomagnesemia in critical illness. Critical Care Clinics of North America 7, 225–252.

Terry, J (1991). The other electrolytes: Magnesium, calcium and phosphorus. Journal of Intravenous Nursing 14(3), 167–176.

Van Hook, K (1991). Hypermagnesemia. Critical Care Clinics of North America 7, 215–223.

Workman, ML (1992). Magnesium and phosphorus: The neglected electrolytes. AACN Clinical Issues in Critical Care Nursing 3(3), 655–663.

CHAPTER 10

· · · · · · · ·
· · · · · · · ·
· · · · · · · ·
· · · · · · · ·
· · · · · · · ·
· · · · · · · ·
· · · · · · · ·
· · · · · · · ·
· · · · · · · ·
· · · · · · · ·

Copper

COPPER BALANCE

Copper is an essential trace mineral. It is a component of various proteins and is required for the formation and development of red blood cells (RBCs). The clinical symbol for this element is Cu. Copper was first identified as an essential mineral when it was noted that children fed formulas containing cow's milk, sucrose, and cottonseed oil developed a classic syndrome of anemia, neutropenia, and hypoproteinemia. These disorders were not corrected by iron supplementation. Further research identified a copper deficiency as the cause of this classic syndrome. The deficiency was also noted in premature infants placed on high-calorie feedings.

Distribution

Approximately 80 to 120 mg of copper is present in the body tissues of a 70-kg adult. The greatest concentrations are in organ tissues, such as brain, liver, heart, and kidney, which contain about one half of the body's copper. The remaining one half is located in bone and muscle, but at lower concentrations than in the organs. Copper is also found in the plasma and blood cells, especially erythrocytes.

The copper protein ceruloplasmin contains about 90% of the plasma copper. The remaining 10% is bound to albumin and amino acids. Copper, once absorbed, is very rapidly transferred to the red blood cells. The ratio of cell copper to plasma copper is about 0.70.

During pregnancy, copper plasma or serum levels increase to almost double that of nonpregnant states. Ceruloplasmin is also increased and both abruptly decrease after birth. Copper changes may be linked with placental insufficiency, intrauterine death, and threatened abortion.

Both serum and urinary copper levels can be assessed. Additionally, copper activity in RBCs, liver, or connective tissue can be assessed as can RBC activity. Evaluating ceruloplasmin may not be the best way to assess copper status. According to the National Research Council (1989), assessment of erythrocyte superoxide dismutase (SOD) activity may be a more useful indicator. Copper uptake studies are still being refined. Plasma copper levels do not correlate with intake.

Sources

Copper is supplied by the diet through foods such as organ meats (heart, kidney, liver), shellfish, nuts, raisins, cereals, and legumes (Mahan & Arlin, 1992) (Table 10–1). The copper content of plant sources varies with the amount of copper in the soil.

The need for adequate levels of copper in the diet of infants and young children raises concerns about the levels of copper in milk sources. Cow's milk is relatively low in copper compared to human milk. The copper level for cow's milk is 0.09 to 0.21 mg/L in winter and 0.17 to 0.50 mg/L in summer, whereas human colostrum contains 0.95 to 1.23 mg/L. Copper levels of second month breast milk decline to 0.60 to 0.95 mg/L and decrease further to about 0.30 to 0.70 mg/L by the ninth month.

TABLE 10–1
COPPER CONTENT OF SELECTED FOODS

Food	Copper Content (mg)
Beef liver, fried, 3 oz	2.4
Cashews, dry roasted, ¼ cup	0.8
Black-eyed peas, dried, cooked, ½ cup	0.7
Molasses, blackstrap, 2T	0.6
Sunflower seeds, ¼ cup	0.6
Chocolate chips, semisweet, ¼ cup	0.5
V-8 juice, 1 cup	0.5
Tofu, firm, ½ cup	0.5
Beans, refried, ½ cup	0.5
Instant breakfast, fortified, 1 envelope	0.5
Cocoa powder, 2T	0.4
Prunes, dried, 10	0.4
Salmon, baked, 3 oz	0.3
Tahini (sesame butter), 1T	0.2
Pizza, cheese, ⅛ of 15″	0.2
Bread, whole wheat, 1 slice	0.1
Milk chocolate, 1 oz	0.1
Milk, 2% fat, 1 cup	0.1

Source: USDA: Composition of Foods. USDA Handbook No. 8 Series. ARS, USDA, Washington, DC, 1976–1986.

Recommended Daily Allowance

According to the National Research Council (1989), no recommended dietary allowance has been established. Instead, a safe and adequate range of dietary intake of copper has been set at 1.5 to 3 mg per day for adults and 0.4 to 0.6 mg per day for infants. Children's needs range from 0.7 to 2.5 mg per day, depending on age. Pregnant women should have 75 micrograms (μg) of elemental copper per day. Daily copper intake in the United States averages between 1 and 2 mg per day (Lutz & Przytulski, 1994).

Implications for Infants and Children

Infants particularly are in great need for copper, especially during the first year of life. Although body stores in healthy newborn infants are generally very high, it is recommended that daily intake for breast-fed infants should be 0.08 mg/kg of body weight. Infants from birth to 6 months should receive a daily dietary copper intake ranging between 0.4 and 0.6 mg per day. Infants 6 months to 1 year should receive 0.6 to 0.7 mg per day.

Implications for Elderly Persons

Absorption is a factor in achieving adequate levels of copper for everyone, regardless of age. In elderly persons, intestinal absorption of all nutrients decreases. However, copper can accumulate with age, producing fatty deposits, cellular necrosis, and pigment changes.

Control

Copper is absorbed from the stomach and small intestine through the process of active and passive transport with the copper bound to the protein albumin as the most likely transport mechanism. Although research regarding the exact process by which copper is controlled is limited, some research indicates that albumin-bound copper is stored in the liver as metallothionein or secreted back into the cells through a process involving ceruloplasmin.

Zinc interferes with copper absorption. High levels of zinc result in an increase in the amount of copper that binds to metallothionein. The bound copper is not transported into the serum and is, instead, lost in the feces. A high-fiber diet may also cause a decrease in copper absorption. Additionally, cadmium, molybdenum, and sulfate alter copper absorption.

The kidneys can conserve copper by the reabsorption in the tubules. Dietary copper that is not absorbed in the gastrointestinal tract is eliminated in the feces. Small amounts of copper are present in the urine, perspiration, and menstrual flow. The amounts of copper that are eliminated are as follows: stool, about 2.5 mg per day; menstrual flow, 0.55; and urine, about 5.8 μg/100 mL.

Functions

Copper is crucial to selected enzymes, known as "cuproenzymes." These are amine oxidases important to the development of protein-based tissue, especially those with collagen and elastin cross-linking. Cross-linking deficiencies can result in muscular defect or bone fragility. Key enzymes utilizing copper are:

- Cytochrome oxidase, a pigment concerned with cellular respiration and a catalyst in the oxidation process
- Lysine oxidase, a basic amino acid essential for growth
- Tyrosinase, used to oxidize melanin, the basic pigment in skin
- Ascorbic acid oxidase, which is involved in the proper utilization of vitamin C

Adequate amounts of copper are needed for cardiovascular and skeletal integrity, central nervous system function, hematologic function, and hair keratinization and pigmentation. Specifically, copper is necessary for red blood cell development, blood clotting (component of factor V), phospholipid synthesis affecting nervous tissue, and melanin (skin pigment) formation. Minute traces of copper greatly facilitate the utilization of iron for hemoglobin synthesis, called the "iron-copper link."

Copper is very important in preventing infection, since insufficient copper can result in leukopenia, a decrease in the white blood cells needed to fight infection. In addition, copper deficiency is related to impaired antibody response, a depressed immune system, and decreased microbicidal activity of granulocytes.

Copper is also involved in the development of collagenous tissue and thus in the maintenance of skin integrity, as well as skeletal and vascular integrity. A copper deficiency results in diseased bones that resemble those seen in scurvy.

COPPER IMBALANCES

Copper imbalances are not common, but when they occur, they can have devastating consequences. A deficit is the most likely imbalance of copper to be seen in the clinical setting.

Copper Deficit

A copper deficit, or hypocupremia, exists when there is an intake of less than 1 mg per day or a blood level less than 85 mg/dL.

Causative Factors

People who have an inadequate oral nutritional intake or who are receiving total enteral feedings or total parenteral nutrition (TPN) without copper supplementation are at risk for copper deficiency. Diseases that cause abnormal copper metabolism, chronic malnutrition, and diarrhea can also result in a copper deficit.

Initially, copper deficiency was not noted with TPN because protein hydrolysate, containing copper, was part of the parenteral nutrition solution. But when crystalline amino acids, which do not contain copper, were used to prepare the parenteral nutrition solution, copper deficiencies resulted. Subnormal values of copper can be seen within weeks of initiating TPN. Chromium and selenium deficiencies may occur in patients receiving long-term TPN as well. These minerals are now added to parenteral nutrition solutions.

Decreased copper plasma levels are also seen with malabsorption diseases, such as celiac and tropical sprues, nephrotic syndrome, and protein-losing enteropathies.

The three syndromes causing copper deficiencies noted in children are syndrome with anemia, syndrome with chronic malnutrition and diarrhea, and Menkes' (kinky hair) syndrome. In the first syndrome, infants have severe anemia related to copper and iron deficiencies. The second syndrome is caused by infant formulas containing inadequate copper or by breast-feeding supplemented with cow's milk, which supplies inadequate amounts of copper. Menkes' syndrome, a sex-linked defect related to dysfunctional copper absorption, is characterized by growth retardation, increased urinary loss of copper, hypothermia, mental deterioration, defective keratinization, and hair pigmentation.

Assessment

Physical Assessment

Copper deficiency develops slowly because copper is stored in the liver in appreciable amounts. The classic syndrome of deficiency is characterized as anemia, neutropenia, and hypoproteinemia. Neutropenia and leukopenia are the best-characterized early signs of copper deficiency in children. As copper deficiency develops, a progression of symptoms appears: a fall in both serum copper and ceruloplasmin levels, failure of iron absorption, neutropenia, leukopenia, bone demineralization, erythropoietic failure, and death. Patients may have subperiosteal hemorrhages and hair and skin depigmentation.

Copper deficit plays an important role in the origin of specific disorders. Thus the occurrence of certain disorders should be viewed as possible manifestations of copper deficiencies. For example, a correlation between low copper levels and abnormal bone formation with spontaneous fractures, reproductive failure, heart failure, and arterial and cardiac aneurysms has been reported. Even mild copper deficiency states can be related to failure to thrive (FTT) phenomena in infants and in prematurity. Table 10–2 summarizes common assessment findings associated with copper deficiency.

Diagnostic Assessment

Copper deficit develops when less than 1 mg of copper is ingested daily. Generally, even poorly balanced diets provide adequate copper to meet the individual's needs. Children need a daily copper intake between 0.05 and 0.1 mg/kg of body weight. Zeman (1991) reports that exact amounts at which a deficiency occurs are not known.

TABLE 10–2
ASSESSMENT FINDINGS
COMMONLY PRESENT
IN PATIENTS WITH
COPPER DEFICIENCY

Anemia
Neutropenia
Hypoproteinemia
Decreased serum copper and ceruloplasmin levels
Decreased iron level
Hypercholesteremia
Bone demineralization
Subperiosteal hemorrhages
Hair and skin depigmentation
Spontaneous fractures
Heart failure
Reproductive dysfunction
Arterial and cardiac aneurysms
Failure-to-thrive syndrome
Premature birth

Because copper works with iron to produce red blood cells, the red blood cell count may be decreased in a person with hypocupremia. Hypercholesteremia may also result from copper deficit in which there is a high zinc to copper ratio. An exceptionally high zinc level impairs the absorption of copper; the resulting hypocupremia leads to high cholesterol levels and coronary heart disease.

Nursing Diagnoses

Common nursing diagnoses for copper deficiency include:

- Nutrition, altered, less than body requirements, related to inadequate intake or intestinal, diaphoretic, or metal losses
- Fatigue, related to anemia
- Skin Integrity, impaired, related to reduced resistance to infection and depigmentation
- Growth and Development, altered, related to the effects of inherited metabolic disorder

Collaborative Management

Medical Treatment

Medical treatment focuses on increasing copper to adequate levels. A daily dose of 0.5 to 1.5 mg copper is recommended for adults; 20 micrograms (µg)/kg per day is recommended for infants and children (Gahart, 1994). Infants and patients on TPN should have the nutritional formula reviewed to be sure adequate copper is included.

The underlying cause of disorders known to be associated with copper deficits should be identified and corrected along with appropriate monitoring and adjustment of copper intake. One milligram of copper sulfate combined with zinc sulfate is recognized as a component in small bowel losses. At least 12 mg/L of zinc are needed to correct the small bowel losses. In diarrhea, 17 mg/L of zinc are needed to correct the copper loss. Replacement fluid should also contain 10 to 12.5 μg of chromium chloride and 0.4 to 0.8 mg manganese.

Adults with malabsorption syndrome also need to be assessed for copper deficiency, along with other multiple fluid and electrolyte disorders, such as saline deficit, and loss of potassium and calcium.

Nursing Interventions

Nursing interventions are focused on proper nutritional intake. Special attention should be given to infants on formulas, especially when cow's milk is used. Breast-feeding should be encouraged when possible. Also, parenteral nutrition formulas and the individual's response to parenteral nutrition need to be closely monitored.

The nurse should teach the patient to consume foods high in copper, although copper supplementation may still be required to meet daily needs initially. Patients also need to be reminded that zinc supplements of more than 1500 mg per day and large amounts of vitamin C can decrease copper absorption and availability. High dietary fiber intake also increases the dietary requirement for copper in the body (Davis & Sherer, 1994).

Copper solutions can be prepared for intravenous administration. They may be given as a single element or combined with other elements as needed. For patients with anemia, attention is directed toward correction of the anemia by restoring both copper and iron to the correct levels.

The nurse helps the anemic patient organize the proper balance of rest and activity periods until the anemia is corrected. If metabolic function is sufficiently compromised, the patient may require oxygen for breathlessness and hypoxia. If the white blood cell (WBC) count is reduced, it is important to pay attention to the prevention of infections and injury, including pathologic fractures. Patients undergoing tube feeding or suffering from malabsorption syndrome need replacement of trace minerals, usually using a standard formula of combined elements.

Copper Excess

Copper excess, or hypercupremia, caused by dietary excess is uncommon. Although the safe and adequate level of copper is regarded to be between 1.5 and 3 mg per day, an adult could probably consume as much as 5 mg per day, and occasionally as much as 10 mg per day, without harm. When hypercupremia does occur, it is generally related to a physiologic dysfunction.

Causative Factors

Copper excess is usually attributed to kidney and liver dysfunction that results in copper being deposited in tissues. Serum concentration levels are increased in patients with infections with unknown etiology, and in patients with pellagra. Ceruloplasmin increases during pregnancy and in women on oral contraceptives and estrogen therapy. Serum copper concentrations in pregnant women are approximately twice the values found in nonpregnant women, so a possible relationship between copper levels and the development of toxemia needs to be considered.

The key disorder associated with copper excess is Wilson's disease, a rare autosomal recessive disorder slightly more common in males than in females. Wilson's disease is due to the genetic absence of the enzyme that synthesizes ceruloplasmin, a copper-binding protein. As a result, copper accumulates in tissue. Clinically, Wilson's disease usually presents between 6 and 20 years of age—rarely before the age of 4 years. Biliary cirrhosis, primary and secondary, can result in excess storage of copper in the liver, kidney, and spleen.

Assessment

Physical Assessment

The primary diagnostic feature in identifying Wilson's disease is Kayser-Fleischer rings, which are greenish-brown pigmented rings at the outer margin of the cornea. Wilson's disease is manifested by degeneration of brain tissue, cirrhosis of the liver, and renal impairment. Copper is excreted in the bile and then deposited in tissues of the liver, kidney, cornea, and brain. As a result, the patient has signs and symptoms of liver and renal disease, including jaundice, ascites, bleeding tendencies, pruritis, and urinary output changes. When copper is deposited in the brain, mental changes, including confusion and personality disorders, can occur.

Prosad (1985) reports that a plasma copper to zinc ratio of greater than 2.0 in chronic alcoholics increases the incidence of hepatitis and cirrhosis of the liver. With an increased ratio, more cases of delirium tremens were observed, as well as a prolonged hallucinatory state.

Diagnostic Assessment

Laboratory results show a decrease in serum ceruloplasmin and copper, along with increased stores in the liver and other tissues, and increased urinary copper excretion. The serum phosphate levels may also be decreased.

Other diagnostic assessment focuses on evaluation of liver and renal function, and determination of laboratory values for blood urea nitrogen, serum creatinine, transaminases, and direct and indirect serum bilirubin.

Nursing Diagnoses

Common nursing diagnoses for patients with a copper excess include:
- Nutrition, altered, more than body requirements, related to the effects of genetic disorder
- Skin Integrity, impaired, related to pruritus

Collaborative Management

Medical Treatment

The objectives of treatment are to relieve the tissue of excess copper and prevent reaccumulation. The treatment of copper excess, especially in Wilson's disease, is to place the patient on a low-copper diet and to administer penicillamine (Cuprimine, Depen). Penicillamine is a chelating agent that binds the copper and allows it to be excreted in the urine. The initial dose is usually 250 mg qid. Doses that produce a urinary copper excretion rate of greater than 2 mg per day should be continued for at least 3 months (Hodgson, Kizior, & Kingdon, 1993). Side effects to watch for with penicillamine are rash, anemia, leukopenia (especially with abrupt reduction of copper), nephrotic syndrome, and a lupus-like syndrome. Pyridoxine may also be administered to prevent pyridoxine deficiency, which can occur with prolonged use of penicillamine treatment.

Dietary restriction includes exclusion of organ meats, shellfish, nuts, legumes, broccoli, whole wheat, chocolate, cocoa, and tea. Tap water may also be a source of copper. Oral administration of potassium sulfide may be used to precipitate copper into the bowel as insoluble copper sulfide; this prevents absorption of copper from the gastrointestinal tract and consequently accumulation of copper from dietary sources.

Biliary cirrhosis, whether the result of copper excess or of jaundice, may be treated with cholestyramine to chelate bile salts and reduce itching. Cholestyramine can cause severe constipation.

Major copper toxicity is treated with edetate calcium disodium, 15 to 25 mg/kg intravenously for 5 days. The treatment course is then interrupted to check the results. Further treatments may be required depending on the severity of the toxicity.

Nursing Interventions

Nursing interventions focus on dietary treatment and skin care. Attention is also directed toward care of subsequent conditions such as kidney and liver disease, including ascites and gastrointestinal problems.

The nurse teaches the patient what high-copper foods to avoid. Information about the therapeutic and adverse effects of drug therapy also needs to be explained. Patients receiving penicillamine or other chelating agents should take the drug on an empty stomach (Table 10–3).

TABLE 10–3

NURSING INTERVENTIONS ASSOCIATED WITH
PENICILLAMINE THERAPY FOR WILSON'S DISEASE

- Give the drug on an empty stomach, preferably 1 hour before or 2 hours after meals, other drugs, or milk.
- Crush tablets; do not crush or break capsules.
- Take with large amounts of water to facilitate excretion.
- Do not give with iron therapy or antacids.
- Monitor elderly patients carefully for adverse effects, especially patients with impaired kidney or liver function.
- Monitor complete blood count and differential.
- Assess for hematuria and proteinuria.
- Observe for pruritic, reddened, and maculopapular rash.
- Teach patients about the adverse effects, as well as the possibility that the drug can cause anorexia and nausea.

CRITICAL THINKING ACTIVITIES

A patient is admitted with a probable diagnosis of Wilson's disease. The initial laboratory test results are as follows:

Blood urea nitrogen (BUN)	62 mg/dL
Serum creatinine	2.4 mg/dL
Aspartate aminotransferase (AST)	58 IU/L
Alanine aminotransferase (ALT)	65 IU/L
Total bilirubin	2.7 mg/dL
Serum ammonia	138 μg/dL

QUESTIONS

1. *Why are the BUN and creatinine levels elevated in this patient?*

2. *Of what significance are the elevations of the AST, ALT, and bilirubin levels (liver function studies)?*

3. *Is the serum ammonia level within the normal range? What would be expected in a patient with Wilson's disease?*

4. *What are the treatment options for this patient?*

REFERENCES AND READINGS

Breslow, R (1991). Nutritional status and dietary intake of patients with pressure ulcers: Review of research literature 1943 to 1989. Decubitus 4(1), 6–21.

Chandra, RK (1985). Trace Elements in Nutrition of Children. New York: Raven Press.

Davis, J, & Sherer, K (1994). Applied Nutrition and Diet Therapy for Nurses. Philadelphia: WB Saunders.

Gahart, BL (1994). Intravenous Medications (10th ed). St. Louis: Mosby–Year Book.

Hodgson, BB, Kizior, RJ, & Kingdon, RT (1993). Nurse's Drug Handbook. Philadelphia: WB Saunders.

Lutz, CA, & Przytulski, KR (1994). Nutrition and Diet Therapy. Philadelphia: FA Davis.

Mahan, LK, & Arlin, M (1992). Krause's Food, Nutrition & Diet Therapy (8th ed). Philadelphia: WB Saunders.

National Research Council (1989). Recommended Dietary Allowances. Washington, DC: National Academy Press.

Pennington, JAT, & Young, BE (1991). Total diet study nutritional elements, 1982–1989. Journal of the American Dietetic Association 91(2), 179–183.

Prosad, AS (1985). Diagnostic approaches to trace element deficiencies. In RK Chandra (ed), Trace Elements in Nutrition of Children. New York: Raven Press.

Yadrick, MK, et al (1989). Iron, copper, and zinc status: Response to supplementation with zinc or zinc and iron in adult females. American Journal of Clinical Nutrition 49(1), 145–150.

Zeman, FJ (1991). Clinical Nutrition and Dietetics. New York: Macmillan.

CHAPTER 11

. . . .
. . . .
. . . .
. . . .
. . . .
. . . .
. . . .
. . . .
. . . .

Iron

IRON BALANCE

Iron is the key mineral responsible for providing oxygen to cells for metabolic processes. When insufficient amounts of iron are available, hypoxia results. In hypoxic states, the patient experiences fatigue and is susceptible to infections. In addition, hypoxia can lead to metabolic acidosis with serious ramifications.

Iron toxic states can also be lethal. According to Hurley (1992), 6 to 12 iron tablets can kill a young person. Among children, iron poisoning deaths caused by iron overdose are second only to deaths related to aspirin overdose. The United States Food and Drug Administration has recently issued a call for improved safety caps on medication packages and bottles containing iron, including multivitamins with iron.

Distribution

Iron exists in the body as bound or free. The amount of iron in the human body is proportional to one's weight and averages 45 mg/kg of body weight. Approximately 60% to 70% of the 4 to 5 g of iron in the human body is in the form of hemoglobin. About one fifth of the total body iron is stored in the liver, spleen, and bone marrow in the form of the iron protein ferritin. When excess iron accumulates in the blood because of rapid red blood cell destruction, the surplus iron is stored in the liver as another compound, hemosiderin. The remaining iron is in the bloodstream, primarily bound to transferrin, a protein that transfers iron in and out of storage depots to the blood plasma (Fig. 11–1).

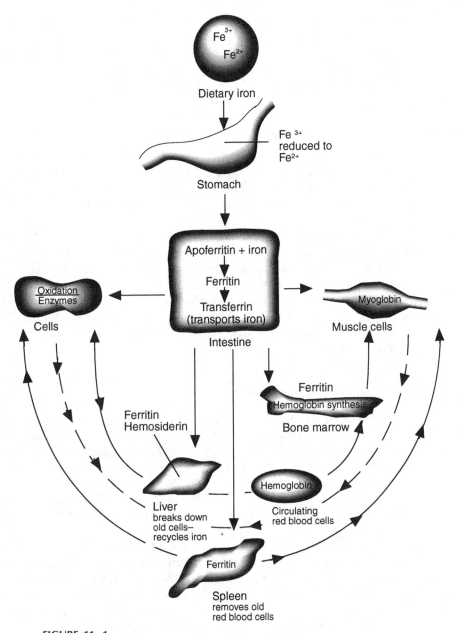

FIGURE 11–1
In the process of iron absorption, iron is absorbed primarily in the small intestine and may be transported or stored to meet the body's needs. (From Lutz, CA, and Przytulski, KR: Nutrition and Diet Therapy. FA Davis, Philadelphia, 1994, p 186.

Large amounts of iron are incorporated in the red blood cells (RBCs). In a woman with a body weight of 70 kg, the RBCs contain about 2180 mg of iron; for a man the figure would be 2750 mg. Bone marrow and muscle mass contain 610 and 520 mg of iron, respectively. Bone marrow releases no more than 5 mg of iron per day, so it is not an adequate buffering source in the event of acute hemorrhage. Blood volume can be provided from the spleen in acute hemorrhage.

Because some body iron is bound to protein and appreciable quantities of iron cannot exist in the body in a free state, a protein deficit state, such as hypoalbuminemia, can contribute to an iron deficiency state.

Sources

Rich iron sources are organ meats, red meats, fish, poultry, grains, dark green leafy vegetables, and prunes. Iron in the form of heme is obtained from meat. Nonheme iron is obtained from vegetables and grains. Bran flakes, prune juice, cooked navy beans, and beef liver are exceptionally good sources.

In the United States, the daily intake of iron in a typical 2500-calorie diet is about 15 mg. The absorption rate for heme iron is high, with an average of about 25% of the heme iron consumed being absorbed. Heme iron consumption is only about 1 to 3 mg per day, however, and accounts for less than 10% of the iron in the daily diet. Nonheme iron constitutes a larger percentage of the daily dietary iron but is less bioavailable, and its absorption is highly influenced by dietary factors. Only about 5% of the nonheme iron is absorbed.

Implications for Infants and Children

Iron sources for infants include breast milk, which is composed of about 0.3 mg of iron per liter of milk, and cow's milk, containing about 0.5 mg of iron per liter of milk. The absorption rates from these sources are 50% and 10%, respectively. This amount is sufficient for the first 3 months, but if milk continues to be the only food source after the 3-month period, an iron deficiency can occur. Iron is typically added to infant foods, including cereal. A meat source is recommended as soon as the infant's gastrointestinal tract is able to digest it.

Several factors affect the ability of the body to absorb iron. Large amounts of alcohol, a high intake of calcium, and vitamin C *enhance* the intestinal absorption of iron. Antacids, phytic acid (phytates) from cereals, and oxalic acid from certain vegetables combine with iron and *decrease* its availability. Excess zinc, manganese, and copper also reduce absorption of iron and each other. Nonheme iron is reduced by tannates found in tea and coffee (Lutz & Przytulski, 1994).

Recommended Daily Allowance

Intake recommendations are based on estimated losses, as well as the amount of estimated gastrointestinal absorption. Absorption is affected by the efficiency of the heme and nonheme irons.

The recommended daily iron intake for infants is 6 mg; for adolescent boys, 12 mg; for women of reproductive age and lactating women, 15 mg; and for pregnant women, 30 mg. These requirements are in contrast to those for men (19 to 51 years), who need about 10 mg per day.

Implications for Elderly Persons

As one ages, gastric acidity decreases, and diminished gastric acidity for any reason impairs the absorption of iron. Nevertheless, iron intake and absorption are adequate in most people until about the age of 75. After this age, elderly people tend to consume less meat, which contains the heme iron that is most readily absorbed. Therefore, elderly people are at risk for iron deficiency.

Control

The biologic levels of most minerals are regulated by increased or decreased excretion. Iron is an exception in that its level is largely controlled by the rate of intestinal absorption. In an iron deficiency state, the percentage of iron absorbed is increased. This mechanism may not be adequate to correct anemia in a person with marginal nutritional intake.

The exact mechanisms controlling iron levels are under continuous study. The iron body pool is the single most important factor. When the pool size decreases, iron absorption increases. The increase in iron absorption when iron storage is low is believed to involve transferrin in the intestinal cells. Vitamin C is essential to the absorption process, as it reduces iron from the ferric to the ferrous oxidation state, which is much more easily absorbed than the ferric.

The role of transferrin in the gastrointestinal tract is to carry iron from the intestinal mucosa to the intestinal cell. Iron, when released by transferrin, combines with the protein apoferritin to form ferritin, an iron compound that can be stored. Iron is stored as ferritin and is enzymatically controlled. However, Zeman (1991) reports that certain iron-sugar complexes may move iron across mucosal cells without binding to ferritin.

Levels of iron show a diurnal variation, with levels 30% higher in the morning than in the evening.

Functions

Iron serves two key functions in metabolic processes: hemoglobin formation and cellular oxidation. Because of the nature of hemoglobin, iron is important to all tissue oxygenation as well as being the vehicle for transporting carbon dioxide from the cells in the catabolic process. Without hemoglobin and cellular iron, hypoxic states would occur.

The hemoglobin molecule consists of four molecules of heme and a protein molecule called globin. Thus, one hemoglobin molecule contains four iron atoms. These iron atoms in hemoglobin comprise 60% to 70% of the body's total iron. Iron, as well as protein, copper, cobalt, vitamin B_{12}, and folate, is crucial to the actual formation of red blood cells. About 20 mg or about 0.5% of the total body store of iron is used for RBC production, with the majority coming from the bone marrow. About 4 g is stored for crucial functions.

Within cells, iron serves to oxidize enzymes. Even though very minute amounts of iron are needed for cellular oxidation, these minute amounts are crucial to the enzyme systems that oxidize glucose. In the classic Krebs cycle, iron is part of the oxidative chains that produce adenosinetriphosphate (ATP), whose hydrolysis produces the energy required for cellular metabolism. Iron is a component in more than 80 enzyme reactions.

Iron is also needed for the hydroxylation of two proteins, proline and lysine, which are essential for collagen synthesis. Thus, it is an important component in wound repair.

IRON IMBALANCE

States of iron imbalance are anemia (deficit) and iron overload (excess).

Iron Deficit

Women, children, elderly persons, and persons with low economic status are at high risk for an iron deficit or deficiency. In the United States, iron deficiency is the most common nutrient imbalance in women and children.

Causative Factors

Iron deficiency can result from inadequate intake or excessive loss of iron. The anemia related to inadequate intake of iron is known as iron-deficiency anemia, also called nutritional anemia. The definition of iron-deficiency anemia is a hematocrit (Hct) of less than 30% and a hemoglobin (Hgb) value of less than 10 g/dL. Hgb values below 6 g/dL categorize severe anemia. Iron-deficiency anemia is common in children and young adults from ages 6 to 24 due to growth needs, limited body stores, and menstruation.

There are three stages in the development of anemia: iron depletion (noted in serum ferritin), iron-deficit erythropoiesis (transferrin saturation reduced), and frank iron-deficiency anemia (low hemoglobin and microcytosis).

Other classifications of anemia are hemorrhagic anemia, which is related to excessive blood loss; pernicious anemia, which is a result of the inability to form hemoglobin in the absence of vitamin B_{12}; and postgastrectomy anemia, which is caused by the inability to reduce iron from absorption because of the lack of gastric hydrochloric acid. When iron blockers are present, the disorder is referred to as malabsorption anemia. Total parenteral nutrition can also play a role in iron deficiency; therefore, patients on total parenteral nutrition need to be assessed for iron needs.

Implications for Pregnant Women

Anemia during pregnancy constitutes a special category of anemia. Because the hemoglobin values decrease in early pregnancy because of the expanded plasma volume, this state is known as physiologic anemia of pregnancy. The drop in hemoglobin, hematocrit, and iron as the pregnancy progresses may be caused not by anemia but by an extracellular volume excess.

The average diet of the pregnant woman in the United States provides only about 10 to 15 mg of iron daily. Of this amount only about 10% to 20% is absorbed, the rest being lost in excreta. Because the common diet contains about 5 to 6 mg of iron per 1000 calories, many women do not consume a sufficient quantity of iron without supplementation. The consequence of anemia in pregnancy, especially severe anemia, can be postpartum hemorrhage, infection, and slow recuperation. Dunnihoo (1992) also lists low birth weight (LBW), fetal distress, and death as major fetal consequences of iron deficiency during pregnancy.

Transcultural Considerations

Groups at especially high risk for iron deficiency are African-American women and members of lower socioeconomic groups. African Americans tend to have lower hematocrit and hemoglobin values when compared to other groups. Some studies suggest genetic factors, but nutritional factors may also contribute. The diet of African Americans tends to be low in iron, calcium, and vitamin A. The diets of Native Americans also tend to be inadequate in these vitamins and minerals (Giger & Davidhizar, 1991).

The most common cause of iron deficiency, second to nutrition, is excessive iron loss due to chronic, usually occult, intestinal or uterine bleeding. There also may be microhemorrhages into pulmonary parenchyma or from intestinal parasites. An iron deficit in adult men and adolescent boys or postmenopausal women suggests blood loss.

Another important contributing factor is malabsorption caused by mucosal disease or after subtotal gastrectomies with gastrojejunostomy. Other metabolic problems that can affect iron utilization are chronic illness and infection, malignancies, and chronic skin diseases. Anemia accompanies chronic renal failure caused by decreased erythropoiesis.

Iron deficiency states may also be associated with other mineral and vitamin deficiencies, most notably zinc deficiency producing anemia, dwarfism and hypogonadism in children; with folic acid deficiency in pregnancy; and with intestinal disease.

Assessment

Physical Assessment

The patient with an iron-deficiency anemia typically complains of fatigue, weakness, and pallor. Assessment findings also include atrophy of papillae of the tongue (atrophic tongue), inflammation of the mucous membranes of the mouth (angular stomatitis), and cracking at the angles of the lips (cheilosis). Fingernails may become fragile, with spoon-shaped heads depressed in the center and raised at the borders. Although rare, pica behaviors, such as eating ice, clays, or starches, may occur in adults, especially African Americans (Giger & Davidhizar, 1991). Signs of acute confusion may occur as a result of hypoxia, especially in the elderly. Splenomegaly is present in 5% to 10% of iron-deficient patients.

In addition to the above symptoms, children with iron deficiency also experience anorexia, decreased resistance to infection, decreased growth, and reversible protein-losing enteropathy. Children may also exhibit pica behaviors.

Diagnostic Assessment

Laboratory tests include a red blood cell (RBC) count, hemoglobin with mean corpuscular volume (MCV), serum iron with iron-binding capacity, and serum ferritin (Table 11–1). In a patient with iron deficiency, the RBCs

TABLE 11–1

COMMON LABORATORY FINDINGS IN
ADULT PATIENTS WITH IRON DEFICIENCY

| Laboratory Test | Normal Value | | Value in Iron Deficiency State |
	Female	Male	
Hemoglobin (Hgb)	12–15 g/dL* (7.1–9.9 mmol/L)	14–17 g/dL* (6.3–9.3 mmol/L)	Decreased
Hematocrit (Hct)	35%–47%*	42%–52%*	Decreased
Serum iron (Fe)	40–150 µg/dL (7.2–26.9 µmol/L)	50–160 µg/dL (8.9–28.7 µmol/L)	Decreased
Serum ferritin	11–122 ng/mL† (11–122 µg/L) 12–263 ng/mL‡ (12–263 µg/mL)	15–200 ng/mL (15–200 µg/L)	Decreased
Serum transferrin	200–400 mg/dL (4.0 g/L)		Decreased
Total iron-binding capacity (TIBC)	250–400 µg/dL (44.8–71.6 µmol/L)		Increased

*Slightly lower in elderly persons.
†Women before menopause.
‡Women after menopause.

are reduced in number and characterized as microcytic (small) and hypochromic (pale). In addition, the following laboratory values are present:

- Serum iron levels are low, less than 50 µg/dL.
- Serum iron-binding capacity is more than 400 µg/dL and transferrin saturation is less than 15%.
- Hgb is less than 10 g/dL, Hct 30% or less, with a mean corpuscular hemoglobin concentration (MCHC) of less than 30 g/dL (low).

Because the majority of the body's iron is in hemoglobin and because the body pool depends on body size and blood volume, males have different laboratory readings than females.

Serum ferritin levels are also measurements of importance when evaluating the storage of iron. The normal serum ferritin level for newborns is 12 to 200 nanograms (ng)/mL, which increases after birth because of destruction of fetal hemoglobin. After about 1 month the serum ferritin levels begin to decline. In men, levels less than 15 mg are associated with iron deficiency and decreased stores. Because ferritin is a key substance in protecting cells from high, potentially toxic concentrations of free iron, a ferritin test is important to establish an adequate presence of ferritin. Each ferritin complex may contain up to 2000 iron atoms.

In iron-deficient states, a bone marrow analysis may show a mild erythroid hyperplasia. As the number of mature RBCs decreases, reticulocytes enter the circulation. Platelets may also be increased.

Serum albumin and serum transferrin values are usually decreased when a patient has an iron deficit. Decreases in protein and iron levels contribute to poor wound healing, especially in the elderly patient.

Nursing Diagnoses

Common nursing diagnoses related to iron deficiency include:

- Nutrition, altered, less than body requirements, related to inadequate iron intake or excessive blood loss
- Fatigue, related to cellular starvation state
- Confusion, acute, related to cerebral and cellular hypoxia
- Activity Intolerance, related to fatigue and hypoxia
- Tissue Perfusion, altered, related to hypoxia

Collaborative Management

Medical Treatment

The key to the treatment of anemia is prevention through proper nutrition. Prophylactic oral supplementation, typically with ferrous sulfate, is the most effective. One 325-mg tablet of ferrous sulfate contains 64 mg of iron. Other preparations include ferrous fumarate and ferrous gluconate. These medications should be given in divided doses after meals and with laboratory monitoring for at least 6 weeks. Further supplementation may be given if the iron stores are also depleted.

For an actual iron deficit, the goal of treatment is replacement of iron. Oral iron is available in a variety of products. High-dose preparations are Mol-Iron (195 mg), TriHemic (115 mg), Trinsicon (110 mg), Iberol (105 mg), Tabron (100 mg), Geritonic Liquid (63 mg), Geritol (50 mg), Oebetinic (38 mg), Femiron (20 mg), Flintstones with iron (18 mg), Unicap Plus Iron (18 mg), Tri-Vi-Sol drops (10 mg), and others (most are combinations of ferrous sulfate and elemental iron). It is recommended that oral iron preparations be used for 1 to 3 months until the hemoglobin level is restored to normal. Possibly, an additional 1 to 3 months may be needed to restore tissue stores. Since the efficiency of iron absorption is decreased by higher body stores, the first month of therapy is usually most effective. Children's doses are 3 mg/kg per day. Oral iron supplements are given in chronic renal failure; however, note that iron should not be administered at the same time as antacids, because these will bind to the iron.

Parenteral iron is administered when anemia is severe or malabsorption is the major cause of deficiency. The intravenous (IV) form of iron is iron dextran, such as Imferon, in saline. The body must be able to reduce the iron to transferrin. Doses are calculated based on body weight and Hgb level. The formula is

$$Fe(mg) = (normal\ Hgb - patient\ Hgb) \times weight\ (kg) \times 2.21 + 1000$$

Intramuscular (IM) iron may be administered. A common dosing schedule is 1 mL (50 mg) each day in a large muscle using the Z-track technique. Intramuscular injections cause pain and skin discoloration as well as pyrexia, headache, palpitations, and metallic taste. Test doses are given to check for anaphylactic response (0.5 mL test before injection). Dunnihoo (1992) reports that about 0.5% of the patients receiving intramuscular injections will have major problems such as phlebitis, serum sickness, or anaphylaxis. Delayed reactions may include fever, arthralgia, splenomegaly, and lymphadenopathy, about 4 to 10 days after treatment. The most common cause of poor response to therapy is noncompliance; other reasons may be poor absorption, continuing blood loss (including occult), and other types of anemia.

The therapeutic goal may be assessed because oral iron will increase the reticulocyte count by 7% to 10% in the first 5 to 10 days of therapy. The hematocrit should rise 5% to 15% in the first month of therapy; the hemoglobin should rise about 2 to 5 g/dL.

In a severe deficit in which the patient's excretory mechanisms are functioning, it is necessary for the patient to have laboratory iron levels of 250 mg or greater in order to have an absorption rate of about 25 mg.

IMPLICATIONS FOR PREGNANT WOMEN • Additional iron is needed during pregnancy because of the expansion in the circulating blood volume and to supply the iron storage need of the developing fetal liver. During pregnancy, the maternal blood volume is greatly expanded, hemopoiesis is increased, and therefore the hemoglobin concentration is increased to sustain about 500 mg of iron. So, about 1 g of ferrous iron (15 mg per day) is required to prevent anemia in the mother during pregnancy. Cravings for pickles and ice cream may be related to the respective iron and calcium contents of these foods.

Patients receiving total parenteral nutrition (TPN) require special attention to iron needs. Woodley and Whelan (1992) recommend that iron requirements be met with monthly parenteral administration (IM or IV) of 1 mL of iron dextran.

Nursing Interventions

Iron requirements vary over the course of the life span, especially from birth through the onset and cessation of menses. It is important to maintain positive iron balance in the newborn to 6 months of age, during the physiologic stress of adolescence, and throughout years of menses. The newborn at birth has about a 3- 6-month supply of iron stored in the liver during fetal development. Because milk is a poor source of iron, it is important to add food supplements to the diet to avoid the classic milk anemia of young children.

The nurse should discuss nutritional concerns and provide health education regarding the many factors that can hinder or enhance the absorption of iron.

Patients also need information about iron medications. Oral iron supplements may lead to constipation. Some patients experience diarrhea and cramping. Patients receiving parenteral iron must be carefully monitored for the possibility of iron overload syndromes and phlebitis at the administration site. The latter may develop if the medication is not properly diluted. Table 11–2 summarizes nursing interventions appropriate for the patient with an iron deficiency.

Iron Excess

Excess is considered to be present when serum levels are greater than 200 μg/dL. The normal range for men is 65 to 200 μg/dL, with an average level for men of 100 μg/dL and for women of 90 μg/dL.

TABLE 11–2
NURSING INTERVENTIONS FOR PATIENTS WITH IRON DEFICIENCY

- Identify high-risk patients, such as elderly and poor persons, African Americans, Native Americans, and patients receiving total parenteral nutrition.
- Monitor laboratory tests (Hct, Hgb, Fe) for changes.
- Monitor for sources of occult bleeding, such as the gastrointestinal tract; check stools for occult blood.
- Teach the patient to consume foods high in iron, especially those containing heme iron (meats).
- Teach the patient about iron supplementation: Take oral iron after meals to prevent gastrointestinal symptoms; drink liquid preparations through a straw to prevent staining of teeth.
- Give intramuscular (IM) iron preparations by deep IM injection using the Z-track technique.
- Encourage rest periods for fatigue associated with anemia.

Causative Factors

Iron overload results when intake is not adequately excreted. Even though iron levels are regulated by intestinal controls, this mechanism cannot cope with a sustained high dietary intake of iron. Excess can also result when massive blood transfusions have been administered for replacement of red blood cells. The most common cause in children is ingesting iron supplements intended for adults. A lethal dose is 3 g in a child. An adult lethal dose is 200 to 250 mg/kg (National Research Council, 1989).

Although iron overload is primarily related to excessive intake, it can also result from excessive absorption, such as in hemochromatosis, cirrhosis, or porphyria. Hemochromatosis is a genetic inborn error of metabolism present in individuals in Northern Europe, North America, and Australia. In this condition, a large amount of iron accumulates in body tissues, including the liver and spleen (hemosiderosis).

Additional clinical syndromes include states in which lysis of cells can raise the serum iron as the red blood cells are destroyed, for example, hemolytic anemia or acute hepatitis. These states are not correlated with increased body iron stores.

Chronic alcoholism can cause iron toxicity by injuring intestinal mucosa and increasing iron absorption. In addition, some alcoholic beverages contain significant amounts of iron, especially inexpensive red wines.

Assessment

Physical Assessment

Iron toxicity, or overload, results in damage to major organs, such as the liver, pancreas, heart, and endocrine glands, as well as the joints. A bronze skin color develops. Severe diabetes mellitus may also occur. Patients with iron toxicity often experience severe headaches, dyspnea, confusion, and lethargy. Excess iron also inhibits zinc absorption.

Diagnostic Assessment

The diagnosis of iron excess is confirmed with a high ferritin level as well as elevated transferrin. A serum iron level greater than 200 µg/dL is expected in patients with iron toxicity.

Nursing Diagnoses

Common nursing diagnoses in patients with iron toxicity include:

- Nutrition, altered, more than body requirements, related to excessive iron intake or regulation defect
- Cardiac Output, decreased, related to cardiomyopathy
- Pain, related to organ and joint damage

Collaborative Management

Medical Treatment

Treatment of iron toxicity focuses on decreasing iron intake and evaluating the causes of increased intestinal absorption. A pint of blood removed once a month will remove about 250 mg of iron and can reverse some tissue damage, especially to the heart and liver.

Chelating agents that bind metals are used in hemochromatosis. One such agent is deferoxamine mesylate, for which the dose is 1 g intramuscularly (IM) every 8 hours as needed. Deferoxamine or diethylenetriamine pentaacetic acid (DTPA) may be given subcutaneously or peritoneally.

Nursing Interventions

Patients receiving iron replacement therapy, especially parenterally, should be monitored carefully for iron toxicity. Frequent laboratory testing can confirm an early diagnosis before organ damage occurs. If the patient experiences iron overload, supportive care to provide comfort and alleviate symptoms is needed.

CRITICAL THINKING ACTIVITIES

An elderly African-American woman is receiving home health care following a total hip replacement. On the second visit after hospital discharge, the nurse notes that the patient seems pale and dyspneic when ambulating with a walker. The patient states that she has felt very tired lately, but attributed her fatigue to rehabilitative therapy. She is also disoriented, a marked change from the first home visit. The nurse takes a blood specimen for laboratory analysis of hemoglobin, hematocrit, and serum iron levels. As suspected, all of these laboratory values are decreased, and the patient is placed on iron supplements three times a day (TID).

QUESTIONS

1. *What risk factors does the patient have that predisposes her to an iron deficiency?*

2. *What signs and symptoms alerted the home care nurse that the patient had this nutrient deficiency?*

3. *When providing patient teaching, what precautions and adverse effects of supplemental iron should the nurse include?*

4. *What foods should the nurse instruct the patient to eat?*

REFERENCES AND READINGS

Beare, PG, & Myers, JL (1990). Wellness: Nutrition and Health. Guilford, CT: Dushkin.

Chernecky, CC, Krech, RL, & Berger, BJ (1993). Laboratory tests and diagnostic procedures. Philadelphia: WB Saunders.

Cook, JD et al (1991). Calcium supplementation: Effect on iron absorption. American Journal of Clinical Nutrition 53(1), 106–111.

Dunnihoo, DR (1992). Fundamentals of Gynecology and Obstetrics (2nd ed). Philadelphia: JB Lippincott.

Giger, JN, & Davidhizar, RE (1991). Transcultural Nursing: Assessment and Intervention. St. Louis: Mosby–Year Book.

Hallberg, L et al (1991). Calcium: Effect of different amounts of nonheme- and heme-iron absorption in humans. American Journal of Clinical Nutrition 53(1), 112–119.

Hunt, JR et al (1990). Ascorbic acid: Effect on ongoing iron absorption and status in iron-depleted young women. American Journal of Clinical Nutrition 51(4), 649–655.

Hurley, JS (1992). Wellness: Nutrition and Health. Guilford, CT: Dushkin

Kokko, JP, & Tannen, RL (1990). Fluids and Electrolytes. Philadelphia: WB Saunders.

Lutz, CA, & Przytulski, KR (1994). Nutrition and Diet Therapy. Philadelphia: FA Davis.

National Research Council (1989). Iron. A report of the Subcommittee on Iron, Committee on Medical and Biologic Effects of Environmental Pollutants, Division of Medical Sciences, Assembly of Life Sciences. Baltimore: University Park Press.

U.S. Department of Agriculture (1987). Nationwide Food Consumption Survey. Continuing survey of food intakes of individuals: Women 19–50 years and their children 1–5 years, 4 days. Hyattsville, MD: Nutrition Monitoring Division, Human Nutrition Information Services.

Woodley, M, & Whelan, A (1992). Manual of Medical Therapeutics (27th ed). Boston: Little, Brown.

Zeman, FJ (1991). Clinical Nutrition and Dietetics. New York: Macmillan.

CHAPTER 12

. . . .
. . . .
. . . .
. . . .
. . . .
. . . .
. . . .
. . . .
. . . .
. . . .

Zinc

ZINC BALANCE

Even though zinc is a mineral that represents 0.2% of the earth's crust, it exists within the human body in minute quantities. However, this microelement plays an important role in maintaining vital body functions.

Distribution

Zinc is an essential microelement present in teeth, bones (bone leukocytes), skin, hair, kidneys, pancreas, lung, liver, eyes, and muscles. It is also important to glandular tissue including the pituitary, adrenals, prostate, and testes, and is found in seminal fluids and spermatozoa. Zinc is present in enzymes, such as carbonic anhydrase, phosphate transferase, and alkaline phosphatase.

About 40% of the plasma zinc is bound tightly to globulins; zinc also binds to albumin. Thus, the level of serum proteins affects the level of serum zinc. Most of the blood zinc is present in the hemoglobin of red blood cells.

Physiologically, there appears to be a rapid turnover in the apparently small body pool of readily available zinc. Although the bones and muscles of the body store relatively large quantities of zinc, these sources are not immediately available when a zinc deficiency occurs. Because absorption rates of zinc are variable, it is recommended that about 12.5 mg of zinc be ingested daily to maintain a positive balance.

Zinc is absorbed through the intestines and stored in the liver. Only about 5% to 10% of the ingested zinc is actually absorbed. Intestinal absorption of zinc can easily be reduced by chelating agents (metal-binding agents such as aluminum hydroxide). Zinc absorption is also reduced when there is a high intake of calcium or phosphates.

Inorganic iron and zinc compete for the same absorption sites. If a person's iron intake is much greater than that of zinc, the absorption of the zinc decreases. Like iron, excessive copper and manganese also reduce zinc absorption (Lutz & Przytulski, 1994).

The normal serum level of zinc is 70 to 150 μg/dL or 10.7 to 23 mmol/L, depending on laboratory measurement technique and method of reporting (Chernecky, Krech, & Berger, 1993).

Control

Zinc is primarily excreted through the feces, to the extent of 2 to 3 mg per day. It is also excreted by the kidneys. Fecal zinc output is related to dietary intake, whereas urinary zinc output is not. Besides inorganic zinc losses through the feces, a major source of loss is through the pancreatic juice. Therefore, patients with pancreatitis and severe diarrhea need to be carefully assessed for zinc depletions. Other routes of excretion are sweat and menstrual flow.

Sources

The best sources of zinc are red meat, shellfish, and liver. Other good dietary sources include milk, eggs, nuts, legumes, and dry yeast. Although zinc also exists in whole grains in significant amounts, it is not easily absorbed by the body. In the United States, meat and meat byproducts provide as much as 70% of the zinc consumed.

Some vegetable sources actually contain zinc-binding agents. Some data are available to indicate that excessive ingestion of phytates found in cereals and whole grains, coffee, tea, and cocoa *increases* the zinc requirement (National Research Council, 1989).

Other factors that affect the viability of certain foods as sources of zinc include mineral deposits in the soil and fertilization practices. Milling of cereals can also decrease the amount of zinc in wheat and other cereal products. Liquid diets are frequently deficient in zinc.

Because zinc is readily available, intake from normal nutrition is generally adequate in both infants and adults except in Third-World countries and other areas where malnutrition exists. Elderly people are prone to malnutrition and therefore are more likely to have zinc deficiency than their younger counterparts.

Recommended Daily Allowance

The recommended dietary allowance for girls and adult women is 12 mg per day; men and pregnant women require about 15 mg per day and lactating women need 19 mg per day. The increased intake for lactating women is especially critical to ensure the daily availability of the 1.6 to 2.6 mg of zinc that is transferred to breast milk.

Implications for Infants and Children

It is recommended that infants receive 3 to 5 mg of zinc per day, especially for the first 6 months because this trace element is important in normal growth, development, and sexual maturation. During rapid growth periods from about 11 years on, young people may require adult levels of zinc (Lutz & Przytulski, 1994).

Functions

The key function of zinc is promotion of the synthesis of genetic code proteins, deoxyribonucleic acid (DNA), and ribonucleic acid (RNA); insulin; and protein. Because it is involved in protein synthesis, zinc is essential for the normal growth and development of children and for proper metabolism in adults.

Further, zinc facilitates achieving normal blood concentrations of vitamin A by mobilizing vitamin A from the liver. Zinc also assists in the carbon dioxide exchange because it is a component of the enzyme carbonic anhydrase.

Recently, the clinical importance of zinc for proper wound healing has been emphasized. Zinc is necessary for the formation of collagen, which is an essential material required for tissue healing and repair. Zinc also provides immunity against disease (Lutz & Przytulski, 1994).

Implications for Elderly Persons

People over 65 years of age are at high risk for zinc deficiency because they may not eat a well-balanced diet complete with adequate amounts of required nutrients. Additionally, intestinal absorption of nutrients is usually decreased in elderly people (Davis & Sherer, 1994). Wounds, especially pressure ulcers, heal more slowly in elderly persons as a result of decreased nutrient absorption.

Vegetarians and people who cannot afford adequate nutrition, such as migrant workers and homeless persons, are also at risk for zinc deficiency.

ZINC IMBALANCES

Zinc deficiency, or hypozincemia, is more common than zinc excess, or hyperzincemia. Excesses occur only in toxic states and total renal failure.

Zinc Deficit

A zinc deficit is defined by a level of less than 70 μg/dL in normal plasma. This finding should direct the healthcare worker to observe for signs of concomitant hypoproteinemia, acute stress, or drugs that affect zinc leads.

Causative Factors

Nutritional deficit is not a common cause of zinc deficiency, except in Third-World countries where malnutrition is common and in high-risk populations such as elderly people, vegetarians, and low economic groups.

Chelation is an etiologic factor in zinc deficiency. Chelation is the formation of insoluble chelates (zinc bound with iron or amino acids, calcium and vitamin D, fiber or cellulose in cereals) that prevent the absorption of zinc from the bowel. Certain medications such as penicillamine, antimetabolites, cisplatin, and estrogens also act as chelating agents, causing zinc to be removed from the blood. Blood loss and increased renal excretion are further etiologic factors contributing to zinc loss. Other causes of zinc deficiency are:

- Malabsorption syndrome
- Blood loss, especially related to parasitism or Crohn's disease
- Increased urinary excretion caused by alcohol consumption
- Increased levels of corticosteroids (steroid medications or Cushing's disease) leading to increased renal excretion
- Altered metabolism, especially of phosphates
- Imbalance between aerobic and anaerobic metabolism, as in acidosis
- Severe burns or other acute catabolic illnesses
- Parenteral nutrition without appropriate supplementation
- Dialysis and diuretics
- Acrodermatitis enteropathic, a childhood disease of hereditary origin
- Diarrhea, existing with gastrointestinal disease or pancreatitis
- Ileostomy
- Rheumatoid arthritis

Transcultural Considerations

Zinc deficiency is common in Mexico and among Mexican-Americans. One of the leading causes of death among children in Mexico and in adjacent Mexican-American communities is malnutrition. For all Mexicans, the two leading causes of death are pneumonia and gastrointestinal disease caused by bacteria and parasites. Although the problems of malnutrition and subsequent infections are primarily economic, lack of education may be another causative factor (Giger & Davidhizar, 1991).

Assessment

Physical Assessment

Zinc disorders may be classified as acute or chronic. Assessment for *acute* hypozincemia focuses on sensory and skin changes. Other findings include anemia, bone deformities, and metabolic malfunctions. Table 12–1 lists the clinical manifestations of acute zinc deficit.

Chronic zinc losses are associated with iron-deficiency anemia, especially in severe cases. Iron-deficiency anemia and the subsequent zinc deficiency are associated with gradual blood loss. Gradual blood loss can occur in such situations as gastrointestinal disorders and parasitism. Signs include fatigue, weakness, dyspnea, and pallor.

Chronic deficiency, especially in children, can lead to bone deformities, hypogonadism, dwarfism, and hyperpigmentation.

Zinc levels have been correlated with both depression and attention deficit disorder. Both of these health problems often improve with supplemental zinc administration.

IMPLICATIONS FOR PREGNANT WOMEN • The results of hypozincemia during pregnancy include in the infant congenital deformities, low birth weight,

TABLE 12–1
CLINICAL MANIFESTATIONS OF ACUTE ZINC DEFICIT

Body System	Clinical Manifestations
Lab and metabolism changes	Low serum albumin (3 g/dL or less); iron deficiency anemia; impaired T-lymphocyte function, i.e., decreased cellular immunity; signs of vitamin A deficiency; in glucose metabolism related to reduced insulin response to glucose
Skin, nails, and hair	Macrovesicular rash; acrodermatitis, red skin with scaling; delayed wound healing; fingernail abnormality (Beau's lines); sparse hair growth, alopecia and desquamation
Neurologic and sensory	Depression and dementia-like symptoms Hypogeusia (decreased taste acuity); hyposmia (decreased olfactory acuity); dysosmia (decreased/unusual smell)

persistent depressed immune function, and early neonatal death. In addition, zinc deficit can cause depressed development and function of the mammary glands in mothers.

Diagnostic Assessment

The finding of a serum zinc level below 70 μg/dL should prompt further assessment for the possibility of hypozincemia, although zinc levels are prone to sudden variations. Decreased zinc levels may also indicate altered phosphate levels caused by changes in enzyme activity (check serum phosphate) and an imbalance between aerobic and anaerobic metabolism (check serum CO_2 levels). An increased serum amylase level indicates pancreatic disease. A complete blood count, serum iron, and total iron-binding capacity are evaluated for patients suspected of having iron-deficiency anemia.

Urinary analysis is not a reliable measurement of zinc levels because of the potential for contamination with feces. Feces is the primary source of excretion of zinc so contamination of the urine with feces would alter the urinary zinc reading. Normal urinary levels are 0.3 to 0.6 mg per day.

Nursing Diagnoses

Common nursing diagnoses associated with zinc deficit include:

- Nutrition, altered, less than body requirements, related to malnutrition or zinc loss
- Skin and Tissue Integrity, impaired, related to inadequate zinc for healing
- Infection, high risk for, related to poor wound healing and desquamation
- Fatigue, related to inadequate tissue perfusion associated with iron-deficiency anemia

Collaborative Management

Medical Treatment

The goal of treatment for the patient experiencing a zinc deficit is to restore the serum zinc level by correcting the underlying causes. Because zinc deficiency is commonly a result of lack of absorption caused by chelation, the recommended treatment is to reduce excessive intake of nutrients that cause chelation. The nutrients to be controlled include combinations of iron, calcium, vitamin D, fiber, and phytates. Drugs that bind metals, such as penicillamine, should be given before meals on an empty stomach. With the decrease of chelation, absorption of zinc should improve.

Chronic iron deficiency, another potential cause of zinc deficit, is treated by locating and correcting possible slow blood loss, such as via the gastrointestinal tract. Once the blood loss is controlled, an increase in iron consumption may be advised. Prescribing nutritional or therapeutic iron supplements and administration of vitamin C, calcium, and vitamin D may also be recommended.

Deficiency caused by inadequate intake of zinc is treated by administering zinc supplements and establishing a balanced diet that includes foods containing zinc, such as eggs, cereal (oatmeal and bran), nuts, red meat, seafood, and legumes. Zinc salts are available as a preventive or therapeutic measure for patients on long-term parenteral therapy, during catabolic illness and rapid growth, or in states of tissue acceleration. Zinc is given in combination with a multiple vitamin supplement to promote healing in elderly patients who have pressure ulcers or other wounds.

Zinc salts are composed of elemental zinc, zinc acetate, zinc gluconate, zinc oxide, and zinc sulfate. As with any nutritional or parenteral therapy involving electrolytes and minerals, caution must be exercised for patients with renal insufficiency.

Nursing Interventions

Nursing interventions are first directed to promotion of overall health and prevention of illness through proper nutrition. A balanced diet for all age groups is paramount. Mild hypozincemia may be treated by increasing foods high in zinc. It is also important to note related conditions in which additional nutrients such as iron, calcium, and vitamins C and D are lost.

If supplements of zinc have been prescribed or recommended, the patient is advised to take the supplements with meals or milk to avoid gastric distress and possible vomiting.

Patients may experience anorexia and need encouragement in order to increase proper nutritional intake. If signs and symptoms of iron-deficiency anemia are present, nursing interventions are directed at relieving fatigue through planning activities and avoiding high-energy activities until the anemia is corrected. Special skin and nail care is needed, especially if the patient has one or more wounds. Delayed wound healing from other disorders may be present, and the patient may require proper wound care using aseptic technique and prescribed treatments. Table 12–2 summarizes the most important nursing interventions for the patient with hypozincemia.

TABLE 12–2

NURSING INTERVENTIONS FOR PATIENTS WITH ZINC DEFICIENCY

- Teach the patient that meats and legumes are very good sources of zinc. If a well-balanced diet is consumed, zinc deficiency is not likely to occur.
- Identify patients at high risk for zinc deficiency, especially elderly and poor persons and those with malabsorption problems.
- Monitor patients receiving zinc supplements for adverse effects, such as vomiting, diarrhea, anemia, lethargy, and renal impairment.
- Remember that zinc supplementation can interfere with iron and copper absorption.
- Give zinc supplements with or after meals.
- For patients with poorly healing wounds or pressure ulcers, monitor zinc levels and obtain a physician's order for zinc salts.
- Collaborate with the dietitian when caring for patients with zinc deficiency or excess.

Zinc Excess

Zinc excess (hyperzincemia) can occur but is extremely rare. Zealous zinc supplementation can block the effect of exercise in increasing high-density lipoproteins (HDLs). A zinc level that is 10 times the recommended daily allowance can affect white blood cells and therefore decrease immune function (Lutz & Przytulski, 1994).

Causative Factors

Hyperzincemia is usually the result of renal failure or renal shutdown. Hyperzincemia has also been associated with intravenous overdose of zinc salts, as well as with the storage of acidic foods in zinc-coated containers. Ingestion of more than 15 mg per day may lead to toxicity. Excessive intake of zinc can interfere with copper absorption and lead to copper deficiency.

Assessment

Physical Assessment

Assessment findings associated with zinc excess include the following symptoms:

- Gastrointestinal: anorexia, nausea, vomiting, diarrhea, gastrointestinal erosions followed by bleeding
- Neurologic: lethargy and vertigo
- Renal: oliguria leading to acute renal failure

Diagnostic Assessment

Zinc can be measured in feces, blood, urine, hair, and nails. Routine serum levels are assessed for patients receiving parenteral nutrition and for those who are receiving zinc supplements. A zinc level of over 150 µg/dL confirms hyperzincemia.

Nursing Diagnoses

Common nursing diagnoses associated with hyperzincemia include:

- Nutrition, altered, more than body requirements, related to decreased renal excretion or excessive zinc supplementation
- Diarrhea, related to the effects of gastrointestinal irritation

Additional nursing diagnoses are relevant if gastrointestinal bleeding or renal failure occurs.

Collaborative Management

Medical Treatment

Prevention of zinc toxicity can be accomplished by monitoring patients who are at high risk, such as those receiving total parenteral nutrition or zinc supplements, and patients with renal insufficiency. Treatment of hyperzincemia is directed to reestablishing renal function by diuresis or dialysis. Even if renal function is not impaired, dialysis may still be needed to rid the body of excess zinc.

Nursing Interventions

If nutritional or parenteral intake has been excessive, especially in patients with compromised renal function, reduction of zinc in the diet is indicated. Other nursing interventions are directed at providing comfort measures and maintaining fluid balance in the patient who experiences diarrhea or gastrointestinal bleeding. Assisting with ambulation and activities of daily living may also be needed if the patient experiences lethargy and vertigo.

CRITICAL THINKING ACTIVITIES

An 87-year-old woman is admitted to a nursing home from the hospital with several stage III and IV draining pressure ulcers. She is severely demented and anemic. As a result, she is being continuously tube-fed and receives a daily multivitamin.

Despite special attention to the pressure ulcers by the staff, her wounds worsen. She remains anemic from chronic lower GI bleeding related to diverticulitis.

QUESTIONS

1. *What two minerals is the patient probably lacking?*
2. *Why is the multivitamin not effective at treating her deficiencies?*
3. *What medications are most likely to be prescribed based on her physical and laboratory assessment?*
4. *What adverse effects should the nurse observe for with these medications?*

REFERENCES AND READINGS

Bogden, JD, et al (1990). Effects of one year supplementation of zinc and other micronutrients on cellular immunity in the elderly. Journal of American College of Nutrition 9(3), 214–225.

Broun, ER, et al (1990). Excessive zinc ingestion: A reversible cause of sideroblastic anemia and bone marrow depression. Journal of the American Medical Association 264(11), 1441–1443.

Chernecky, CC, Krech, RL, & Berger, BJ (1993). Laboratory Tests and Diagnostic Procedures. Philadelphia: WB Saunders.

Davis, J, & Sherer, K (1994). Applied Nutrition and Diet Therapy for Nurses. Philadelphia: WB Saunders.

Doenges, ME, & Moorhouse, MF (1991). Pocket Guide: Nursing Diagnoses with Nursing Interventions. Philadelphia: FA Davis.

Dore, F, Duggy, P, Peterson, M, Catalanotto, F, Marlow, S, Ostrom, M, and Weinstein, A (1990). Zinc profiles in rheumatoid arthritis. Clinical and Experimental Rheumatology 8(6), 541–546.

Fosmire, GJ (1989). Possible hazards associated with zinc supplementation. Nutrition Today 24(3), 15–18.

Fosmire GJ (1990). Zinc toxicity. American Journal of Clinical Nutrition 51(2), 225–227.

Giger, JN, & Davidhizar, RE (1991). Transcultural Nursing: Assessment and Intervention. St. Louis: Mosby–Year Book.

Kokko, JP, & Tannen, RL (1990). Fluids and Electrolytes. Philadelphia: WB Saunders.

Lutz, CA, & Przytulski, KR (1994). Nutrition and Diet Therapy. Philadelphia: FA Davis.

McLouglin, IJ, & Hodge, JS (1990). Zinc in depressive disorders. Acta Psychiatrica Scandinavica 82(6), 451–453.

Moser-Veillon, RB (1990). Zinc: Consumption patterns and dietary recommendations. Journal of the American Dietetic Association 90(8), 1039–1093.

National Research Council (1989). Recommended Dietary Allowances. Washington, DC: National Academy Press.

Sandstead, HH (1991). Zinc deficiency. A public health problem? American Journal of Diseases in Children 145(8), 853–859.

Yadrick, MK, et al (1989). Iron, copper, and zinc status: Response to supplementation with zinc or zinc and iron in adult females. American Journal of Clinical Nutrition 49(1), 145–150.

CHAPTER 13

· · · · · · · · ·
· · · · · · · · ·
· · · · · · · · ·
· · · · · · · · ·
· · · · · · · · ·
· · · · · · · · ·
· · · · · · · · ·
· · · · · · · · ·
· · · · · · · · ·
· · · · · · · · ·

Trace Minerals

The chief minerals or elements that function as electrolytes in the body are sodium, chloride, potassium, calcium, phosphorus, and magnesium. Other elements—nitrogen, carbon, and hydrogen—are discussed in Chapter 3. Included in the present chapter are elements that, although small in quantity, are important for healthy human function. These trace elements include chromium, cobalt, fluorine, iodine, manganese, molybdenum, and selenium.

CHROMIUM

Chromium is an important mineral that activates several enzymes. It is considered an essential mineral even though the low levels of chromium in the body make it difficult to determine body requirements. The symbol for chromium is Cr. Like iron, chromium has two different oxidation states. The +3 oxidation state is the form in living systems. A complex of chromium with nicotinic acid is known as the glucose tolerance factor (GTF).

Chromium Balance

Distribution

Most of the absorbed chromium is excreted through the urine. Even though concentrations of chromium in the body tend to decrease with age, the lungs are a site where chromium continues to accumulate.

Sources

The best sources of chromium are brewer's yeast and calves' liver. American cheese and wheat germ are also appropriate sources of chromium because they contain a chromium-dinicotinic acid-glutathione complex with high bioavailability. Other sources include spices, corn oil, and mushrooms. Secondary sources of chromium include potatoes with skin, whole-grain bread, beef, chicken legs, and cheese. Fresh fruit, fish, and seafood are fairly good sources.

Whereas whole-grain products are considered a secondary chromium source, foods based on refined grains are not good sources. The milling process used to refine grains causes a loss of chromium from the grain. Rice and sugar are very poor sources. In fact, a person "loading up" with white sugar and white flour actually experiences a chromium loss because these highly processed foods increase insulin production. When insulin production is high, chromium is lost. Phosphates in milk may also cause a loss of chromium by binding with the chromium to form insoluble salts that are lost in feces.

Recommended Daily Allowance

Because information on the level of chromium in foods is limited, the recommended daily intake level is tentatively established at 50 to 200 μg for persons aged 7 years and above. Infants need only 10 to 60 μg daily, whereas toddlers and children up to 6 years old need 30 to 120 μg (Table 13–1). Safe and adequate consumption of chromium is best achieved by maintaining a well-balanced diet.

Control

When the daily intake of chromium is 40 μg, approximately 0.5% of the intake is absorbed. The absorption percentage increases when chromium intake drops below 40 μg per day. About 1% to 2% is excreted in the urine and some is excreted in the feces as phosphate salts. Oxalates increase absorption and phytates (from grains and cereals) decrease absorption. Hyperglycemia and hyperinsulinemia both increase chromium losses in the urine.

TABLE 13–1
RECOMMENDED DAILY
ALLOWANCE FOR CHROMIUM

Category	Age (years)	Chromium (μg)
Infants	0–0.5	10–10
	0.5–1	20–60
Children and adolescents	1–3	20–80
	4–6	30–120
	7–10	50–200
	11+	50–200
Adults		50–200

Chromium is absorbed in the small intestine with varying degrees of efficiency, ranging from about 1% for simple salts to as high as 10% to 25% for other food sources. It is currently estimated that the normal amount being absorbed is between 50 and 100 μg per day.

Functions

Research on chromium function is focused on the glucose tolerance factor. The glucose tolerance factor is a biologically active complex of elements tentatively identified as chromium, nicotinic acid, and glutathione. The glucose tolerance factor appears to play a significant role in the effectiveness of insulin utilization and in the normal metabolism of carbohydrates and lipids. This factor increases the removal of excess glucose from the blood and maintains serum glucose levels. Because chromium increases the affinity of the cell membrane receptors for insulin, it is considered a cofactor with insulin. When the chromium level is low, insulin will bind less tightly to receptor sites, and glucose clearance will be reduced. Cofactors with insulin are crucial to intracellular burning of glucose.

Chromium's key function in lipid metabolism involves the proper release of free fatty acids. Chromium may also be linked to improved wound healing.

Chromium Imbalance

Chromium Deficit

Because it is very difficult to measure body chromium, the best indicators of imbalances come from studies in which significant improvement of the patient's condition results from the supplementation of chromium. Deficiency states have most commonly been linked to a diabetes-like syndrome and to growth failure in children with protein-calorie malnutrition.

Assessment

Chromium deficit is related to overloading of glucose, excessive consumption of milk, and excessive consumption of refined flours and other processed foods. Chromium deficit may occur in the early states of type II diabetes mellitus. Chromium deficiencies may also be present in patients receiving total parenteral nutrition. A deficiency is manifested by fasting hyperglycemia and abnormal glucose levels. Deficits have occurred with protein-calorie malnutrition, type I diabetes, multiparity, maldevelopment, and old age. Impaired tolerance tests, weight loss, and peripheral neuropathy appear to be related to chromium deficit. Patients may also develop metabolic central encephalopathy.

Serum chromium levels vary from 1 to 3 μg/mL. Chromium levels tend to increase rapidly with insulin level increases, and decrease in patients experiencing infection. Urine excretion is about 5 to 10 μg per day.

Collaborative Management

The key treatment is intake of chromium in the amount of 75 to 250 μg daily. In deficit states related to total parenteral nutrition, Kokko and Tannen (1990) recommend 10 to 15 μg of intravenous chromium chloride daily for the first few days, followed by 10 μg weekly.

Nursing interventions are primarily to facilitate the achievement of a balanced diet. If glucose intolerance develops, meal spacing, similar to the practice followed in diabetic meal planning, is recommended.

Chromium Excess

Toxicity related to trivalent chromium has not been reported to date in humans. A study by the International Programme on Chemical Safety correlated chronic exposure of occupational workers to dusts containing chromates, which contain the +6 oxidation state of chromium, to increased incidents of cancer of the bronchi.

COBALT

Cobalt is the trace element necessary for vitamin B_{12} formation. The symbol for cobalt is Co. The trace element discussed in this chapter is the naturally occurring nonradioactive isotope cobalt 59, which should not be confused with cobalt 57, a radioisotope of cobalt used in the Schilling test for diagnosis of pernicious anemia, or cobalt 60, a radioisotope used in radiation therapy.

Cobalt Balance

Distribution

Cobalt as a component of vitamin B_{12} is stored in the liver. The normal blood level, which is a reference for both the element as a free substance and the element that is in the red blood cell is about 1 μg/dL.

Sources

Although cobalt is synthesized from foods in the gastrointestinal tract of animals, microorganisms that exist in the human gastrointestinal tract have minimal capacity for performing this synthesis in areas where vitamin B_{12} can be absorbed. Thus, in humans, the most reliable source is consumption of the muscle and organ meat of animals in which the animals have already transformed the cobalt to cobalamin. Sources include beef, eggs, fish, milk products, organ meats, and pork.

Cyanocobalamin, a commercially available form of vitamin B_{12}, is also a source for cobalt.

Recommended Daily Allowance

The National Research Council has not established a recommended dietary allowance for cobalt. Although the actual quantity required is unknown, as little as 0.045 to 0.09 μg daily maintains bone marrow function in anemia patients with pernicious imbalances.

Mahan and Arlin (1992) suggest that following the dietary requirement for vitamin B_{12} will ensure adequate levels of cobalt. The dietary recommendation for vitamin B_{12} is 3 μg per day.

Control

Absorption rates for cobalt vary. Cobalt is excreted in the urine and unabsorbed quantities are excreted in feces.

Functions

Cobalt's only known function is to form vitamin B_{12}. This vitamin is necessary for the intrinsic function factor from the gastric mucosa in order to produce red blood cells. Vitamin B_{12} is essential for the maturation of red blood cells, and is a factor in the normal functioning of all cells.

Cobalt Imbalance

Cobalt Deficit

Cobalt deficiency with normal dietary intake is rare and is related to vitamin B_{12} deficiency. Vitamin B_{12} deficiency, and consequently cobalt deficiency, are uncommon in the United States. When these do occur they are rarely related to inadequate diet. Approximately 95% of vitamin B_{12} deficiency is related to inadequate absorption rather than inadequate diet.

Strict vegetarians may become deficient in vitamin B_{12}. Such a deficiency usually occurs only when a vegetarian diet is followed over a period of 3 to 6 years.

Assessment

The development of pernicious anemia is the key manifestation related to cobalt deficiency. Pernicious anemia is characterized by fatigue, megaloblastic anemia, and mental depression. The vitamin B_{12} deficiency may also manifest with glossitis, diminished reflexes, and amenorrhea.

The laboratory test specific for confirming pernicious anemia is the Shilling test, which uses a tracer dose of radioactive cobalt (^{57}Co). Without the intrinsic factor in the gastric juices, only about 10% to 20% of the dietary vitamin B_{12} is absorbed. Additionally, achlorhydria is characteristic of the disorder.

Collaborative Management

The treatment for cobalt deficit is to treat the vitamin B_{12} deficiency by administering 15 to 30 μg of vitamin B_{12} daily and by increasing dietary intake of cobalt during the initial treatment. Thereafter, about 30 μg of vitamin B_{12} is recommended every 30 days. Food sources recommended to increase cobalt consumption are liver, kidney, lean meat, and dairy products.

Cobalt Excess

Excessive levels of cobalt caused by normal dietary consumption in humans do not appear to be a common problem.

FLUORIDE

Because of the conflicting reports as to the nature of fluoride in relationship to human growth and development, fluoride is not classified as an essential element. However, because of the significance of the role fluoride, along with calcium, plays in the development of strong teeth and bones, fluoride is considered one of the important trace elements. Fluoride is the term used for the ionized form of the element fluorine. These two terms, in practice, are considered interchangeable.

Fluoride, in addition to iodine, has received significant attention over the years because of its relationship to the formation of teeth. Health policy in the United States has been concerned with the incidence of dental caries and the potential fluoride has to reduce them. As a result of this concern, stannous or sodium fluoride is placed in many toothpastes to reduce caries. Mahan and Arlin (1992) report that dental caries have decreased by 50% in the last 15 to 20 years as a result of water fluoridation.

Fluoride Balance

Distribution

Fluoride is located primarily in the teeth and bones. The average skeleton contains about 2.5 mg.

Sources

The primary source of fluoride is drinking water, which may contain fluoride naturally or as a result of supplementation. Fluoridation of public water supplies is usually aimed at producing a concentration of 1 part per million or 0.7 mEq/L. Other dietary sources of fluoride include tea, marine fish containing edible bones, and gelatin. One cup of tea can contain as much as 0.1 mg of fluoride. Beef liver is also a good source. Food cooked in Teflon cookware contains fluoride leached out of the Teflon; in contrast, aluminium-surfaced cooking utensils can reduce the amount of fluoride in foods.

Food processing involving the use of fluoridated water also increases the dietary intake of fluoride. The free fluoride available in water is more readily available to the body than the protein-bound fluoride in foods. Sodium fluoride in aqueous solution (fluoridation of water) is estimated to be 100% absorbed.

The American Dental Association recommends fluoride supplements until about 13 years of age or until the adult teeth are fully formed. Dentifrices used to brush teeth also make additional fluoride available for absorption by tooth enamel.

Recommended Daily Allowance

Although studies have measured the daily intake of fluoride in various substances, very little data are available to document levels of absorption. For children, the overall recommended amount is 0.08 mg/kg per day. The estimated safe range for adults is 1.5 to 4.0 mg per day (Mahan & Arlin, 1992).

Control

There is a direct relationship between the level of consumption of fluoride and the levels of fluoride in the plasma. Plasma levels show diurnal variation. Fluoride is primarily excreted through the kidneys, and urine contains approximately 90% of the excreted fluoride.

Functions

Fluoride is primarily involved with the formation and strengthening of the tooth enamel and, thus, reducing the incidence of dental caries. It is also believed to be related to the utilization of calcium in bone formation.

Fluoride replaces the hydroxyl (OH) grouping in the calcium phosphorus salts that compose the teeth and bones. This replacement results in the formation of fluorapatite, which creates a stronger, smoother surface enamel of teeth prior to eruption. Fluoride-rich enamel increases the tooth's resistance to "acid attack" generated by plaque bacteria. Plaque hardens and forms a complex material known as calculus or tartar. Fluoride is most effective on smooth surface enamel and less effective on teeth with pits and fissures, such as premolars and molars.

Some studies suggest that the interaction between fluoride and calcium phosphorus salts also has a positive impact on osteoporosis; however, further studies are still ongoing. Sodium fluoride may be given to patients to treat or prevent osteoporosis.

Fluoride Imbalance

Fluoride Deficit

The impact of fluoride deficit has been assessed primarily through studies comparing the incidence of dental caries in geographic regions where fluoride occurs naturally in the water supply with those in regions where fluoride in the water supply is low. The results of these studies indicate that the key symptom of fluoride deficit is the development of dental caries.

Assessment

Other assessment of fluoride deficit is confirmed by radiologic examination for dental pathology and bone matrix changes. In therapeutic doses, fluoride promotes the maintenance of newly synthesized bone matrix and inhibits bone reabsorption.

Collaborative Management

Treatment is the fluoridation of water at 1 part per million and dental care as prescribed.

Fluoridation of community water has been implemented in more than 5000 communities in the United States; thus, it is estimated that about 100 million people receive fluoridated water. Fluoridation of public drinking water can reduce the incidence of dental caries by 65%. Such a reduction means a decrease in the need for and cost of multiple-surface fillings, crowns, and extractions.

Prophylactic care, in addition to fluoride supplements and fluoridation, includes avoiding cariogenic foods. Foods with a high sugar content are particularly cariogenic. Thus it is important to rinse and brush the teeth after consuming these foods. Ideally, it is recommended that between-meal snacks be restricted to raw fruits and vegetables because the act of chewing stimulates saliva flow. Acid foods, such as apples, should not be eaten late in the day because of the reduced saliva flow. The key feature of a healthy dental program is regular brushing, flossing, and professional dental care.

Nursing care is directed to helping the client to develop a self-care regime, a prophylactic dental care program, good nutrition, and proper eating and chewing habits (Table 13–2).

Fluoride Excess

Fluoride excess, known as fluorosis, results from excessive, chronic ingestion of fluorides. Mild fluorosis can appear at doses of 0.1 mg/kg per day with maximal level of 2.5 mg. Two and one-half to 8 mg/kg in drinking water or food, leads to mottling of the dental enamel, especially in children. The American Dental Association also advises against providing dietary fluoride supplements to children in communities with adequately fluoridated water or with water supplies that have a natural fluoride content of 0.7 parts per million or higher. Caution is also recommended to prevent small children under the age of 6 years from swallowing too much fluoridated mouth rinse.

Toxic levels of fluoride can produce nausea, vomiting, abdominal pain, diarrhea, and tetany. The latter can cause death from cardiovascular failure. Toxic levels are usually the result of poisons, such as insecticides, or of years of daily consumption in the range of 20 to 80 mg. Some therapies for osteoporosis have recommended an intake of 50 mg per day for 3 months. These patients must be monitored carefully because the potential for toxicity is present.

TABLE 13–2
PATIENT EDUCATION FOR PREVENTION OF FLUORIDE DEFICIENCY

- Drink plenty of fluoridated water if available.
- Use a fluoride supplement for children until the eruption of the third molars (about the age of 16 years).
- Brush teeth at least once in the morning and once at night, being sure to brush along the gum line.
- Use toothpaste and mouthwashes that contain fluoride.
- Ensure that children have regular dental checkups and fluoride treatments (usually about every 6 months).
- Avoid aluminum antacids, which decreased fluoride absorption.
- Avoid calcium supplements if taking fluoride supplements because absorption of both elements will decrease.

IODINE

Iodine is a micromineral responsible for the formation of thyroid hormones, which regulate basal metabolic rate and cellular metabolism. The symbol for iodine is I.

Iodine Balance

Distribution

Iodine is contained in three pools in the body: a circulating inorganic pool; a circulating organic pool, which includes that bound to serum protein; and, an organic pool contained within the thyroid. The circulating pools include the thyroid hormones, primarily thyroxine (T_4) and triiodothyronine (T_3). Iodine is concentrated in the thyroid (organic form) and in salivary glands (inorganic and mineral form). It is absorbed in the form of iodide.

The body contains about 15 to 20 mg of iodine, of which 60% to 80% is concentrated in the thyroid. The thyroid gland must absorb about 60 μg of iodine per day to maintain a proper hormone level and to provide sufficient iodine resources to compete for distribution balance with the kidneys and the salivary and gastric glands. The thyroid gland secretes two hormones, T_3 and T_4, that require iodine for their manufacture. The third thyroid hormone, calcitonin, does not contain iodine and is involved with calcium metabolism (see Chap. 7).

Sources

Geographic region has an impact on the availability of iodine. Costal regions are more likely to have sufficient natural iodine than mountainous or noncoastal regions, both in foodstuffs and in the environment. Iodine is a mineral obtained from food and water; and, in coastal regions, from the iodine-laden costal mist. Water content varies from 0.1 to 2 μg/L in goiter belts to 2 to 15 μg/L in nongoitrogenic areas. Foods containing iodine are seaweed (such as kelp) and seafood (such as halibut liver and cod and fish liver oils, 18 μg/oz), dairy products (3 to 4 μg/oz), eggs (4 to 10 μg/each), and meat (5 μg/oz). The iodine content of meat depends on the iodine content in the animal's foodstuffs. Fruits and vegetables are generally low in iodine (1 μg/oz).

The primary source of iodine in the United States is iodized table salt (76 μg/g). Processed bread is another source of iodine because iodates are used as dough oxidizers in the bread-making process (500 μg/100 g). Multivitamins are also a source of iodine.

Recommended Daily Allowance

Because it is somewhat difficult to evaluate intake of iodine, excretion of iodine is used as the criterion to assess whether levels are adequate. An excretion of more than 50 μg of iodine per gram of creatinine indicates that iodine levels are adequate for normal functions. If the level drops to between 25 and 50 μg/g creatinine, the risk of hypothyroidism increases. Although an intake of 50 to 70 μg per day is deemed sufficient to maintain an adequate excretion level of iodine, the recommended daily allowance for adults is 150 μg.

Implications for Pregnant Women and Children

Recommended daily allowances of iodine are increased by 25 μg for pregnant women and by 50 μg for lactating women. This requirement is related to increased needs of the fetus and infant rather than any loss by the pregnant woman. The recommended dietary allowance for newborns is 40 μg per day, with older infants needing 50 μg per day. Children's needs range from 70 to 120 μg depending on age.

Control

Iodine is regulated by passive distribution mechanisms in which most iodine is retained by the thyroid. Iodine is easily absorbed by the gut. Like chloride, and unlike many of the other trace elements, it passes through membranes by diffusion. Iodine passes into body reserves in the three pools mentioned previously: plasma, organic; plasma, inorganic; and thyroid (organic). The thyroid form is degraded by the liver and the iodine is conserved.

Intake of iodine occurs daily through ingestion of foodstuffs and water containing iodine. Iodine output is primarily through urinary excretion, the average excretion being 50 μg or more per day. Some iodine is excreted through perspiration and in feces from bile. Because it is readily absorbed, iodine is not one of the trace minerals whose levels are jeopardized by malabsorption syndromes.

Functions

Iodine's key function is to contribute to the production of organic thyroid hormones. Thyroxin and other thyroid hormones are essential for normal basal metabolism, that is, the metabolism of all cells. Thyroid hormones, and the role of iodine in sustaining these hormones, are important to the proper growth and development of children into healthy adults. Thyroid hormones are necessary not only for growth and development but also for maintaining proper basal metabolic rate.

Thyroxine is 65% iodine by weight. When the iodine level is insufficient, the thyroid gland enlarges in an attempt to produce sufficient thyroxine.

Iodine Imbalance

Iodine Deficit

Iodine deficiency results in decreased cellular oxidation and decreased basal metabolic rate (BMR) and leads to hypothyroidism and, in children, to growth retardation.

Iodine deficiencies have only one major cause, lack of intake. The cause may be specific to the individual or endemic to a region. The World Health Organization has estimated the world incidence of goiter is approximately 200 million (Mahan & Arlin, 1992). Endemic goiters exist in at least 12 European countries as well as many Third-World countries.

The association between iodine deficiency and cretinism in goitrogenic regions was made almost 40 years ago. The development of neonatal cretinism depends on interaction between the nutrient deficiency, susceptibility to goitrogenic factors, and genetic characteristics.

There is some indication that certain foods contain goitrogens that block absorption or utilization of iodine. Some foods naturally containing goitrogens are cabbage, turnips, peanuts, cassava, and soybeans. Local water supplies may also contain goitrogens (Mahan & Arlin, 1992).

Assessment

PHYSICAL ASSESSMENT • The clinical manifestations of iodine deficit are related to cellular hypoxia and lowered basal metabolic rate (BMR): fatigue, weakness, asthenia, and slowed mental and neurologic responses. Cardiovascular responses included hypotension, bradycardia, and pretibial edema. Other general manifestations related to lowered BMR include temperature intolerance, cold hands and feet, dry hair, irritability, nervousness, and obesity. Decreased peristalsis may cause constipation. Check of the Achilles tendon may detect slow deep-tendon reflexes.

The disorder gives rise to myxedema in adults and cretinism in children. Myxedema is characterized by anemia, edematous facies, large tongue, slow speech, edema of the feet and hands, alopecia, mental apathy, drowsiness, and sensitivity to cold. Cretinism is manifested by severe mental deficiency, spasticity, deaf mutism, dysarthria, shuffling gait, shortened stature, and hypothyroidism. Less severe manifestations may include deaf mutism, mental retardation, and neuromotor abnormalities. Severe iodine deficiency during pregnancy and the early postnatal period results in cretinism.

In addition to arrested development, hypothyroidism leads to dystrophy of bones and soft tissue. The patient may also have decreased peristalsis and slow deep-tendon reflexes. Thyromegaly, that is, development of an enlarged thryoid (goiter) may result.

DIAGNOSTIC ASSESSMENT • Iodine is measured as both inorganic and organic forms. All the forms occur in the serum in very low concentrations. The level for free inorganic iodine is about 0.08 to 0.16 μg/dL. Most of the thryoid hormone that is measured is T_4. The amount of T_3 is much smaller; however, T_3 is more potent than T_4 so its levels are important. Organic iodine measured in the serum reflects the level of free thyroid hormone (thyroxine, or T_4). Thyroxine levels for adults range from 5 to 12 μg/dL, or 65 to 155 nanomoles (nmol)/L (Chernecky, Krech, & Berger, 1993). In pregnancy or with use of birth control pills this level increases as much as 14.5 μg/dL. Iodine binds to protein, so a protein-bound iodine level (PBI) can also be measured.

The level of triiodothyronine (T_3) for adults ranges 80 to 230 nanograms (ng)/dL from or 1.2 to 3.5 nmol/L (Chernecky, Krech, & Berger, 1993). These levels increase in pregnancy and with the use of oral contraceptives. Slightly higher values are also found in children.

Iodine can also be measured with the radioisotopes iodine 123 (^{123}I) or iodine 131 (^{131}I). With ^{123}I, a scan is taken 24 hours after oral ingestion. ^{131}I can be scanned after periods of 2, 6, and 24 hours. Because iodine is used directly by the thyroid gland to manufacture its hormones, changes in these levels may reflect primary disease of the thyroid. On the other hand, because the thyroid functions under the direction of the anterior pituitary gland, thyroid disorders may be secondary to a master gland problem. Thyroid-stimulating hormone (TSH) is assessed to determine whether pituitary failure is a cause of hypothyroidism. Assessing TSH makes it possible to confirm or reject a diagnosis of hypothyroidism caused by lack of nutritional intake of iodine (Table 13–3).

IMPLICATIONS FOR ELDERLY PERSONS • Thyroxine levels decrease with aging, especially as plasma protein levels decrease.

Collaborative Management

The treatment for iodine deficiency is dietary replacement. Because iodine is a component of many multivitamin and mineral preparations these afford protection from iodine deficiency; however, the most basic societal protection against mass deficiency is the manufacture of iodized salt. The usual intake of 2 g per day of iodized salt easily provides the necessary iodine. Mandatory production of iodized salt has been adopted by many nations, including Canada, but it is not policy in the United States.

The average daily intake is 230 to 740 μg per day. Iodized salt contains about 76 μg of iodine per gram of salt. It is highly recommended that iodized salt be used in noncoastal areas or in goiter belts. Coastal regions, in addition to the usual practice of frequent eating of seafood, have an additional source of iodine in the atmosphere. In very severely iodine-deficient areas of the world, injections of iodized oil have shown to provide protection for 2 to 4 years (Mahan & Arlin, 1992).

Nursing interventions include promotion of proper diet. If the patient is on a low-salt diet, it may be appropriate to encourage the patient to make use of other sources of iodine. Further, assistance with planning daily activities to include a balance of rest and activity levels is an important nursing activity. (See Chap. 20 for further nursing care related to thyroid disorders.)

TABLE 13–3
LABORATORY VALUES FOR ASSESSMENT
OF THYROID FUNCTION

Laboratory Test	Normal Range	SI Units
Triiodothyronine (T$_3$)		
Adult	80–230 ng/dL	1.2–3.5 nmol/L
Infant	110–280 ng/dL	1.70–4.31 nmol/L
Child	83–213 ng/dL	1.28–3.28 nmol/L
Thyroxine (T$_4$)		
Adult	5.0–12.0 μg/dL	65–155 nmol/L
Pregnancy	9.1–14.0 μg/dL	117–181 nmol/L
Elderly	5.5–10.5 μg/dL	71–135 nmol/L
Thyroid-stimulating hormone (TSH)		
Adult	<10 μIU/mL	<10 mIU/L
Elderly		
Female	2.0–16.8 μIU/mL	2.0–16.8 mIU/L
Male	2.0–7.3 μIU/mL	2.0–7.3 mIU/L

Data from Chernecky, CC, Krech, RL, & Berger, BJ: *Laboratory Tests and Diagnostic Procedures*. WB Saunders, Philadelphia, 1993.

Iodine Excess

Iodine excess is very rare. It is related to excessive intake of iodine, usually exceeding 2000 μg per day.

MANGANESE

The nutrient manganese is needed only in minuscule amounts. However, in spite of the small quantity needed, manganese is considered essential for central nervous system functioning, metabolism, human reproduction, and normal bone structure. It also plays a part in the utilization of vitamins and the activation of enzymes. The symbol for manganese is Mn.

Manganese Balance

Distribution

It is not possible to measure body stores (Mn^{2+}) of manganese as a divalent ion. About 10 to 20 mg is present in adults, mostly in the liver, pancreas, and kidneys, with small amounts in the eyes, bones, and salivary glands. The serum level is very low, about 1.4 μg/L.

Sources

Broadly speaking, whole grains and cereals are the best sources of manganese, followed by fruits and vegetables. Animal tissue, seafood, and dairy products are considered poor sources. Manganese is found in nuts, whole grains, and legumes; and to a lesser degree in bananas, egg yolks, green leafy vegetables, prunes, and liver. It is also present in coffee and tea. The average diet provides about 3 to 9 μg daily.

Recommended Daily Allowance

Lack of a practical method of assessing manganese status has lead to difficulty in establishing safe and adequate dietary intakes. Because there is a lack of documented deficiencies or excess among the general population, current dietary intake appears to be adequate. The National Research Council has set a provisional daily dietary intake level at 2 to 5 mg for adults and adolescents of both sexes. The suggested levels for children are set at an average intake of 1.4 mg per day. Levels for pregnancy are not yet known.

Control

Manganese is absorbed in the small bowel. The absorption rate decreases when there are excesses of calcium, phosphate, or iron. Manganese is one of the trace elements that is considered to be poorly absorbed, thus large amounts are eliminated directly into the feces. The amount that is absorbed binds with serum globulins and is transported to cells for utilization. The absorbed manganese is quickly released into the bile and is excreted. The selective excretion process through the bile controls the tissue levels of manganese. Inappreciable amounts are excreted in the urine.

Implications for Elderly Persons

Studies to date reported by the National Research Council do not show decrease or increase in manganese accumulation with aging.

Functions

Manganese is needed for central nervous system function, reproduction, and bone-building functions. It is an important element in the activation of enzymes involved in both fat and carbohydrate metabolism. Manganese, as a component of metalloenzymes, functions as an enzyme activator, especially for kinases, esterases, peptidases, and various types of carbohydrate enzymes. Tissues containing mitochondria also contain high concentrations of manganese. Some important chemical reactions associated with manganese include:

- Activation of enzymes that form urea
- Prevention of ammonia toxicity
- Activation of amino acid metabolism as a component of proteinases that split specific amino acid bonds
- Glycolysis

Manganese is a component in the activation of lipoprotein lipase, an enzyme that breaks down fat in the serum. Manganese also plays a role in the utilization of vitamins B and B1 and vitamin E.

Manganese Imbalance

Manganese Deficit

The occurrence of manganese deficit is unlikely in individuals consuming a balanced diet. Low-sodium and low-protein diets may supply inadequate amounts of manganese.

Assessment

Inadequate intake of manganese results primarily in neuromuscular disorders and manifestations, such as vertigo, hearing disturbance or loss, and ataxia. Symptoms of weight loss, transient dermatitis, and slow hair growth may be present.

Collaborative Management

Treatment for deficit is directed to replacement of manganese through improved nutrition.

Nursing care is directed towards teaching the patient about a balanced dietary regime, especially if the patient is on a therapeutic low-sodium or low-protein diet. Because of sensory perceptual alterations, attention needs to be given to preventing injury, as well as assisting with ambulation and movements.

Manganese Excess

The potential for manganese excess appears to be fairly minimal. The only observed excesses have occurred as a result of occupational exposure in such industries as mining and steel production, in which workers have inhaled manganese dust or fumes. Excess has not been demonstrated to be a consequence of excessive nutritional intake, although concentrations are increased in the serum as a result of liver disease.

Assessment

Manganese excesses accumulate in the liver and central nervous system and produce extrapyramidal symptoms, much like those of Parkinson's disease. These symptoms include tremors, including pill-rolling; gait change or lack of facial expression due to muscular rigidity; and hypokinesia. Increased salivation may also be noted. Furthermore, if dust inhalation leads to toxicity, psychiatric disorders may result. An occasional daily intake up to 10 mg is still considered safe.

Collaborative Management

Prevention via occupational health monitoring is the key intervention in avoiding an excess of manganese.

MOLYBDENUM

Molybdenum is a micromineral that is an essential component of the enzymes xanthine oxidase, aldehyde oxidase, and sulfite oxidase. As with many of the trace elements, most research on this trace mineral has been conducted with animals. An interrelationship of molybdenum, copper, and sulfate absorption has been demonstrated. The symbol for molybdenum is Mo.

Molybdenum Balance

Distribution

Molybdenum is found in minute quantities in humans. It is absorbed from the gastrointestinal tract and excreted through the urine.

Sources

The levels of molybdenum vary according to the soil in which the food was grown. Food sources include milk, organ meats, legumes, whole grains, and dark-green leafy vegetables. The amount available in most drinking water supplies varies between 2 and 8 μg per day. Table 13–4 lists the main sources of trace minerals by food groups.

Recommended Daily Allowance

Most diets contain sufficient quantities of molybdenum, and supplements are not recommended. The amount of mineral needed is estimated to be 75

TABLE 13-4

MAIN SOURCES OF TRACE MINERALS BY FOOD GROUPS

Mineral	Mixed Dishes/ Miscellaneous	Meats	Milk	Fruits/ Vegetables	Grains
Iron	Fortified infant formulas, iron cookware	Meat (esp. liver), eggs	Seafood	Dark-green leafy, nuts, legumes	Selected cereals, whole grains
Iodine	Fortified table salt	Saltwater fish	Milk*		Bakery products*
Fluoride	Fluoridated water, supplements, tea	Fish			
Zinc		Shellfish, red meat			
Copper		Organ meats, shellfish		Nuts, legumes	
Selenium		Lean meat, seafood	Milk products	Legumes	Whole grains
Chromium	Thyme, black pepper	Meat (esp. liver), oysters	Milk products		Brewer's yeast, whole grains
Manganese				Legumes, seeds, nuts green leafy	Wheat bran
Cobalt		Meat, eggs	Milk, cheese		
Molybdenum		Organ meats		Legumes	Grains

*Contaminant Iodine

From Lutz, CA, & Przytulski, KR: Nutrition and Diet Therapy. FA Davis, Philadelphia, 1994.

to 250 μg per day. Infants and children up to the age of 7 years need a little less.

Control and Function

Molybdenum is a key component in a number of metabolic enzymes. Xanthine oxidase is a catalyst for the oxidation of an end-product of protein catabolism, xanthine, to form uric acid. The liver enzyme aldehyde oxidase is responsible for the catabolism of aldehydes to carboxylic acids. Molybdenum is also a key component of the flavoproteins, enzymes important to cellular respiration.

Molybdenum Imbalance

Molybdenum Deficit

Naturally occurring deficits are not well documented at this time. In rare cases, patients on total parenteral nutrition may be vulnerable to molybdenum deficits.

Molybdenum Excess

Excess molybdenum is very uncommon. High levels of environmental molybdenum have been documented in a few cases of excess. Dietary intake in excess of 10 to 15 mg per day may result in gout-like symptoms associated with increased blood levels of molybdenum, uric acid, and xanthine oxidase.

Elevated levels of molybdenum result in loss of copper through the urine. Thus, an excess of molybdenum can result in a copper deficit.

SELENIUM

The amount of selenium in the soil varies from area to area. Little is known about the effects of this trace mineral on humans, but it is considered to have an important role in preventing the breakdown of polyunsaturated fatty acids. The symbol for selenium is Se. Animal research on selenium has been extrapolated to humans with regard to its function and effects. Selenium is a major factor in enzyme systems, and along with vitamin E, acts as an antioxidant to protect cells.

Selenium Balance

Distribution

Selenium deposits occur in the liver, heart, kidney, and spleen. Selenium is also located in plasma, platelets, and erythrocytes, and is excreted in the urine.

Sources

The best sources of selenium are seafood, organ and muscle meats, and poultry. Selenium levels in plant sources are very unpredictable because they depend on selenium soil levels. Whole grains can be an adequate source depending on soil conditions, but milling of grains increases the loss of this mineral. In some regions, selenium levels in humans coincide with selenium levels in livestock, which, in turn, depend on the amount of selenium in the soil.

Transcultural Considerations

People who have a subsistence farming diet based largely on milk and whole grains have a low rate of cancer. This diet, where most of the farm produce is kept for home use rather than marketed, is rich in selenium, mangesium, fiber, and other anticarcinogenic substances. This type of diet is consumed in the Appalachian region of the United States (Giger & Davidhizar, 1991).

Recommended Daily Allowance

About 50 to 70 µg of selenium per day is needed to maintain body stores and replace losses. This assumes an absorption rate of 80% and is not recalculated for variations in vitamin E intake. Selenium needs increase with age and during pregnancy (65 μg per day) and lactation (75 μg per day).

Implications for Infants

Infants from birth to 6 months need 5 to 10 μg per day. For nursing infants, 750 mL of breast milk daily supplies 13 μg of selenium. In the second 6 months of life, 15 μg per day is recommended.

Control

Very little specific information is available regarding the physiologic control of selenium. According to Mahan and Arlin (1992), selenium is metabolized in anionic form. It is absorbed in the small intestine, with the highest rate of absorption in the duodenum. Selenium is transported initially by albumin and subsequently by α_2-globulin.

Functions

The major functions of selenium involve immune mechanisms, fat metabolism, and cellular protection. It serves as an antioxidant along with vitamin E and ATP synthesis. Selenium may function in the immune system by aiding in the production of interferon, the body's natural antiviral and anticancer substance. The remedial actions of selenium and vitamin E as antioxidants may reinforce each other. The result of this protection is that the cell membranes, cell nuclei, and chromosomes are less likely to be damaged by the constant barrage of environmental carcinogens.

Selenium has slowed cancer growth in mice without affecting healthy cells. The demonstrated ability of selenium to lower fat in animals may point to a possible role for selenium in the reduction of breast cancer, because there is a positive correlation between a high percentage of body fat and breast cancer.

Selenium works with both vitamins A and E. Vitamin E may act to stabilize selenium so that it is more readily accessible for utilization by the body.

Selenium Imbalance

Selenium Deficit

In spite of a wide range in the amounts of selenium consumed, the human body seems to be able to maintain an adequate balance of selenium. Thus, selenium deficits are rare. However, the potential role of selenium in reducing the incidence of cancers has focused increasing interest on selenium deficits. Vitamin E also has a role in maintaining selenium balance.

Causative Factors

Selenium deficits have been documented primarily through geographic identification of diseases that occur when selenium consumption from the diet is low. A relationship between selenium deficit and poor health is suggested by the correlation between low levels of soil selenium, and thus low bioavailable selenium, and high rates of such diseases as breast cancer. Low blood levels of selenium have in fact been linked to increased cancer rates (Davis & Sherer, 1994). Another example of correlation between low selenium consumption and disease has been documented in China. Two key disorders, called Keshan disease and Keshan-Beck disease, have occurred in some regions of China. One disorder linked to this regional selenium deficiency is manifested as a severe cardiomyopathy in children. The second disorder, also related to cardiomyopathy, occurs in women of childbearing age.

Livestock, also a source of selenium, have been studied extensively in connection with a disorder called white muscle disease, which occurs when a selenium deficit exists. However, the condition may be caused by a concomitant deficiency of vitamin E.

The risk for deficiency occurs when the vitamin E amounts vary or when the levels of polyunsaturated fatty acids are too high. This occurs because vitamin E acts as an antioxidant, as fatty acid metabolism tends to increase the production of oxidants.

Assessment

Cardiomyopathy and muscular weakness and fatigue can result from a selenium deficit. Dietary history is not reliable as a means of validating selenium consumption because selenium content in foods varies widely with the selenium content in the soil.

Most serum selenium is bound to protein, and the laboratory results are significantly affected by serum protein levels. Normal blood levels are 0.1 to 0.34 µg/mL, with red blood cells containing greater amounts. Evaluation of erythrocytes is more representative of the long-term status of selenium than is evaluation of the plasma. In addition, animal studies suggest that the results obtained by evaluation of plasma alone do not correlate with the selenium content of other body pools. Hemolysis can also alter the serum amounts. Hair analysis values are yet to be established.

According to the National Research Council, evaluation of glutathione peroxidase activity measures biologically active selenium. The advantage of measuring glutathione peroxidase to assess selenium is that it is easier to do. The disadvantage is that glutathione peroxidase activity plateaus when selenium levels are high; thus, it is only an effective measure of selenium in populations with low selenium levels. Sensitive biochemical indicators of selenium toxicity are not known.

Collaborative Management

The usual treatment for selenium deficiency is dietary intervention. From 100 to 200 µg per day are recommended until the deficient state is corrected. A daily maintenance dose of 50 to 70 µg is recommended for adults.

Of special concern are clients on total parenteral nutrition (TPN). Replacement of trace minerals is required for severely malnourished patients and for those expected to receive TPN for more than a month. Kokko and Tannen (1990) recommend infusion of selenomethionine to improve the cardiomyopathic state.

Nursing interventions include collaboration with the dietitian to perform nutritional assessments noting a diet that is sufficient in daily recommended allowances. In severe depletion, low cardiac output and concomi-

TABLE 13–5
MINERAL ELEMENTS REQUIRED BY THE BODY

Major Minerals (> 100 mg or more/day)	Trace Elements (< 100 mg daily)	Possible Essential Trace Elements (No RDA)	Trace Elements (Unknown Function or Contaminants)
Calcium (Ca)	Iron (Fe)	Nickel (Ni)	Aluminum (Al)
Phosphorus (P)	Copper (Cu)	Tin (Sn)	Barium (Ba)
Sodium (Na)	Zinc (Zn)	Silicon (Si)	Strontium (St)
Potassium (K)	Manganese (Mn)	Vanadium (V)	Mercury (Hg)
Magnesium (Mg)	Iodine (I)	Boron (Bo)	Silver (Ag)
Chlorine (Cl)	Molybdenum (Mo)	Cadmium (Cd)	Gold (Au)
Sulfur (S)	Fluorine (Fl)	Arsenic (As)	Antimony (Sb)
	Selenium (Se)	Lead (Pb)	Others
	Chromium (Cr)		
	Cobalt (Co)		

From Davis, JR, & Sherer, K: Applied Nutrition and Diet Therapy for Nurses (2nd ed). WB Saunders, Philadelphia, 1994. Reprinted with permission.

tant hypoxia and fatigue must be addressed by appropriate cardiac care. Most relevant for hospitalized patients is keen observation of those receiving total parenteral nutrition, especially malnourished patients. The nurse should also be aware of two situations which further increase the risk of selenium deficits: reduced vitamin E intake and increased intake of polyunsaturated fatty acids. Both of these factors increase the risk of selenium deficiency because the production of oxidants is increased.

Selenium Excess

Toxic levels in humans have not yet been established. In animals, levels exceeding 3 μg/g of diet are toxic.

OTHER TRACE ELEMENTS

Many other trace elements exist in the body, but their precise roles and requirements have not yet been established. Boron is needed for bone development and boron deficiency can contribute to osteoporosis. Boron is found in fruits, vegetables, nuts, and legumes (Davis & Sherer, 1994). Table 13–5 lists the mineral elements in the body.

CRITICAL THINKING ACTIVITIES

A 7-year-old girl is brought to the emergency department of a small rural hospital complaining of right ear pain and a fever. As part of the physical examination, she is noted to have severe dental caries and gum disease. The family lives on a small farm and drinks well water. The patient has never been to a dentist and does not brush her teeth regularly.

QUESTIONS

1. *What trace element deficiency does the patient have?*

2. *What risk factors made her prone to this mineral deficiency?*

3. *What treatment does she need for the deficiency?*

4. *What patient and family education is required to prevent further health problems related to the mineral deficiency?*

REFERENCES AND READINGS

Chernecky, CC, Krech, RL, & Berger, BJ (1993). Laboratory Tests and Diagnostic Procedures. Philadelphia: WB Saunders.

Davis, J, & Sherer, K (1994). Applied Nutrition and Diet Therapy for Nurses. Philadelphia: WB Saunders.

Giger, JN, & Davidhizar, RE (1991). Transcultural Nursing: Assessment and Interventions. St. Louis: Mosby–Year Book.

Guyton, AC (1991). Textbook of Medical Physiology. Philadelphia: WB Saunders.

Kokko, J, & Tannen, R (1990). Fluids and Electrolytes (2nd ed). Philadelphia: WB Saunders.

Lutz, CA, & Przytulski, KR (1994). Nutrition and Diet Therapy. Philadelphia: FA Davis.

Mahan, LK, & Arlin, M (1992). Krause's Food, Nutrition, and Diet Therapy. Philadelphia: WB Saunders.

PART TWO QUESTIONS

• • • • • • • • • • •

1. What is a common cause of hypokalemia, especially in elderly persons?
   a. Excessive potassium supplements
   b. Metabolic acidosis
   c. Diuretic therapy
   d. End-stage renal disease

2. Which assessment finding is consistent with hypokalemia?
   a. Muscle weakness
   b. Increased reflexes
   c. Decreased thirst
   d. Kussmaul's respirations

3. Which of the following electrocardiogram (ECG) findings is typical in a patient with hypokalemia?
   a. Flat P waves
   b. Elevated ST segments
   c. U waves
   d. Prolonged P–R intervals

4. Which of the following nursing interventions is not necessary for the patient receiving a liter of IV fluids with 40 mEq of potassium added?
   a. Monitor serum potassium levels.
   b. Place the patient on a cardiac monitor.
   c. Monitor the site for redness and pain.
   d. Place the infusion on an electronic infusion device.

5. A patient has the following serum electrolyte values:

Potassium	3.9 mEq/L
Chloride	93 mEq/L
Sodium	142 mEq/L

   Which of the following electrolyte imbalances does the patient have?
   a. Hyperkalemia
   b. Hypernatremia
   c. Hypokalemia
   d. Hypochloremia

6. Which of the following patients is at the greatest risk for hip fracture related to osteoporosis?
   a. A 45-year-old African-American woman
   b. An 85-year-old Caucasian woman

c. A 75-year-old African-American man

d. An 85-year-old Caucasian man

7. Which of the following assessment findings is not associated with hypocalcemia?

a. Irritability

b. Trousseau's phenomenon

c. Muscle cramps

d. Muscle weakness

8. Which of the following nursing interventions would be appropriate for a patient with a serum calcium of 13 mg/dL?

a. Strain all urine.

b. Restrict fluids.

c. Increase dietary calcium.

d. Give thiazide diuretics as ordered.

9. What endocrine gland regulates both serum calcium and phosphorus levels?

a. Pituitary

b. Pancreas

c. Parathyroid

d. Thyroid

10. Which of the following electrolyte imbalances typically accompanies hyperphosphatemia?

a. Hyperkalemia

b. Hypocalcemia

c. Hypernatremia

d. Hypochloremia

11. What nursing intervention is necessary for a patient with hypomagnesemia?

a. Fluid restriction

b. Blood transfusion

c. Intravenous fluids

d. Seizure precautions

12. Which of the following drugs may be used for a patient with acute hypermagnesemia?

a. Calcium gluconate

b. Sodium bicarbonate

c. Potassium chloride

d. Glucagon

13. Which of the following foods are rich sources of iron?

a. Milk

b. Citrus fruits

c. Red meats

d. Cheese

14. When assessing a patient for zinc deficiency, which of the following patients would be at the highest risk?

a. Adolescents

    b. Elderly persons

    c. Pregnant women

    d. Preschool children

15. Which of the following information regarding fluoride should be included in health teaching for children?

    a. Use a fluoridated toothpaste.

    b. Drink plenty of well water.

    c. Use a Water-Pik when cleaning teeth.

    d. See the dentist every other year.

PART THREE

TECHNIQUES AND PROCEDURES FOR MAINTENANCE OF FLUID AND ELECTROLYTE BALANCE

CHAPTER 14

· · · · · · · ·
· · · · · · · ·
· · · · · · · ·
· · · · · · · ·
· · · · · · · ·
· · · · · · · ·
· · · · · · · ·
· · · · · · · ·
· · · · · · · ·

GASTROINTESTINAL INTUBATION AND ENTERAL NUTRITION

Gastrointestinal intubation and enteral nutrition (tube feeding) affect fluid and electrolyte balance because they either substract from or add to the body's fluids and electrolytes. In either of these processes, acid-base balance may also be altered.

GASTROINTESTINAL INTUBATION

Gastrointestinal (GI) intubation is most commonly used for decompression, to remove gastric or intestinal secretions by suction; or for enteral feeding when oral intake is inadequate. Three major types of tubes can be used — the nasogastric tube, the nasointestinal tube, and the gastrostostomy tube, all of which are described later in this chapter.

Occasionally, GI intubation (Sengstaken-Blakemore tube) is used for compression — the internal application of pressure to treat GI hemorrhage from bleeding esophageal varices (a complication of cirrhosis). Either a three-lumen or four-lumen tube with two balloon attachments is available. One lumen is used for GI suction, the second is used to inflate the gastric balloon, the third is used to inflate the esophageal balloon, and the fourth lumen is used to aspirate secretions.

Purposes

When used for decompression, intubation connected to suction relieves abdominal distress caused by lack of peristalsis, which sometimes follows surgery, or by reverse peristalsis, which sometimes accompanies an intestinal obstruction. Without adequate peristalsis, gaseous and liquid substances build up in the stomach and cause pain and abdominal distention. A nasogastric tube evacuates the stomach of gastric secretions, swallowed saliva, and air. It is also used for patients with conditions such as pancreatitis, cholecystitis, postoperative ileus, and nonoperative management of peptic ulcers.

Another purpose of gastric intubation is to "pump" the stomach when a toxic substance, such as a drug overdose, has been introduced into the stomach. In this procedure, a large-bore tube is inserted and suction is applied to remove the offending substance. This technique is also referred to as lavage.

Gastric intubation may be used for feeding via the oral route when the patient has difficulty ingesting food. Tube feedings administer calories and nutrients, especially the protein necessary for tissue healing. Conditions in which patients may require tube feeding include severe dementia, unconsciousness, mental illness, extreme weakness caused by chronic debilitating illnesses, difficulty in swallowing, severe anorexia, and oral surgery. Feeding by intubation is not a preferred method, but it is sometimes necessary to maintain the patient's nutritional state until other problems can be improved or remedied.

Types of Tubes

The most commonly used gastrointestinal tube is the nasogastric tube, which is usually inserted through the nose and advanced into the stomach. In infants, the tube may be inserted through the mouth and into the stomach.

Nasogastric Tube

The most common and familiar nasogastric (NG) tube, the Levin tube, is a single-lumen plastic tube. In adult sizes, it usually ranges from 12 to 18 French (12F to 18F). The more flexible and pliable small-bore feeding tubes are 8F to 12F. The smaller the number, the smaller the tube diameter. Infants and children require very small tubes.

All NG tubes are made of radiopaque material to facilitate the examination of placement by x-ray. Double-lumen NG tubes, such as the Salem sump, have a sump vent open to air ("blue pigtail") that partially suspends the tube to keep it from coming in close contact with the gastric mucosa. The sump vent helps equalize the outside pressure with the pressure inside the stomach. In this way, it prevents damage to the gastric mucosa, especially during suction (Fig. 14–1).

Both Levin and Salem sump tubes are used for decompression. The Levin tube may also be used for *short-term* feeding, but if a feeding solution is given through a Salem sump tube, it will leak out of the sump airway onto the patient or bed.

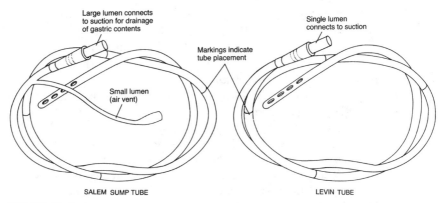

FIGURE 14-1
Two types of nasogastric tubes. (From Ignatavicius, DD, Workman, ML, and Mishler, MA: Medical-Surgical Nursing: A Nursing Process Approach, ed 2, WB Saunders, Philadelphia, 1995, p 416. © 1992 M. Linda Workman. Reprinted with permission.)

For *long-term* feeding, softer, more pliable small-bore tubes made of silastic or polyurethane are used in most cases. Most of these tubes have a small weight on the distal end so that the tube advances into the duodenum instead of remaining in the stomach. Use of these nasoduodenal (ND), or nasoenteral, tubes decreases the incidence of dislodgment and aspiration. Examples are the Dobbhoff and Keofeed tubes.

Nasogastric Tube Insertion

The NG tube may be inserted through the nares or the oral cavity to the stomach. Nasal insertion is usually preferred over oral insertion because it is less likely to stimulate the gag reflex. NG insertion is usually used if the tube is to be left in place over a long period. The orogastric approach is used when the nares are structurally abnormal or when intubation will be used for only a short time. NG insertion is also contraindicated in the presence of nasal trauma with bleeding or when the tube is too large to pass through the nares.

If possible, the patient should be in an upright position (high Fowler's position). The patient, who must remain supine, should be positioned on the right side. These positions use gravity to facilitate the passage of the tube.

The nurse determines the length of tubing needed, before insertion, by the traditional method or the Hanson method. In the traditional method, measurement is made from the tip of the patient's nose to the earlobe, and from the earlobe to the lower end of the sternum. The average measurement for an adult is 18 to 22 inches (45 to 55 cm), and is an estimate of the length of tube needed to reach the stomach. This method of measurement is controversial in infants. Studies have shown that the tube may not be in the stomach, but rather lodged in the esophagus if this measurement technique is used (Whaley & Wong, 1995). In using the Hanson method, the nurse notes the first 50-cm (20-inch) point on the tube, then carries out the traditional measurement. Tube insertion should be to the midway point between the 50-cm and the traditional measurement marks.

Unless specimens for cytology are to be obtained, the tube is well lubricated with water-soluble jelly. (Lubricants interfere with the visibility of the cells.) Jelly, water, or saline can be used for lubrication. Water-soluble material should be used in case the tube enters the trachea. Oil-based lubricants are irritating to the respiratory tract and may cause pneumonitis.

To insert the tube into the nostril, the tube is grasped about 3 inches from the lubricated end. Holding the tube close to the end makes the tube stiffer and facilitates entry. The nares are held upward while the tube is advanced forward and downward. The upward position of the nares straightens the passageway that the tube must follow. The tube should not be forced if there is resistance, because using force could cause trauma to the mucosa and break the blood vessels. If no resistance is felt, the tube is passed to the back of the oropharynx, which should be about the premeasured distance from the tip of the nose to the earlobe. The patient's eyes may tear because passage of the tube stimulates the tear glands and ducts.

The patient is then asked to flex the head so that the chin rests on the chest. This position encourages the tube to enter the posterior pharynx and then the esophagus. The patient should now begin swallowing to help move the tube into the esophagus rather than the trachea. Deep breathing, on the other hand, tends to draw the tube into the trachea. At this point, the nurse asks the patient to open the mouth (if able) and determines whether or not the tube is circling the mouth. Peristalsis of the esophagus also helps advance the tube. If the patient is unable to swallow, stroke the throat upward, toward the chin. The stroking causes swallowing, which closes the epiglottis and increases pharyngeal contraction and esophageal peristalsis.

If excessive coughing, cyanosis, or inability to speak occurs, the tube should be withdrawn. These are signs that the tube has passed into the larynx and is stimulating the cough reflex in the trachea.

Because advancement of the tube often stimulates the gag reflex, it may be necessary to stop advancement for a short time until gagging is controlled. To help decrease the psychologic aspects of gagging, the conscious patient can pant until the gagging sensation passes.

Again, the tube is not advanced if resistance is encountered. Resistance may indicate the presence of an obstruction or abnormality. This problem should be reported to the physician.

Verifying Tube Placement

After tube insertion, the patient is asked to talk, if able. If a large-bore tube was passed through the vocal cords, a patient who was previously able to talk will not be able to speak. Inability to speak indicates improper tube placement.

The most reliable method for validating correct tube placement is x-ray examination, which should be performed after any GI tube has been inserted.

Placement may also be checked by aspirating gastric contents using a large piston syringe. A return of gastric secretions usually means that the tube is in the stomach. Shifting the patient's position so that he or she is lying on the left side will sometimes facilitate a return of gastric secretions. Another commonly used method of verifying the tube's position involves injecting about 30 mL of air into the tube with a syringe. A stethoscope is placed over the cardia of the stomach to listen for a swooshing sound as the air enters the stomach. Although this auscultatory method may seem effective, it is *not* reliable, especially for patients with the newer, small-bore feeding tubes. For these patients, testing the pH of the aspirate has been suggested as a way to verify placement (Fater, 1995). The pH of aspirates from NG tubes ranges from 1 to 4 because of the high acidity of gastric secretions; the aspirates from nasointestinal tubes (described later) are more alkaline, with pH values of 6 or higher.

Once the tube is inserted, it is taped to the nose. A good taping method is to split a 2- or 3-inch long tape about half way down, leaving two tails. The unsplit portion of the tape is placed on the nose, and the two tails are wrapped around the tube. This method of taping prevents the tube from jiggling, which could stimulate the gag reflex or irritate the nares. If the tape is placed tightly against the opening of the nares, a pressure sore can develop.

Gastric Decompression

If the tube is used for decompression, it is attached to wall or Gomco suction. The gastric suction setting for a Levin tube is low intermittent suction. High pressure may traumatize the gastric mucosa. For a Salem sump tube, the suction should be set on low continuous or high intermittent suction. If low intermittent suction is used, the GI secretions will leak out of the blue pigtail vent. A special valve is available for the blue pigtail to prevent leakage of secretions.

Nursing Interventions During Decompression

Proper oral care is extremely important during intubation. The tube stimulates the mucosa and can become an irritant. Patients with gastric tubes tend to breathe through the mouth, which dries the membranes. The mouth needs frequent cleaning either with a solution of normal saline and hydrogen peroxide solution or by brushing the teeth (if possible). Some physicians allow patients to keep the mucous membranes moist by sucking on ice chips. Large amounts of ice should not be used, however, because water is not isotonic. A glass of cool water for rinsing the mouth can be soothing, but the patient should not swallow the water.

Because there is no food to stimulate the salivary glands, they may not empty, and parotitis (inflammation of the parotid glands) may develop. Sucking on hard, sour candy may eliminate this problem. This intervention may be contraindicated, however, because it also stimulates the gastric juices, which may promote ulcer formation. Excessive gagging may be a problem for some patients. Gagging can be controlled to some degree by administering an antacid to coat the pharynx. If the antacid is ineffective, an anesthetic agent that decreases the sensations of the pharynx may be used. Anesthetics commonly used for this purpose are lidocaine spray, lidocaine viscous, and Cetacaine. A nonalcohol-based mouthwash should be made available to the patient to alleviate some of the bad tastes and odors that accompany intubation.

In addition to oral measures, the nurse must maintain patency of the tube during decompression by checking for kinks and observing gastric drainage. To promote gastric drainage, the tube can be irrigated with 30 mL of normal saline (isotonic solution) every 4 hours and as needed. Hypotonic solutions, such as water, should never be used because they can cause electrolyte imbalances. The amount of irrigant must be measured so that it can be subtracted from the drainage output to give an actual drainage output measurement. The drainage should also be observed for clarity, consistency, color, and odor. Typically, gastric contents are greenish-brown, with flecks of mucus. There should not be an extremely foul odor or any blood. If the presence of blood is suspected, the drainage can be tested.

At all times, the head of the bed should be raised to prevent aspiration and facilitate drainage. Comfort measures include care of the nares as well as oral care. The opening of the nares tends to accumulate dried secretions. These secretions should be removed at least daily with a cotton swab soaked in peroxide. The tape should also be changed every day because the adhesive becomes gummy and leaves a residue. The nose can become very sore if these preventive measures are neglected.

Another important aspect of nursing care is to monitor for fluid, electrolyte, and acid-base imbalances. The patient with nasogastric decompression is at high risk for fluid deficit, hyponatremia, hypokalemia, hypochloremia, and metabolic alkalosis from loss of gastric acid secretions. The nurse monitors electrolyte and arterial blood gas values carefully and reports significant changes immediately to the physician. Each of these imbalances is discussed in Parts I and II of this book.

Removal of Nasogastric Tube

For removal, the end of tube is clamped to prevent gastric secretions from dropping into the trachea. The tape securing the tube is gently removed. The patient is instructed first to inhale deeply, then to exhale slowly while the tube is being removed. Deep breathing relaxes the muscles and prevents aspiration of gastric secretions into the trachea. The tube should be pulled out in one continuous motion.

Because many patients are nauseated by the sight of the tube, it should be wrapped in a towel or blue pad as it is removed, and kept out of sight. The patient needs oral care to prevent irritation and nausea from the tube. The tube is discarded in the infectious waste container as indicated by agency policy. Lozenges or warm, saline rinses may help relieve the soreness of the throat.

Nasointestinal Tubes

Nasointestinal (NI) tubes are used for decompression of the intestinal tract or as a presurgical aid, although they are less commonly seen today than years ago. Intestinal decompression is the removal of both gaseous and liquid substances from the intestines. Although it may be necessary to remove the liquid contents of the intestines, it is extremely important to maintain fluid and electrolyte balance. Losses most frequently associated with removal of gastrointestinal fluids are those of sodium and chloride. Potassium losses may be significant when they are from the lower gastrointestinal tract. Intestinal intubation relieves the pressure in the intestines caused by flatus. Flatus or gas is usually a result of slowed or absent peristalsis caused by anesthesia, or by surgery in which the intestines have been manipulated. Peristalsis may also be upset by an intestinal obstruction, which may actually cause the intestines to contract in a *reverse* direction in an attempt to move the secretions out of the body. This is the reason that nausea and vomiting, possibly projectile, are often seen in the patient with an intestinal obstruction.

The most commonly used intestinal tubes are the Miller-Abbott and the Cantor tubes. The Miller-Abbott tube is a double-lumen tube. It is 10 feet long and most frequently 16F in diameter. The lumina are independent of each other. One lumen is used for inflation of the bulb at the end of the tube. The balloon and metal tip assist forward movement. The other lumen is for the aspiration of gas and secretions and is connected to suction, which pulls the gas and secretions out by vacuum.

Unlike the NG tube, which can be inserted by the nurse, the NI tube is usually inserted by the physician through the nose into the stomach in the same manner as the Levin tube. Before insertion, the tube must be measured to determine the correct length. The markings on the tube can be used for an estimate.

Unlike the NG tube, the NI tube is *not* taped to the nose after insertion. The tube must be free so that the action of peristalsis can draw it into the intestines. If the tube is anchored, it cannot move into place in the duodenum. The tube may be coiled and placed in a plastic sack attached to the patient's gown. This enables the tube to move and keeps the patient from becoming entangled in the tubing. Varying the position of the patient often aids the movement of the Miller-Abbott tube. Allowing a time lapse between each shift, the patient is first turned to the right side, then to the back, and then to the left side. X-ray examination will verify the placement of the tube.

The progress of advancement can be monitored by checking the scale marked on the tube. The physician may order the nurse to advance the tube manually into the stomach, designating the times and distances of the advancement.

The Cantor tube is a 10-foot, single-lumen tube. An 18F diameter is most commonly used. Its main characteristic is a mercury-filled bag at the end of the rubber tubing to facilitate gravity flow insertion. The physician will insert a specific amount of mercury (usually 5 mL) into the bag with a needle and syringe. Because of its density, the mercury will not leak out of the rubber bag. As with the Miller-Abbott tube, the physician is usually responsible for insertion. Advancement of the tube may be ordered, or the physician may choose to advance the tube under fluoroscopy to allow the movement of the tube to be observed with the help of x-ray visualization.

Agency policy determines whether removal of the tube is the responsibility of the physician or the nurse. It is often the physician's responsibility.

Gastrostomy Tube

Whereas both the NG and NI tubes are inserted through the nose into the GI tract, the gastrostomy tube (G-tube, or GT) is placed directly through the skin into the stomach. The G-tube is *not* used for decompression, but is used exclusively for long-term feeding.

The GT can be inserted laparoscopically or surgically through a small incision. The percutaneous endoscopic gastrostomy (PEG) tube is becoming very popular because it does not require a surgical incision. It can be inserted at the bedside under local anesthesia and conscious sedation.

Jejunostomy Tube

For patients who do not have a stomach or have had a large amount of the stomach damaged or removed, a jejunostomy tube (J-tube, or JT) can be inserted in the same manner as the G-tube. Like the G-tube, it is used only for feeding and is not very commonly used. Table 14–1 lists the types of feeding tubes and indications for each one.

ENTERAL NUTRITION

Enteral nutrition is the introduction of nutrients into the gastrointestinal tract via a tube. Also called tube feeding or gavage, it is indicated when the patient is unable to maintain adequate nutrition by other means but still has patent gastric and intestinal passageways and normal digestive patterns. In collaboration with the dietitian, the physician prescribes the amount, frequency, and kind of feeding for each patient.

A number of commercial formulas are available, depending on the patient's needs. Most formulas provide 1 calorie per milliliter of solution (e.g., Ensure, Jevity, and Osmolite), but Ensure Plus has 1.5 calories/mL and TwoCal HN has 2 calories/mL. It is important that formulas not be substituted for one another unless they provide the same nutrients.

The prescribed formula is usually diluted when the feeding begins in order to help the patient's GI tract to adapt gradually to the feeding. Additional water is also provided with the tube feeding, to meet the body's water needs.

Like parenteral nutrition, enteral nutrition may be partial or total. Par-

TABLE 14–1
TYPES OF FEEDING TUBES: INDICATIONS AND PRECAUTIONS

Type of Tube	Indications	Precautions
Nasogastric (NG) tube	Short-term use	Watch for aspiration. Observe for irritation of nares and throat.
Nasoduodenal (ND) tube	Short-term use (but can be used longer than traditional NG tube)	Observe carefully for signs of incorrect placement; x-ray is the only reliable method for verifying placement. If aspirate obtained, check pH.
Gastrostomy tube (G-tube, GT)	Long-term use	Check incision site for infection. Check for correct placement.
Jejunostomy tube (J-tube, JT)	Long-term use when stomach not available	Same as above.

tial enteral nutrition (PEN) means that the patient can consume some nutrients orally, but that supplemental nutrition is needed to meet the body's requirements. Some patients, for instance, have an order for tube feeding at night if they consume less than a designated amount of their meals during the day. This regimen is particularly common for elderly patients. Total enteral nutrition (TEN) implies that the patient must depend entirely on the tube feeding for nutrition.

The caloric requirements are based on the patient's weight and body needs. Calories are administered primarily for their protein-sparing action. If the protein supply is not adequate, the body cannot make and repair tissues and may eventually burn body proteins for energy. Nutrients burned for energy are consumed in the following order: carbohydrates, then fats, and then proteins. The nurse should be alert to inadequate intake and utilization and correct these by whatever method will improve the patient's nutritional status.

Tube Feeding Methods

There are basically three methods of tube feedings used today—bolus, continuous, and cyclic. All methods accomplish the same goal, and for all methods the patient must be in a sitting position or have the head of the bed raised up by no less than 30 degrees during and after feeding.

Bolus Intermittent Feeding Method

Bolus tube feeding is the oldest method and is used far less commonly today than the other two methods. A specified amount of formula and water are administered at specified times during a 24-hour period, usually every 4 hours. This method can be administered through a large syringe or by using a mechanical pump or controller.

Before every bolus feeding, the placement of the tube is checked, as described earlier, to ensure that it is in the stomach. A syringe is used to withdraw any nutrients remaining in the stomach from the previous feeding. This residual is measured and returned into the stomach. The physician usually specifies what to do if the residual is too high (e.g., over 100 mL).

The formula is placed into the syringe or bag threaded through the pump and attached to the tube. If a syringe is used, gravity controls the flow of the liquid through the tube. The rate of flow is controlled by the height at which the syringe is placed. If the solution does not flow easily, the feeding may need to be thinned with water. In some cases, the flow can be started by applying a small amount of pressure generated by the bulb of the syringe or plunger. Rapid administration must be avoided because it tends to increase gastric discomfort and to stimulate the gastrocolic reflex. Feedings high in calories may precipitate diarrhea. At no time should air be allowed to enter the tubing.

After completion of the feeding, the tubing should be flushed with 50 to 75 mL of water. This may be specified by the physician. The water need not be sterile. The amount of water may need to be adjusted according to the fluid needs of the patient and depending on the amount of flushing needed to prevent clogging of the fluid in the tube. The intakes of both formula and water are recorded. After the feeding, the tube is clamped to prevent leaking.

Continuous Feeding Method

Probably the most popular method for tube feeding is the use of a small-bore feeding tube and a food pump to introduce formula into the stomach, duodenum, or jejunum. Unlike the previous method, the mechanical food pump maintains a continuous flow of nutrients at a constant rate. The other methods do not provide this accuracy in maintaining a constant flow.

The food pump connective tubing is attached to the gastric tube by a plastic connector. Pulleys, usually on the front side of the pump, control the rate of flow, which is usually regulated in terms of milliliters per hour. The rate is determined by the feeding volume and the time period specified by the physician. Because continuous feeding is required, the liquid should never be allowed to clog in the tubing. In the hospital setting, no more than 4 to 8 hours of formula is placed into the feeding bag at one time to prevent bacterial growth. In the home or nursing home setting, a closed system may contain more than 8 hours of formula. The bag must be labeled with the type, amount, and rate of feeding, as well as the time the formula was added. Each feeding administration set (bag and tubing) can usually be used for 24 hours.

Although the feeding is continuous, the nurse must check for placement (using aspirate pH or residual), measure and record the amount of residual feeding, and flush the tube with the prescribed amount of water every 4 hours. Accurate intake and output are essential.

If diarrhea occurs, the feeding may need to be diluted or the formula may need to be changed. Some formulas are hyperosmolar, pulling fluid into the gut and causing diarrhea.

During continuous feedings, the patient must be monitored to be sure that the feeding is not being administered too rapidly. Rapid feeding may incite nausea and vomiting from the high-caloric intake, or distention, which may cause spasms of the diaphragm exhibited by hiccoughs (singultus). Slowing the feeding down or discontinuing if for a short time will aid the patient.

If the patient is on a ventilator or has a cuffed endotracheal or tracheostomy tube, the cuff should be kept inflated to prevent aspiration.

Cyclic Feeding Method

Cyclic tube feeding is similar to continuous feeding except that the infusion is stopped for a period of time each day, usually 4 to 8 hours, to allow the GI tract to rest. This "down time" is typically scheduled for the morning while the patient is receiving care, treatments, and so forth. The care for the patient is the same as that described above.

Complications of Tube Feeding

Complications of tube feeding can be divided into those related to the tube and those related to the formula used for feeding.

Complications Related to the Tube

With any GI tube, but especially the NG tube, the risk of vomiting and aspiration is present. The NG tube inhibits the effectiveness of the gag and cough reflexes. In addition, the large-bore tubes like the Levin tube cause irritation to the throat and nose. Erosion of these tissues can occur with prolonged use. Using small-bore, flexible feeding tubes avoids this complication.

The biggest complication associated with tubes is obstruction. Because some of the formulas are thick, the tube diameter is small, and medications may be administered through the tube, the tube may become clogged. Table 14–2 lists interventions that can help maintain a patent tube.

Gastrostomy tubes pose a different kind of problem in that they may be dislodged, or even fall out of the G-tube opening. Sometimes a Foley catheter is used as the G-tube so that it can be changed easily by the nurse. Others must be replaced by the physician in a hospital setting. Another complication associated with surgically inserted G-tubes is the risk of infection in the surgical incision. A sterile dressing should be maintained around the tube and changed every day. The nurse should observe the skin for redness, warmth, and purulent drainage—all signs of possible infection. If purulent drainage is present, it should be cultured and sent to the lab.

Complications Related to the Formula

Potential complications associated with the tube-feeding formula include fluid, electrolyte, and nutritional imbalances.

Fluid Imbalances

Fluid volume overload is more common than fluid volume deficit in patients receiving tube feedings. For example, the volume of the formula added to the volume of the supplemental water may be too much for the elderly patient who has compromised renal or cardiac function. In addition, the osmolarity of most formulas is greater than that of the plasma, due to their high protein and sugar content. These substances act to pull water from the intracellular and interstitial spaces into the intravascular compart-

TABLE 14–2
TIPS ON MAINTAINING FEEDING TUBE PATENCY

- Do not put crushed pills through the tube; try to use liquid medications instead.
- Flush the tube before *and* after every feeding and every 4 hours.
- Use warm water for flushing, after checking residual.
- If the tube becomes clogged frequently, consider having the formula changed.
- If the tube is clogged, use 30 mL of warm water and apply gentle pressure on the syringe piston.
- Avoid the use of carbonated beverages to unclog a tube *unless* water is ineffective.

ment (plasma). This fluid shift can lead to congestive heart failure and pulmonary edema, which are life-threatening complications.

For patients with poor cardiac or renal function, more concentrated calorie formulas, such as those that supply 2 calories/mL, may be used. Low-sodium formulas are also available.

When hyperosmolar formulas are given too rapidly, the pulling of water into the intestines causes diarrhea. In elderly patients or in infants and small children, dehydration can result from fluid losses. To help prevent this potential problem, the formula is usually diluted initially and the rate is slow at the beginning. The strength and rate can be gradually increased over a few days to a week, until the patient's intestinal tract adjusts to the therapy. Diarrhea may also be caused by bacterial contamination (Vines et al, 1992).

Electrolyte Imbalances

Sodium, potassium, phosphorus, magnesium, and zinc imbalances may result from long-term tube feeding. Dilutional hyponatremia due to the shifting of fluid into the plasma is more common today than hypernatremia. In the elderly, hypernatremia with fluid deficit can occur when very hyperosmolar formulas are used. In this situation, the kidneys are not able to excrete sufficient sodium because there is not enough water to facilitate its removal from the body.

When patients are started on enteral feedings, potassium, phosphorus, and magnesium shift from the extracellular fluid compartment into the intracellular fluid compartment, resulting in serum deficits. Hyperkalemia can occur if the formula contains a large amount of this electrolyte. Hyperkalemia and hyperphosphatemia tend to occur in tube-fed patients with renal disease.

Zinc deficiency has received a great deal of attention because elderly patients receiving tube feedings have a high incidence of pressure ulcers. Most patients who are tube-fed should receive zinc supplements (see Chap. 12).

Nutritional Imbalances

Enteral formulas typically contain a large amount of sugar, which can lead to hyperglycemia and weight gain. Weight gain may also occur as a result of fluid retention because of high carbohydrate and sodium content.

More commonly, the frail elderly, debilitated patient on tube feedings loses weight despite attempts to increase consumption of calories. Serum albumin levels remain low and the patient is prone to skin breakdown. The exact cause for this phenomenon is not known.

Nursing Interventions Related to Tube Feeding

Nursing interventions are directed towards ongoing fluid, electrolyte, and nutritional assessments while the patient is receiving enteral nutrition. Weights, along with intake and output, are monitored carefully. The nurse observes for complications and takes measures, as discussed earlier, to prevent them. Table 14–3 highlights the major interventions needed for tube-feeding maintenance and care.

TABLE 14–3

NURSING INTERVENTIONS FOR CARE AND MAINTENANCE OF TUBE-FED PATIENTS

- Monitor and record the patient's intake and output while on tube feeding, unless the patient is stabilized.
- Monitor laboratory values, especially BUN, creatinine, glucose, albumin, and electrolytes.
- Ensure that the correct formula and rate are used.
- Keep the head of the bed up or have the patient in a sitting position during tube feeding and 1 hour after tube feeding (bolus method).
- In the hospital setting, hang only 4 to 8 hours of formula (depending on policy) in the feeding bag, unless a closed bag system is used (formula comes already packaged in bag or plastic bottle).
- Change the feeding administration bag or plastic bottle every 24 hours.
- Label the bag or bottle with the formula, rate, strength, and time.
- Check for placement and residual formula every 4 hours; return residual formula into stomach.
- If long-term enteral feeding is expected, use flexible, small-bore nasoenteral tube or gastrostomy tube.
- Flush the tube with the prescribed amount of water as ordered at least every 4 hours.
- Check the feeding pump for proper operation.
- Maintain tube patency using methods listed in Table 14–2.

REFERENCES AND READINGS

Bockus, S (1991). Troubleshooting your tube feedings. American Journal of Nursing 91(5), 24–30.

Bockus, S (1993). When your patient needs tube feedings—Making the right decisions. Nursing93 23(7), 34–42.

Edes, TE, Walk, BE, & Austin, JL (1990). Diarrhea in tube-fed patients. American Journal of Medicine 88(1), 91–93.

Eisenberg, P (1991). Pulmonary complications from enteral nutrition. Critical Care Nursing Clinics of North America 3(4), 641–649.

Eisenberg, P, Metheney, N, & McSweeney, M (1989). Nasoenteral feeding tube properties and the ability to withdraw fluid via syringe. Applied Nursing Research 2(4), 168–172.

Fater, KH (1995). Determining nasoenteral feeding tube placement. MEDSURG Nursing 4(1), 27–32.

Metheney, N (1993). Minimizing respiratory complications of nasoenteral tube feedings—State of the art. Heart and Lung 22(3), 324–329.

Metheney, N, et al. (1990). Effectiveness of the auscultatory method in predicting feeding tube placement. Nursing Research 39(5), 262–267.

Metheney, N, et al (1993). How to aspirate fluid from small-bore feeding tubes. American Journal of Nursing 93(5), 86–88.

Vines, SW, et al (1992). Research utilization: An evaluation of the research related to causes of diarrhea in tube-fed patients. Applied Nursing Research 5, 164–173.

Whaley, LF, & Wong, DL (1995). Nursing Care of Infants and Children. St. Louis: Mosby–Year Book.

CHAPTER 15

. . . .
. . . .
. . . .
. . . .
. . . .
. . . .
. . . .
. . . .
. . . .

Intravenous Therapy and Parenteral Nutrition

Intravenous (IV) therapy is the most common type of infusion therapy used to treat patients with a variety of diseases and health problems. It is the practice of administering fluids, nutrients, or drugs directly into the bloodstream through a venous opening called a venipuncture. A small peripheral vein or large "central" vein may be used for IV access. As with gastrointestinal intubation and tube feeding, IV therapy influences fluid and electrolyte balance.

PURPOSE OF INTRAVENOUS THERAPY

Intravenous therapy has several purposes. It replaces fluids and nutrients more rapidly than oral administration. It is used to sustain patients who are unable to take substances orally. Another important use is to administer drugs that must be instantly available to the body, such as those administered to patients with acute infection.

In contrast to conventional intravenous therapy, which supplies fluids and drugs, parenteral nutrition is the IV infusion of nutrients in a highly concentrated solution. The purpose of parenteral nutrition is to provide fluids, nutrients, and electrolytes for patients who are severely debilitated and unable to achieve homeostasis through normal ingestion of fluids and food. It is also used for patients who are undergoing major surgery and those who will have long-term recovery. This special type of IV therapy is described in more detail later in this chapter.

SELECTION OF INTRAVENOUS FLUIDS

It is important for the nurse to be familiar with the types of fluids (solutions) used and the procedures for their safe administration. Before its administration, the IV solution should be checked against the physician's orders for type, amount, percent of solution, and rate of flow. The nurse should be aware of situations that contraindicate particular IV solutions. For example, a patient with congestive heart failure is usually not given a saline solution because this type of fluid encourages the retention of water and would therefore exacerbate heart failure by increasing the fluid overload. A diabetic patient does not typically receive dextrose (sugar) solutions. Close evaluation of the type of fluid administered is also required for patients with renal and liver diseases, and for elderly and very young persons because these groups of people cannot tolerate an excessive fluid volume.

SELECTION OF SITE

Intravenous therapy can be broadly classified by whether the site is peripheral (peripheral venous access) or central (central venous access). Nurses can perform venipuncture (placing the access device into a peripheral vein) for patients receiving peripheral therapy. Physicians generally insert central venous access devices, although certified nurses can insert peripherally inserted central catheter lines (see later discussion under Central Venous Access).

When selecting a venipuncture site, a number of factors are considered: the characteristics of the site, the purpose and duration of the infusion, the kind and amount of fluid administered, the condition of the vein, and the patient's overall comfort and condition.

PERIPHERAL VENOUS ACCESS

The available sites for peripheral infusion are the superficial veins of the extremities, or of the scalp in infants. The most common veins used include (in order of common use):

- Veins of the forearm (basilic, cephalic)
- Veins around the cubital fossa (antecubital, cephalic, basilic)
- Veins in the radial area
- Veins in the hand
- Veins in the thigh (femoral, saphenous)
- Veins in the foot
- Veins in the scalp (infants and elderly persons)

The most frequently used sites are the veins of the forearm, because the bones of the forearm act as a natural support and splint. The veins of both arms should be inspected very closely before selecting a site. The nurse determines the patient's dominant side and selects the other side for a venipuncture site. IVs should not be started in an arm that is weak, traumatized, or paralyzed. In addition, the arm on the same side as a mastectomy (breast removal) should not be used.

The most accessible vein is not necessarily the most ideal site. The antecubital space, for example, has a large vein more easily accessible than veins in a distal location; however, use of the antecubital area restricts movement of the arm. Bending the elbow may easily obstruct the flow of solution, causing infiltration (fluid seeping into the interstitial tissue), which could lead to thrombophlebitis. If infiltration occurs from the antecubital vein, the lower veins cannot usually be used for further puncture sites. It is preferable to start the infusion *distally* to provide the option of proceeding up the extremity if the vein is ruptured or infiltration occurs. Leg veins are not usually used by the nursing staff for venipuncture because of the risk of phlebitis (vein inflammation) and thrombophlebitis (clot formation with vein inflammation), a danger that is greater in the legs than in the arms. Thrombophlebitis can lead to the release of blood clots (emboli) that can lodge in the lungs or other major organ.

In considering use of the fossa area, the close proximity of the artery to the vein should be noted in order to avoid puncturing it. Acute, severe pain in the arm and hand upon venipuncture usually indicates an arteriospasm caused by arterial puncture.

The size and condition of the veins must be considered in relation to the amount and type of fluid to be administered and the duration of therapy. Veins in the back of the hand are small, roll more easily, and are often difficult to anchor. This problem is especially evident in the elderly. In contrast, the veins on the back of a child's hand are often prominent, convenient, and easily accessible using a small butterfly-wing needle. Because the condition of the vein is also a consideration, veins that are reddened or tortuous or that roll are not suitable sites. If the patient is hypovolemic, the distal peripheral veins have a greater tendency to collapse than the proximal veins.

A thrombosed vein can be detected by palpation because it will feel hard and cordlike to the touch. Use of a thrombosed vein only increases the possibility of thrombophlebitis. Stagnation of blood, as in varicose veins, in conjunction with the added trauma of venipuncture, enhances the possibility of thrombophlebitis.

Edema in a potential site may make it difficult to locate the vein. Applying pressure in the local area may force the edema from the adjacent interstitial spaces and allow the vein to be palpated.

The purpose of the infusion influences the selection of a site. If the purpose is the replacement of blood, a large vein should be used because of the relatively high tonicity of the solution. The more viscous solution requires a larger vein and needle caliber to sustain an adequate flow. The peripheral vascular pressure in the patient may be increased enough to make it difficult for the blood to flow freely into the vein. In fact, because of its viscosity, the blood may have to be mechanically pumped into the patient's vein. Infusion of large quantities requires a larger vein.

When hypertonic or irritating solutions are to be administered, use of a larger vein should be considered. Both of these solutions produce an inflammatory reaction in the blood vessel, causing some local edema that narrows the diameter of the vessel. There may not be a sufficient blood flow in

TABLE 15–1
TONICITY OF COMMONLY
USED INTRAVENOUS SOLUTIONS

Solution	Tonicity
0.45% saline (½NS)	Hypotonic
0.9% saline (NS)	Isotonic
5% dextrose in water (5%D/W)	Isotonic
10% dextrose in water (10%D/W)	Hypertonic
5% dextrose in 0.9% saline (5%D/NS)	Hypertonic
5% dextrose in 0.45% saline (5%D/½NS)	Hypertonic
5% dextrose in 0.225% saline (5%D/¼NS)	Isotonic
Lactated Ringer's solution (RL)	Isotonic
5% dextrose in lactated Ringer's solution	Hypertonic

the smaller veins to dilute a hypertonic solution, and further irritation may develop. Table 15–1 lists commonly used IV solutions and their tonicity.

The duration of therapy is another factor influencing the selection of the site. For example, when the duration is longer, a central venous access is usually chosen. Larger vessels should be used when prolonged therapy is anticipated, such as use of the subclavian vein for parenteral nutrition. For short-term therapy, the smaller veins of the lower arm are more appropriate. The preservation of veins becomes critical during a prolonged course of intravenous therapy; thus, it is important to perform the initial venipuncture distally, performing each succeeding venipuncture more proximally. Using alternate arms for infusion will also help to preserve the veins.

Implications for Elderly Persons

The veins in the hand of an elderly patient are frequently fragile, difficult to stabilize, and surrounded by such thin skin that blood can easily bleed into the interstitial spaces (subcutaneous tissue). The hands should also be avoided in an elderly person because movement and self-care are restricted. Table 15–2 summarizes individualized nursing interventions for elderly patients receiving IV therapy.

Selection of Equipment

Because infusion equipment varies among companies, it is necessary to become familiar with the equipment available at a specific institution. However, there are some basic choices to make among the various types of equipment.

Cannula Selection

Peripheral IV therapy is administered through a cannula, a small, hollow plastic (used most often) or metal tube (Fig. 15–1). The gauge of cannulas ranges from approximately 14 to 27. The smaller the gauge number, the larger the outside diameter of the cannula. Larger gauges allow a higher fluid rate than smaller ones.

TABLE 15–2

NURSING INTERVENTIONS FOR ELDERLY PATIENTS RECEIVING INTRAVENOUS THERAPY

- Do not use regular tourniquets; use a blood pressure cuff instead.
- Use the shortest cannula or needle possible.
- Use the smallest gauge cannula or needle possible, unless blood is being administered.
- Do not tap the vein to "bring it up."
- Do not use the veins of the hand.
- Avoid large, tortuous superficial veins.
- Place the IV in the patient's lower forearm on the nondominant side unless the limb is weakened, traumatized, or paralyzed.
- Anchor the IV cannula with a transparent dressing unless the skin is exceptionally thin or the patient is taking corticosteroids.
- Use dry sterile gauze and wrap arm loosely with gauze or flexible netting if the skin is thin and fragile; keep the puncture site exposed for assessment.
- Avoid restraints on the arm with the IV.
- Use electronic infusion devices or volume-controlled sets to prevent accidental fluid overload.
- Assess the IV site every 2 to 4 hours for complications.
- Maintain meticulous intake and output record.

In selecting a cannula, the gauge and length should be considered. The gauge size is determined by the amount, concentration, and duration of therapy. If it is anticipated that an infusion of blood will follow an intravenous fluid, a larger-gauge cannula, such as a No. 16 or 18, should be used. The gauge should be smaller than the vein chosen to lessen the amount of trauma to the vein and to facilitate adequate dilution of the fluid. If the gauge of the needle closely approximates the size of the vein, the limited amount of dilution of the intravenous fluid while it is running increases the irritation to the vein.

The length of the cannula also affects the flow rate. The longer the shaft of the cannula, other factors being equal, the slower the flow. Shorter shafts lessen the chance of infiltration from movement of the cannula. Common

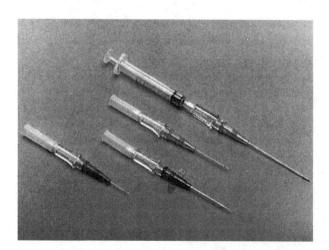

FIGURE 15–1
Example of a peripheral cannula. (From Booker, MF, and Ignatavicius, DD: Infusion Therapy: Techniques and Medications. WB Saunders, Philadelphia, 1996, p 102, with permission.)

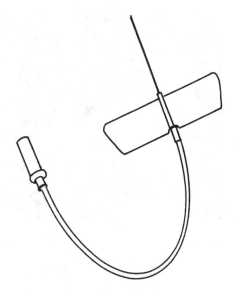

FIGURE 15–2
Butterfly winged needle.

sizes used for adults are gauges 20, 21, and 23, and lengths of 1 to 1½ inches.

In general, larger-gauge cannulas are used for adults; smaller-gauge cannulas are used for infants, children, and elderly persons. IV therapy in these groups may require a butterfly needle, also referred to as a scalp needle. As seen in Figure 15–2, the plastic "wings" help to hold this short needle during insertion and assist in stabilizing it once inserted.

Other Equipment

Other necessary equipment may include an IV administration set, extension tubing, filter, and an electronic infusion device. IV therapy uses either a closed or an open system. A closed system consists of a plastic bag that collapses as the IV solution is infused. Because the bag collapses, there is no need to vent with outside air. The second system is called an open system. The open system uses rigid containers, usually glass bottles. Because the container is rigid, outside air must be vented into the bottle to replace the fluid as it is infused. Venting encourages the fluid to move out of the intravenous bottle. Plastic bags, that is, the closed system, are the preferred system.

Both systems use hydrostatic pressure to aid infusion. An increase in the height at which the intravenous container is placed increases hydrostatic pressure, thereby increasing the rate of infusion.

Additional containers can be connected to either system. This is called a piggyback procedure. Generally, additional containers are used to administer medication. The needle from the second container is inserted into the self-sealing rubber cap of the first bottle after cleaning the site with alcohol. Most facilities that have a large volume of patients receiving IV therapy use a type of needleless piggyback system to prevent accidental needle stick and subsequent risk for blood-borne diseases, such as hepatitis B and human immunodeficiency virus (HIV) infections (Fig. 15–3).

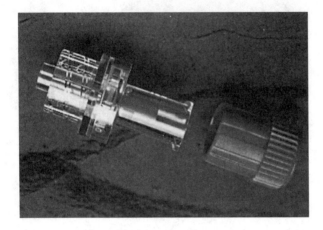

FIGURE 15–3
Example of a needleless system. (From Booker, MF, and Ignatavicius, DD: Infusion Therapy: Techniques and Medications. WB Saunders, Philadelphia, 1996, p 67, with permission.)

Infusion Administration Sets

The basic administration set includes tubing, a section for inserting the tubing into the bottle, and an adapter for connecting the tubing to the needle (Fig. 15–4). The drip chamber and tubing regulate the size of drop or the amount that can be infused at one time. Each manufacturer specifies the *drop factor* for each type of administration set. For example, some sets deliver 10 drops (gtt) per milliliter (i.e., 10 gtt = 1 mL). Some deliver 15 or 20 drops/mL. These sets are referred to as macrodrip sets.

Unlike macrodrip sets, all microdrip sets deliver the same number of drops per milliliter, or 60 in 1 mL. The smaller drop size may be preferable to facilitate counting drops when slow drip rates are ordered. This is partic-

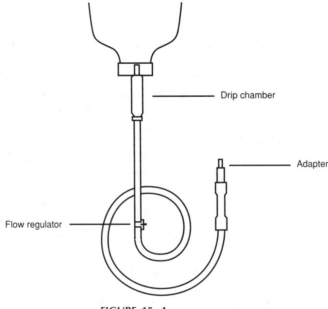

FIGURE 15–4
Basic administration set.

ularly true in administering intravenous fluids to infants, the elderly, or patients on restricted fluid intake. Cardiac patients on a keep vein open rate (KVO) and patients with renal or liver failure require close intravenous and fluid monitoring. The KVO rate is determined by hospital policy, usually approximately 30 to 50 mL/h.

Another type of infusion set that is used for infants and children to deliver small amounts of fluid has an additional volume control chamber (Fig. 15–5). Examples of these are the Volutrol, Soluset, and Buretrol sets. The volume control chamber of these infusion sets holds a maximum of 100 to 150 mL of fluid. The chamber is calibrated in milliliter increments.

Any of the infusion sets may be used with an additional filter, if a filter

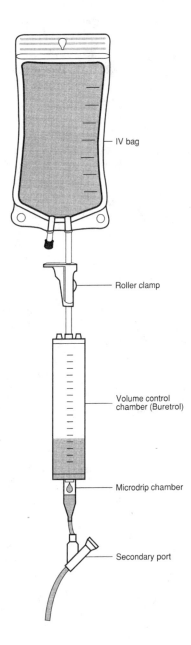

IV bag

Roller clamp

Volume control chamber (Buretrol)

Microdrip chamber

Secondary port

FIGURE 15–5
Buretrol administration set. (From Lipsey, SI, and Ignatavicius, DD: Math for Nurses. WB Saunders, Philadelphia, 1994, p 320, with permission.)

is not provided, and with an extension tubing to allow more flexibility in movement of the patient. The use of filters is increasing. They are particularly useful in eliminating microscopic particles. The incidence of thrombophlebitis is now greatly reduced by the use of filters. Some tubing, such as in the blood administration set, comes with a large filter as part of the Y-infusion set.

Electronic Infusion Devices

Electronic infusion devices are designed to provide a consistent flow rate. These devices can be broadly classified as controllers and pumps. A controller is a gravity-run system. The desired rate is set on the device and an alarm sounds if a problem, such as kinked tubing, occurs. Most controllers can deliver up to 400 mL/h. A disadvantage of this type of device is that the flow rate is limited and the alarms can be a nuisance.

A pump pushes the fluid into the vein at a faster rate than gravity. Most models can deliver up to 999 mL/h or more. The advantage of this device is that large volumes of fluid can be delivered if needed; however, the pump may continue to push fluids into the patient even if the access device is not in the vein.

Electronic infusion devices that are used in an institution such as a hospital are stationary. These devices are usually mounted on an IV pole on wheels and the patient can ambulate or be moved by pushing the pole. Home infusion therapy has revolutionized IV equipment, including pumps. Ambulatory pumps tend to be smaller and easily transportable. Most pumps are programmable to deliver a set amount of fluid or medication. As for the stationary pumps, since each pump is different, the nurse must refer to the manufacturer's instructions about their use.

Venipuncture

The technique of inserting a cannula or other access device into a vein is called venipuncture. Venipuncture is a skill that must be practiced repeatedly to achieve proficiency. Procedures for venipuncture can be found in skills books and health-care agency policies and procedures. Gloves must be worn at all times to protect against exposure to blood.

Occlusive Dressings

After the cannula is inserted and anchored, the site is covered with a sterile occlusive dressing to prevent infection. A dry dressing using a sterile gauze pad and tape or a transparent semipermeable membrane (TSM) dressing (such as Op-Site or Tegaderm) may be used. Dressings are changed every 48 to 72 hours, per agency policy, or when they are contaminated.

CENTRAL VENOUS ACCESS

Central IV therapy is administered through a central venous catheter (CVC). This route is selected for patients on long-term fluid or drug therapy, total parenteral nutrition, or multiple infusions of hyperosmolar solutions. The physician usually inserts venous access devices, such as tunneled catheters and implanted ports. Certified nurses may insert the peripherally inserted central catheter (PIC catheter, or PICC).

Most central catheters are inserted through the skin into the subclavian or internal jugular vein at the bedside. These are used for up to 4 weeks. These devices typically have more than one lumen (opening) and are made of flexible material. When venous access is needed for longer than 4 weeks, a tunneled or implanted catheter may be inserted in the operating room.

The tunneled CVC is most commonly inserted through the skin into the subclavian or internal jugular vein and advanced into the superior vena cava or right atrium. The proximal end is then tunneled through a subcutaneous pocket of the chest wall and brought out below the nipple line (Fig. 15–6).

Implanted ports have no external exit. A catheter is inserted into the subclavian or internal jugular vein. Then, a plastic or metal port is implanted in a subcutaneous pocket of the chest wall where it is connected to the catheter. The port cannot be seen but presents as a bulge under the skin that is accessed by a special needle, such as the Huber needle shown in Figure 15–7. Other types of central catheters are available, but discussion of each type is beyond the scope of this book.

In contrast to peripheral IV therapy, central access devices are associated with life-threatening complications. For example, when the catheter is advanced, pneumothorax, hemothorax, or perforation of the vessel are pos-

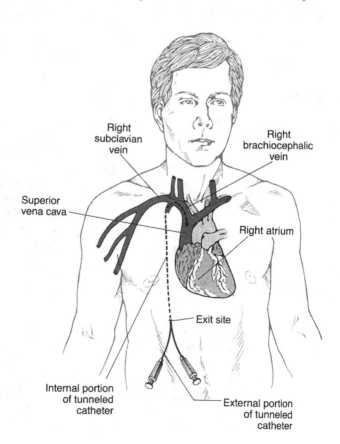

FIGURE 15–6
The usual anatomic position of tunneled catheters. (From Ignatavicius, DD, Workman, ML, and Mishler, MA: Medical-Surgical Nursing: A Nursing Process Approach, ed. 2. WB Saunders, Philadelphia, 1995, p 277, with permission.)

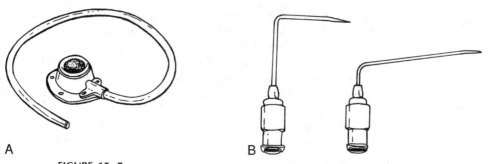

FIGURE 15–7
A, An implanted port and B, Huber (right-angle) needles. (From Ignatavicius, DD, Workman, ML, and Mishler, MA: Medical-Surgical Nursing: A Nursing Process Approach, ed. 2. WB Saunders, Philadelphia, 1995, p 279, with permission.)

sible complications. The catheter can become dislodged, even if it is sutured into place. Venous thrombosis and endocarditis are other less common potential problems.

Like the peripheral IV site, the central IV site is covered with an occlusive dressing. The procedure for changing the dressing is more complex and individualized, depending on agency policy.

COMPLICATIONS OF INTRAVENOUS THERAPY

There are three main types of complications from conventional IV therapy: infection, tissue damage, and fluid and electrolyte imbalances.

Infection

Venipuncture interrupts the integrity of the skin, the first line of defense against infection. The longer the therapy continues, the greater is the risk of infection. Patients who are already immunocompromised from diseases like cancer or treatments like chemotherapy are at especially high risk for infection. These patients have an altered or lowered white blood cell count that predisposes them to infection. The elderly are also at high risk because aging alters the effectiveness of the immune system. Frequent nursing assessments of the puncture site are essential. Table 15–3 lists nursing interventions to help prevent or detect infection in patients receiving IV therapy.

Tissue Damage

Tissues most commonly damaged by IV therapy include the skin, veins, and subcutaneous tissue. Tissue damage is uncomfortable and can have permanent negative effects.

Skin damage can occur from using the tourniquet, tapping the skin over the vein (a practice that should not be done), and penetrating the skin with the cannula. Ecchymosis (bruising) results when skin is damaged. Skin can also be damaged from tape or dressing adhesives, which can give rise to allergic reactions or abrasion.

TABLE 15–3

NURSING INTERVENTIONS TO PREVENT INFECTION IN PATIENTS RECEIVING INTRAVENOUS THERAPY

- Wash hands thoroughly before inserting a venous access device (VAD); wash hands before working with IV for maintenance.
- When inserting a VAD, clean the skin with an antimicrobial solution using an inner-to-outer circular motion.
- Use sterile technique when inserting a VAD.
- When changing the dressing over the IV site, use sterile technique.
- Change the IV dressing every 72 hours, when the dressing is contaminated, or as specified by agency policy.
- Change the IV tubing every 24 to 48 hours depending on agency policy; tubing for medication administration may be used for as long as 72 hours.
- Label tubing, dressing, and solution bags clearly, including the date and time when changed.
- Do not let an IV bag or bottle hang for more than 24 hours.
- Do not allow the IV tubing to touch the floor.
- Swab access ports before adding medications or solutions with 70% alcohol (alcohol prep) or other equally effective solution.

Phlebitis, an inflammation of the vein, can occur from either mechanical or chemical trauma. The Intravenous Nursing Society Standards of Nursing Practice (Corrigan et al, 1990) recommend a postinfusion phlebitis scale to assess the signs and symptoms associated with phlebitis (Table 15–4). If this problem occurs, the nurse removes the cannula immediately.

Infiltration, also called extravasation, is another form of tissue damage. It is defined as seepage of intravenous fluid out of the vein into the surrounding interstitial spaces. This may occur when an access device (e.g., cannula) has become dislodged or perforates the wall of the vein. The result is edema, pain, and/or coolness or warmth. The possibility of infiltration can be evaluated by occluding the vein proximal to the intravenous site. If the IV fluid continues to flow, the cannula is probably outside the vein. If infiltration has occurred, the IV device should be removed immediately. It is important not to rub the area, which might cause development of a hematoma. Compresses (warm or cool, depending on the solution) should be applied over the affected area. Documentation of the infiltration and its effects is critical.

TABLE 15–4

POSTINFUSION PHLEBITIS SCALE

Score	Defining Criteria
0	No skin or vein abnormalities
1+	Pain at site, erythema present, possible edema
2+	Pain at site, erythema present, possible edema, red streak along vein
3+	Pain at site, erythema present, possible edema, red streak along vein, vein cord palpable

Data from Corrigan, A, et al: *Intravenous Nursing Standards of Practice.* Intravenous Nurses Society, Belmont, MA, 1990.

Fluid and Electrolyte Imbalances

In contrast to the mostly localized problems described in the previous sections, fluid and electrolyte imbalances are systemic and can be life-threatening. Fluid overload is a major complication of IV therapy. Infants, children, and the elderly are at particularly high risk for this complication. Chapter 2 of this book describes the nursing assessment and interventions for patients experiencing fluid volume excess.

Electrolyte balance can also be affected by IV therapy. If the patient receives a solution that is hypertonic or hypotonic, sodium and potassium balance can easily be altered. Overzealous replacement of electrolytes can result in any number of electrolyte imbalances. The chapters in Part Two of this book discuss electrolyte imbalance, both assessment and interventions.

Nursing Interventions

There are many nursing interventions related to IV therapy. Of paramount importance are the following:

- Know the types of intravenous fluid, their content, and the medications and fluids with which they are compatible and incompatible.
- Be familiar with intravenous equipment. Know the sizes and types of cannulas and the situations for which they are recommended.
- Develop skill in locating veins and performing venipunctures. A clean, confident thrust into the vein will minimize the insult to the vein and decrease the pain to the patient.
- Calculate and monitor drip rates. Record the amount of fluid infused.
- Secure cannulas and tubing to prevent dislodgment and subsequent complications, and write on the tape or IV label crucial information, such as date of insertion, size of the needle, and flow rate.
- Observe for complications and report any signs of vein irritation. Follow the policies and procedures of the agency.
- Be aware of and sensitive to the individual needs of each patient.

PARENTERAL NUTRITION

When a patient is unable to use the gastrointestinal tract for nutrition, parenteral nutrition is used to provide the nutrients that are needed to support life. Conventional IV solutions containing 5% dextrose provide less than 200 calories/L; therefore, more concentrated dextrose solutions with proteins, fats, vitamins, and minerals are necessary for adequate nutrition.

Parenteral nutrition, sometimes called hyperalimentation or "hyperal," may be partial, administered by peripheral venous access; or total, administered by central venous access. It may be administered continuously 24 hours a day or may be cycled. Cycled parenteral nutrition is especially useful in the home or in nursing homes. In this method, the patient's infusion is interrupted for several hours a day to allow for activities, treatments, and so forth (Viall, 1995a). The patient on a cycled feeding, however, must be able to tolerate an increased flow rate to receive the necessary amount of daily nutrients.

Partial Parenteral Nutrition

As the name implies, partial parenteral nutrition (PPN) does not meet all of the patient's nutritional needs. It is typically given to patients who can take some oral nutrition, but not enough to meet the body's needs. When people are ill, they require additional carbohydrates, proteins, and healing elements to improve their health.

Partial parenteral nutrition is commonly delivered via a large peripheral vein through a cannula or catheter. Solutions that are typically used are amino acid–dextrose (protein-dextrose) solutions (usually 10% dextrose) and lipid (fat) emulsions (usually 10% or 20%). Most lipid emulsions are isotonic, but the amino acid–dextrose solutions are generally hypertonic, ranging from 300 mOsm to nearly 1200 mOsm. Vitamins and minerals are added to this solution to provide the daily recommended requirements.

A newer product available for PPN is a mixture of lipids, amino acids, and dextrose, referred to as a 3:1 solution, total nutrient admixture (TNA), or triple-mix solution (Weinstein, 1993). Most often the mixture is available in 3-L bags and is delivered through an electronic infusion device. Before administering lipids, the patient should be assessed for allergy to eggs, a contraindication for lipid infusion.

An in-line filter is required for parenteral nutrition, to remove crystals from the solution. For standard solutions, a 0.22-micron (μ) filter can be used. If TNA or lipid emulsion is administered, a filter of at least 1.2 μ is needed to allow fat particles to pass (Viall, 1995a).

Total Parenteral Nutrition

Total parenteral nutrition (TPN) can be thought of as a complete meal in a bottle. It is used for patients requiring extensive nutritional support for a prolonged period of time and is administered by central venous access, as described earlier in the chapter.

Total parenteral nutrition solutions have a much higher dextrose concentration than that used for PPN, often as high as 50%. The concentrations of synthetic amino acids are also high, usually 3% to 5%. The high concentrations of the TPN solutions mean that they are hyperosmolar, with osmolarities about 3 to 6 times greater than that of blood. The amino acid–dextrose solution is a base to which the pharmacist can add specific amounts of vitamins, electrolytes, and other minerals, based on each patient's specific needs. Insulin may be added to prevent high serum glucose levels that result from the high concentration of dextrose. TPN is always delivered via an electronic infusion device, preferably a volumetric pump.

Complications of Parenteral Nutrition

Because both PPN and TPN require venous access, some of the complications of parenteral nutrition are the same as those discussed earlier in this chapter as complications of IV therapy. Infection is more likely with TPN than with conventional IV therapy or PPN because the concentration of dextrose is higher in TPN. Dextrose and amino acids provide an excellent medium in which bacteria can reproduce, so vital signs, especially temperature, must be measured frequently. To help prevent infection and solution incompatibility, IV medications and blood are not given through the TPN line.

In addition to those problems, several other complications specific to parenteral nutrition can occur. These complications include fluid and electrolyte imbalances; acid-base imbalances; hyperglycemia; and hyperglycemic, hyperosmolar nonketotic syndrome (HHNS). In children, gastrointestinal complications may occur. Any of these complications can be life-threatening. In general, complications or parenteral nutrition can be minimized or prevented by initiating the therapy slowly, monitoring the patient and laboratory values carefully, and tapering the therapy slowly when discontinued.

Fluid and Electrolyte Imbalances

Patients receiving parenteral nutrition are at a very high risk for fluid imbalance. In addition to delivering extra fluid to the body, the solutions used are extremely hyperosmolar and stimulate fluid shifts between fluid compartments. The dextrose and amino acids can "pull" water from the interstitial and intracellular spaces into the intravascular compartment (plasma). The resulting plasma expansion together with hyperglycemia (high sugar level) can cause osmotic diuresis, then dehydration and hypovolemic shock. If not treated promptly, renal function may be compromised, as evidenced by an elevated blood urea nitrogen (BUN) and creatinine.

If the patient has renal or cardiac dysfunction, the extra fluid can cause fluid overload, resulting in congestive heart failure and pulmonary edema (see Chapter 19).

Any number of electrolyte imbalances can occur in patients receiving parenteral nutrition, in particular, decreases in potassium, phosphate, and magnesium. Most often these deficiencies are the result of the shifting of electrolytes from the extracellular fluid compartment into the intracellular compartment, especially if there is concurrent administration of insulin.

Hypo- or hypercalcemia may be seen in patients on parenteral nutrition. Hypocalcemia can occur with hypomagnesemia or decreased serum protein, since half of the serum calcium is bound to albumin. Hypercalcemia can occur in patients receiving parenteral nutrition for an extended period of time. Part II of this book discusses each of these imbalances in detail.

Acid-Base Imbalances

Metabolic alkalosis or acidosis can be seen in patients on parenteral nutrition. Alkalosis occurs because the acetate in the solution is converted to bicarbonate, a base. Acidosis most likely results from metabolism of amino acids, causing excess hydrogen ion production. Both of these imbalances are described in Chapter 3 of this text.

Hyperglycemia

Hyperglycemia results from high concentrations of dextrose in the solutions that are used for PPN and TPN, and perhaps from the stress of illness or injury. It is the most common metabolic problem associated with this therapy. Hyperglycemia may be prevented by careful monitoring of the blood glucose and the administration of regular insulin.

Hyperglycemic, Hyperosmolar Nonketotic Syndrome

If hyperglycemia is not controlled, hyperglycemic, hyperosmolar nonketotic syndrome (HHNC) can result. The increased serum glucose results in osmotic diuresis and hypertonic dehydration. The mortality rate from this problem is very high, especially in the elderly. Fortunately, HHNS is less frequent today thanks to careful monitoring of patients receiving parenteral nutrition. Chapter 20 describes this complication and its treatment.

Gastrointestinal Complications

Liver disease is the most important gastrointestinal complication that is seen in children who receive TPN (Whaley & Wong, 1995). Although the exact cause is not known, it occurs more in premature infants who have been started on TPN at an early age. Children with liver disease can develop hepatic necrosis, cholelithiasis (gallstones), and in advanced disease, cirrhosis or liver failure.

Nursing Interventions

The interventions for patients receiving PPN or TPN include those discussed earlier in the general section on nursing interventions. Because the patient receives nutritional support with hyperosmolar solutions, additional interventions are required. These interventions are listed in Table 15-5.

TABLE 15-5
NURSING INTERVENTIONS ASSOCIATED
WITH PARENTERAL NUTRITION THERAPY

- Carefully check each bag or bottle of solution for accuracy against the physician's order.
- Check the TPN solution for "cracking" (i.e., separation of the fluid into layers); do not use if cracked.
- Monitor the accuracy of the electronic infusion device by marking the solution container in hourly increments.
- If the solution is "behind," do not increase the rate to get it back on schedule.
- Check the IV puncture site carefully for signs of infection and placement.
- Start TPN slowly and taper it slowly to prevent major changes in serum glucose levels; a recommended flow rate is one half of the prescribed rate for 1 hour before discontinuing.
- Weigh the patient every day early in the morning.
- Monitor laboratory values carefully, especially BUN, creatinine, electrolytes, and glucose, which are typically drawn daily for the first 3 to 7 days, then twice a week.
- Monitor liver enzymes every 3 to 4 days, especially in children.
- Record the patient's intake and output carefully.
- Observe for signs and symptoms of complications, such as crackles in patients with congestive heart failure.
- Monitor blood glucose and vital signs every 4 to 6 hours or as specified by the agency or physician.
- Administer vitamin K as ordered for clotting because TPN solution does not contain vitamin K.
- Change the IV administration set according to agency policy, usually every 24 to 72 hours.
- If new TPN solution is not available, administer dextrose 10% until the next TPN container is ready.

REFERENCES AND READINGS

A.S.P.E.N. (1993). Section II: Rationale for adult nutrition support guidelines. Journal of Parenteral and Enteral Nutrition 17(4), 5SA–62A.

Booker, MF, & Ignatavicius, DD (1996). Infusion Therapy: Techniques and Medications. Philadelphia: WB Saunders.

Camp-Sorrell, D (1992). Implantable ports: Everything you always wanted to know. Journal of Intravenous Nursing 15(5), 262–273.

Corrigan, A, et al (1990). Intravenous Nursing Standards of Practice. Belmont, MA: Intravenous Nurses Society.

Coulter, K (1992). Intravenous therapy for the elder patient: Implications for the intravenous nurse. Journal of Intravenous Nursing 15(Suppl), S18–S23.

Dudek, S (1993). Nutrition Handbook for Nursing Practice. Philadelphia: JB Lippincott.

Hadaway, L (1991). IV tips. Geriatric Nursing 12(2), 78–81.

Holder, C, & Alexander, J (1990). A new and improved guide to I.V. therapy. American Journal of Nursing 90(2), 43–47.

Ignatavicius, DD, Workman, ML, & Mishler, MA (1995). Medical-Surgical Nursing: A Nursing Process Approach. Philadelphia: WB Saunders.

Jurf, JB (1994). Evaluating needleless products. MEDSURG Nursing 3(3), 176–180.

Lipsey, SI, & Ignatavicius, DD (1994). Math for Nurses. Philadelphia: WB Saunders.

Lorenz, B (1990). Are you using the right IV pump? RN 53(6), 31–36.

Marcoux, C, Fisher, S, & Wong, D (1990). Central venous access devices in children. Pediatric Nursing 14, 123–133.

Millam, D (1990). Electronic infusion devices. Nursing90 20(8), 65–68.

Millam, D (1993). How to teach good venipuncture technique. American Journal of Nursing 93(7), 38–41.

Roundtree, D (1991). The PIC catheter: A different approach. American Journal of Nursing 91(8), 22–26.

Smith, R (1993). A nurse's guide to implanted ports. RN 56(3), 48–53.

Viall, C (1990). Your complete guide to central venous catheters. Nursing90 20(2), 34–41.

Viall, C (1995a). Taking the mystery out of TPN, Part 1. Nursing95 25(4), 34–43.

Viall, C (1995b). Taking the mystery out of TPN, Part 2. Nursing95 25(5), 56–59.

Vonfrolio, LG (1995). Would you hang these IV solutions? American Journal of Nursing 95(6), 37–39.

Whaley, LF, & Wong, DL (1995). Nursing Care of Infants and Children. St. Louis: Mosby–Year Book.

Weinstein, SM (1993). Plumer's Principles and Practice of Intravenous Therapy (5th ed). Philadelphia: JB Lippincott.

Wickham, R (1990). Advances in venous access devices and nursing management strategies. Nursing Clinics of North America 25, 345–364.

Wood, L, & Gullo, S (1993). IV vesicants: How to avoid extravasation. American Journal of Nursing 93(4), 42–46.

CHAPTER 16

. . . .
. . . .
. . . .
. . . .
. . . .
. . . .
. . . .
. . . .
. . . .
. . . . . .

Peritoneal Dialysis

Peritoneal dialysis is one of two methods of dialysis used for the treatment of patients with acute or chronic renal disease. The other type is hemodialysis (see Chapter 17).

Peritoneal dialysis is a process by which undesired substances are removed from the blood by instillation of a sterile solution into the peritoneal cavity (Fig. 16–1). Renal failure is a malfunction of the kidneys in which metabolic waste products (e.g., urea and creatinine) are not sufficiently excreted. Peritoneal dialysis has two main objectives: (1) to remove toxic substances, such as drugs, from the body; and (2) to remove excess fluid, electrolytes, and other waste products, such as urea. Dialysis can temporarily restore fluid, electrolyte, and acid-base balance.

Peritoneal dialysis can be used alone or as a supportive measure between extracorporeal hemodialysis procedures. Before the 1970s, peritoneal dialysis for chronic renal failure was used only when the patient could not tolerate hemodialysis or when hemodialysis was not available; however, subsequent improvements in technique and equipment have made peritoneal dialysis a viable alternative to hemodialysis for many patients with chronic renal failure.

PRINCIPLES OF DIALYSIS

Peritoneal dialysis is accomplished by osmosis and diffusion. A sterile dialysate solution is instilled into the abdominal cavity where it is retained until the toxic substances diffuse from the bloodstream across the peritoneal membrane into the dialysate (Fig. 16–2). The fluid is then drained from the body, taking excess fluid and toxic or waste products with it. Excess fluids

221

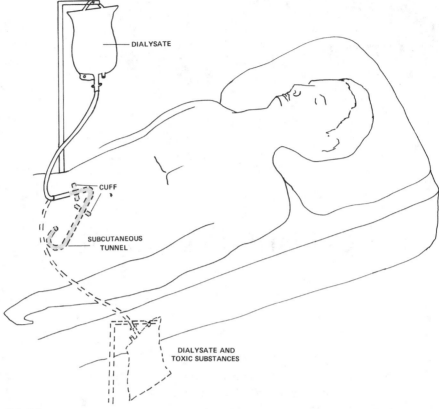

FIGURE 16–1
Peritoneal dialysis. Dialysate is infused into the peritoneum using a closed system plastic bag. Waste products diffuse across the peritoneal membrane to the dialysate; dialysate and toxic substances are evacuated into the same plastic bag. The bag is discarded after one use.

are removed by osmosis, whereas the larger molecular waste products, such as urea and creatinine, are removed by diffusion. Blood cells and molecules too large to pass through the peritoneal membrane are retained in the bloodstream.

TYPES OF DIALYSATE

The composition of the peritoneal dialysis solution, called the dialysate, is essentially the same as normal serum with three exceptions: first, bicarbonate is replaced with 40 mEq/L of lactate in the dialysate solution because lactate is more stable; second, potassium is omitted or added separately as the physician determines the patient's need; and third, glucose is added in various strengths to facilitate osmosis.

The patient's normal serum contains 0.1% glucose (1 g/L), whereas peritoneal dialysate solutions contain 1.5% glucose (15 g/L), 2.5% glucose (25 g/L) or 4.25% glucose (42.5 g/L). The most common dialysate solution is 1.5% glucose. The 2.5% glucose solution offers an intermediate step in both ultrafiltration (removal of water) and volume, between the 1.5% and

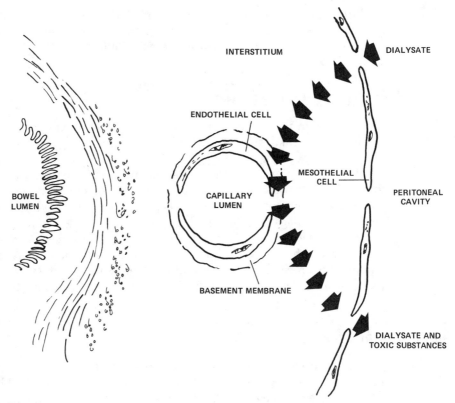

FIGURE 16–2

Cross section demonstrating peritoneal dialysis. Waste products (e.g., excess creatinine, acids, potassium) inside the capillary cell (intracellular fluid compartment) cross the cell membrane into the interstitium (extracellular fluid compartment) where the dialysate picks up toxic substances and then moves back across the peritoneal membrane for removal from the body.

the 4.25% solutions. A 4.25% glucose dialysate solution is used to remove excess amounts of fluid from the bloodstream. This may cause the patient to experience intermittent symptoms of distention because of the increased movement of substances from the plasma into the peritoneal cavity where dialysis takes place.

Use of dialysate containing lactate is contraindicated when the patient is unable to metabolize the lactate. Infants, for example, are frequently hypoxic and, therefore, cannot oxidize lactate. As a result, the lactate moves from the peritoneal cavity into the extracellular fluid, causing bicarbonate to shift into the peritoneal cavity where it is eliminated with the dialysate. The patient then develops progressive metabolic acidosis characterized by a high anion gap with an accumulation of lactate ions in the extracellular fluid. When administering a dialysate containing bicarbonate rather than lactate, calcium should be eliminated from the dialysate (because of precipitation) and administered intravenously.

Medications may be added to the dialysate using the peritoneal membrane, which accommodates two-way movement. Typical examples of additives are sodium heparin to prevent clot formation in the catheter or tubing, potassium chloride for replacement, and antibiotics to treat suspected or confirmed peritonitis (peritoneal infection). Additives must be checked for compatibility with the dialysate. Any contamination of the dialysate by the health team or the patient increases the risk of infection.

ADMINISTRATION MODALITIES

Administration modalities can be divided into two broad groups: (1) non-machine, or manual; and (2) machine (automated), using an automatic peritoneal dialysis cycling machine (Table 16–1).

In an institutional setting, dialysate may be administered from 2-L flexible bags. The physician may prescribe a series of single-bag exchanges or use of the automatic peritoneal dialysis cycling machine with multiple exchanges of 2-L bags (Fig. 16–3).

During continuous dialysis, the dialysate remains in a constant state of exchange as substances move across the peritoneal membrane. An intermittent dialysis procedure is one in which no dialysate is in the peritoneal cavity for an extended period of time, that is, 10 to 14 hours during the day.

PROCEDURE

Patients who are having insertion of a peritoneal dialysis catheter should void before the procedure to prevent accidental puncture of the bladder. A mild sedative may help anxious patients to better tolerate catheter insertion.

Peritoneal dialysis is performed through a surgically implanted catheter in the patient's peritoneal cavity. The catheter consists of a cuffed Silastic tubing, connector, and disposable rubber cap. A single-cuffed catheter may be used for acute conditions. A double-cuffed catheter, anchored in the subcutaneous tissue, is used for chronic peritoneal dialysis. The cuff is anchored by fibroblastic growth embedded in the Dacron felt cuffs. This growth is usually well established in 3 or 4 days and is complete in 10 to 14 days after implantation.

TABLE 16–1
METHODS OF PERITONEAL DIALYSIS

Type	Use	Usual Setting
NONMACHINE		
Single-bag series (manual method)	Acute conditions	Medical care unit
CAPD — continuous ambulatory peritoneal dialysis	Chronic conditions	At home
MACHINE		
IPD — intermittent peritoneal dialysis	Acute conditions	Medical care unit
CIPD — chronic intermittent peritoneal dialysis	Chronic conditions	Renal unit; at home
CCPD — continuous cycling peritoneal dialysis	Chronic conditions	At home

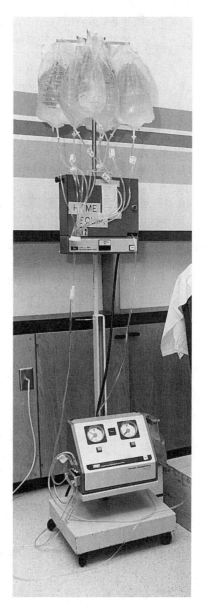

FIGURE 16–3
AMP automatic cycling machine is used for intermittent peritoneal dialysis (IPD), chronic intermittent peritoneal dialysis (CIPD), and continuous cycling peritoneal dialysis (CCPD). (Courtesy of St. Francis Regional Medical Center, Wichita, Kansas.)

The double cuff permits mechanical fixation and creates a double barrier against bacterial invasion of the peritoneal cavity through the catheter tract. Nevertheless, sterile technique must continue to be used to control the possibility of infection. Many patients experience an inflammatory response to the insertion of the catheter.

The success of peritoneal dialysis correlates with the ability of the patient and the health team to control infection. Contamination, when it occurs, is usually through the lumen of the dialysis catheter. The peritoneum has a high resistance to infection; however, the presence of the catheter provides a nidus for infection, allowing organisms and pathogens not normally dangerous to become a significant threat. The incidence of peritonitis has been greatly reduced through the use of indwelling catheters and closed-system automated equipment. Should infection occur, it can usually be controlled through antibiotic therapy.

Because the nurse plays a major role in the dialysis procedure, it is essential to have a working knowledge of electrolyte and fluid balance.

ACUTE CONDITIONS

Acute conditions are often treated by hemodialysis, described in the next chapter. However, some patients who are not candidates for hemodialysis have peritoneal dialysis instead.

The more common acute conditions in which peritoneal dialysis could be indicated are intoxication from dialyzable poisons or drugs; acute renal failure caused by shock; hemolysis caused by mismatched blood transfusion; chronic renal failure involving a sudden loss of therapeutic control in primary renal disease; transplant preparation; electrolyte, acid-base, or fluid imbalance; hemorrhagic pancreatitis; and hypothermia.

Conditions in which peritoneal dialysis could be contraindicated include abdominal adhesions causing functional impediments; severe abdominal trauma; previous abdominal surgery; diffuse bowel disease, for example, Crohn's disease; perforated bowel or colostomy; aortic grafts; and the presence of coagulation defects.

Single-Bag (Nonmachine) Administration

Administration of solution via the single 2-L bag (manual) modality remains a method for treating acute problems. The dialysis cycle, known as an exchange, has three phases—an inflow period, in which the dialysate enters the peritoneal cavity; a dwell time, in which the processes of osmosis and diffusion bring about equilibration (homeostasis); and a drain time, in which the dialysate (effluent, or outflow) exits the peritoneal cavity into a drainage container. The physician prescribes the length of each phase of the dialysis cycle, which is usually between 15 and 20 minutes: for example, inflow, 20 minutes; dwell, 20 minutes; and drain, 20 minutes.

The automatic cycling machine may be used continuously or intermittently. An average continuous cycle lasts for 3 days. An intermittent cycle varies in length of dwell time and number of cycles based on the physician's orders.

Intermittent Peritoneal Dialysis (Machine)

Intermittent peritoneal dialysis (IPD) is an established method for treating acute renal failure; however, the advantage of hemodialysis over peritoneal dialysis for patients with acute renal failure is questionable.

Complications of Peritoneal Dialysis for Acute Conditions

Problems that may be associated with treatment of acute conditions are mechanical (fibrin clots or sludging in tubes); infectious (usually caused by contamination); and medical (for example, dysrhythmias). Complications of peritoneal dialysis are described in more detail later in the chapter.

Nursing Interventions

The nurse is responsible for maintaining an accurate record on the flow sheet of the precise time, composition, and amount of dialysate solution for intake and output on all exchanges. (See Fig. 16–4 for a sample flow record sheet.) The patient undergoing peritoneal dialysis requires continuous and skillful care. Prevention of contamination to avoid infection is paramount because of increased medical risks to the patient. Every effort should be made to keep the patient as comfortable as possible.

ACUTE PERITONEAL DIALYSIS FLOW SHEET

	PRE	POST		
Date	1-2-96			
Time	0900	1540		
Weight	115	113	CATHETER EXIT	Clean
B.P.	130/70	130/72	SITE STATUS	Erythematous
Temp	98.4	98.6		Infected (pus)
Pulse	100	96	EFFLUENT	Clear — blood-tinged
Heparin (units/2L)	500 u/2L	as		Cloudy — Hemorrhagic
KCL (meq/2L)	3-4 mEq/L	ordered	Total No. of Exchanges 7	Slightly sanguinous
Other additives:	Antibiotics	per doctor	Dialysate 1.5 Total No. 6	
	insulin		Dialysate 4.25 Total No. 1	
			Post Dialysis	Intraperitoneal
Inflow time:	20 minutes		Heparinization 3000 Units	Catheter
Dwell time:	20 minutes		Final Fluid Balance + 100	
Outflow time:	20 minutes		Method	Cycler manual

TIME	Exch. No.	Volume IN	OUT	Fluid Balance	Previous	Present	Nurses Notes:
0900	1	2000	1800	+200			Time:
1000	2	2000	1700	+500			0900 Placed on P.D. via sterile
1100	3	2000	2400	+100			technique. Area occlusively
1200	4	2000	2000	+100			dressed c̄ Betadine ointment & 4x4s.
1300	5	2000	2200	−100			1340 Drainage slowed — catheter adjusted
1400	6	2000	1900	0			1440 Abdominal pain during drainage —
1500	7	2000	2100	+100			5 cc 2% Procaine into
	8						Catheter
	9						
	10						
	11						
	12						
	13						
	14						
	15						
	16						
	17						
	18						
	19						
	20						
	21						

FIGURE 16–4
Example of an acute peritoneal dialysis flow sheet.

Meals should be scheduled when the abdomen is empty of dialysate fluid. The patient is usually more comfortable if the backrest or head of the bed is elevated. Changing the patient's position often facilitates drainage of the dialysate solution. Vital signs should be taken before the procedure for baseline comparison, and as often thereafter as the patient's condition indicates, the recommended frequency being every 8 hours. The nurse should weigh the patient, before the dialysate is infused and when the dialysate has been drained off, to determine if fluid is being retained or depleted. The returned dialysate should be observed for cloudiness, shreds of fibrin or mucus, blood, or fecal contents. This effluent drainage should be clear and light yellow.

The nurse as well as the patient should be alert to signs and symptoms of infection, such as abdominal pain on inflow, cloudy dialysate return, low-grade fever, general malaise, and anorexia. This information should be recorded.

CHRONIC CONDITIONS

Common chronic conditions in which peritoneal dialysis could be indicated are end-stage renal disease (ESRD) from diabetes with retinopathy and neuropathy; serious bleeding problems; poor vascular access; blood transfusions refused for religious reasons; elderly patients; and unstable cardiovascular systems characterized by severe coronary artery diseases, cardiomyopathy, severe dysrhythmias, or inadequate cardiac output.

Continuous ambulatory peritoneal dialysis (CAPD), chronic intermittent peritoneal dialysis (CIPD), and continuous cycling peritoneal dialysis (CCPD) are all variations of peritoneal dialysis that promote patient independence, are safe, and are less expensive than traditional peritoneal dialysis. All of these techniques can be performed at home after the patient has been screened for physiologic and psychologic stability, taught the procedure, and shown how to keep a daily log. The patient and family or other caregivers are instructed to take vital signs and weight before starting each dialysis procedure. It is suggested that a dialysis nurse be on call 24 hours a day to assist the patient with nursing concerns.

The patient should be tested for peritoneal membrane transport properties following insertion of the catheter. If transport properties are impaired, the patient will need some residual renal function for successful peritoneal dialysis because there will be a decrease in maximum urea clearance. Continuous ambulatory peritoneal dialysis is the preferred method if the patient has impairment. Uremic symptoms can develop if the patient has a drastically reduced peritoneum mass transfer ability. In such cases, the exchange schedule will need to be altered, or hemodialysis may need to be considered. Good results have been obtained using peritoneal dialysis for patients who have average ultrafiltration rates and peritoneal transport characteristics.

Metabolite transport in children differs from that in adults. Scaling of dialysate by the child's weight rather than by body surface correlates more accurately with adult values.

Chronic Intermittent Peritoneal Dialysis (Nonmachine)

Chronic intermittent peritoneal dialysis (CIPD) is an intermittent process using an average of 12 to 16 cycles with 2-L bags, 5 to 6 times per week. The patient's weight and blood pressure determine the percentage composition of dialysate solution to be used.

Chronic intermittent peritoneal dialysis requires 40 hours or more per week because of the slow clearance rate of the peritoneal membrane. It is usually performed at home 3 to 6 times a week at night and discontinued during the day. CIPD may be performed in a renal center during the daytime. Hemodialysis requires 9 to 15 hours per week, but CIPD has the advantage that it can be carried out during sleep time. No matter which method of dialysis is used, dialysis time is increased as renal function declines.

Children and elderly patients are good candidates for CIPD because the automated equipment is relatively simple to use. CIPD is particularly attractive to those caring for small children with end-stage renal disease, because it does not require vascular access. It can be performed at home during the night, which minimizes the disruption of daily routine and school; however, reports concerning the use of CIPD with children have not been positive. Pediatric patients frequently exhibit hypertension and hyperphosphatemia with development of severe renal osteodystrophy. Growth rate is poor.

Chronic intermittent peritoneal dialysis is effective in situations in which hemodialysis is not well tolerated. Patients with unstable cardiovascular systems characterized by severe coronary artery disease, cardiomyopathy, severe dysrhythmias, or inadequate cardiac output are reported to benefit from CIPD.

Problems of inadequate dialysis may occur because of the slow clearance rate associated with CIPD. Inadequate dialysis may cause malnutrition, shrinking of body mass, and uremia. If the loss of clearance and a decline in residual renal function becomes significant, continuous cycling peritoneal dialysis (CCPD) should be considered.

Continuous Ambulatory Peritoneal Dialysis (Nonmachine)

Continuous ambulatory peritoneal dialysis (CAPD) is a self-administered process in which 2 L of dialysate solution are instilled into the peritoneal cavity and remain there for a predetermined time. The dialysate is drained, replaced with fresh solution, and the cycle is repeated. The CAPD cycle consists of 10 to 15 minutes inflow, 4 to 5 hours dwell, and 10 to 15 minutes drain. The dialysate exchange is repeated 4 to 5 times daily; morning, noon, mid-afternoon, evening, and bedtime. The bedtime exchange remains in the peritoneal cavity overnight. Adequate dialysis can usually be maintained via the exchanges, using a 2.5% and/or a 4.25% glucose solution for one or two of the exchanges. The remaining exchanges should use a 1.5% glucose solution. The selection of dialysate is based on the individualized needs of each patient.

Self-dialysis using CAPD is an alternative to hemodialysis and CIPD for patients with the interest and capability to perform the process. The procedure requires the patient to have good hand-eye coordination, the ability to perform aseptic technique, a willingness to adhere to the daily routine, and the availability of adequate facilities to exchange the dialysate. It is imperative that adequate personnel be available 24 hours per day for patients on CAPD.

Because this is a self-administered procedure and because it is extremely repetitious, the patient may become lax in maintaining asepsis or doing the prescribed number of exchanges per day. There is a direct correlation between maintenance of asepsis and the control of peritonitis. Reinforcement by the dialysis team regarding the use of aseptic technique by the patient is imperative.

Continuous ambulatory peritoneal dialysis provides many advantages to the patient over other forms of dialysis. The patient will experience an increased energy level accompanied by more normal taste and appetite. There are no major restrictions in diet as long as the patient eats a moderate, well-balanced diet, avoiding an excess of sodium. Incidents of pernicious anemia are reduced, because CAPD provides less clearance of vitamin B_{12}, which is essential to the production of red blood cells. A return of general well-being occurs. Children respond well to CAPD, with an improvement in growth rate due to the steady removal of waste products.

The advantages of CAPD over hemodialysis are continuous biochemical and fluid control; patient mobility and freedom from a machine; reduced dietary restrictions; simplicity of operation; elimination of the need for routine blood access; stabilized fluid level, reducing stress to the heart and blood vessels; and maintenance of lower blood urea nitrogen (BUN).

The advantage of CAPD is that it is usually superior to CIPD in maintaining appropriate biochemical parameters.

Continuous Cycling Peritoneal Dialysis (Machine)

Continuous cycling peritoneal dialysis (CCPD) is a nocturnal, automated, prolonged-dwell dialysis system. The dwell time on the first three night-time exchanges is 3 to 4 hours or less. Before discontinuing, 2 L of dialysate solution is instilled into the peritoneal cavity. The dwell time for this fourth and final exchange extends through the day until the next night's dialysis. This procedure takes advantage of the kinetics of long dwell time, and avoids the limitations associated with daytime dialysis.

Each of the four exchanges uses one 2-L bag of dialysate solution. The three exchanges used during the night are usually with 4.25% solution, and the fourth exchange is with 1.5% solution. These percentages were selected because analysis of clearance rates indicates that 4.25% glucose solution reaches maximum clearance in under 3 hours, whereas 1.5% solutions continue to provide clearance for approximately 7 hours.

Continuous cycling peritoneal dialysis reduces the number of times the system is opened per day, as it requires access to the system only twice a day.

Complications of Peritoneal Dialysis for Chronic Conditions

Problems associated with the treatment of chronic conditions using CIPD, CAPD, and CCPD are mechanical (fibrin clots or sludging in catheter) or infectious and medical (for example, cardiovascular, peripheral vascular, gastrointestinal, or musculoskeletal diseases). Infection occurs most frequently with CAPD. CCPD has a lower incidence of peritonitis. CIPD provides the best control of infection. Peritonitis is the most serious problem related to chronic peritoneal dialysis, because scarring of the peritoneum decreases peritoneal clearance rates.

Pain and respiratory distress can result from the peritoneal dialysis procedure, especially during the first few exchanges. Both of these problems should subside after a week or two (Chambers, 1995). In the meantime, the dialysate can be warmed to increase comfort.

Nutrition can also be a problem. Data on protein loss during dialysis are inconclusive. A protein intake of 1 to 1.5 g/kg of body weight per day is considered adequate for patients with a good appetite and no infection. Dietary adjustments must be made when the catabolic rate increases, as with infection or administration of steroids. Diet is least restricted with CAPD, moderately restricted with CCPD, and most restricted with CIPD. Caloric intake may need to be controlled if there is excess weight gain due to absorption of glucose from the dialysate solution.

Psychologically, patients are most likely to prefer CAPD because of the convenience. Patients are less likely to accept CIPD and CCPD.

Nursing Interventions

Nursing interventions include ongoing assessments for potential complications and fluid and electrolyte imbalances. In addition to the interventions described earlier, the nurse is responsible for checking the flow of fluid. Constipation is the major cause of inflow and outflow complications (Chambers, 1995). The physician usually orders a bowel preparation, high-fiber diet, and daily stool softener before inserting the peritoneal catheter. Other causes of flow problems include tubing or catheter kinks, fibrin clot formation, and migration of the catheter. The drainage bag must be placed below the level of the peritoneum to facilitate drainage.

During the course of treatment, the nurse weighs the patient every morning on the same scale and monitors vital signs frequently. Signs and symptoms indicative of peritonitis must be reported to the physician immediately. These manifestations include fever, abdominal pain, rebound abdominal tenderness and distention, nausea and/or vomiting, general malaise, and cloudy outflow drainage fluid.

The nurse is also responsible for changing the dressing around the catheter, using sterile technique. When removing the old dressing, signs of infection, such as swelling, redness, or discharge, should be noted and reported.

REFERENCES AND READINGS

Baer, CL (1990). Acute renal failure. Nursing90 20(6), 34–40.
Baer, CL, & Lancaster, LE (1992). Acute renal failure. Critical Care Quarterly 14(4), 1–21.
Chambers, JK (1995). Interventions for clients with chronic and acute renal failure. In DD Ig-

natavicius, ML Workman, & MA Mishler. Medical-Surgical Nursing: A Nursing Process Approach (2nd ed, pp 2111–2156). Philadelphia: WB Saunders.

Dunetz, P (1992). If your med/surg patient is on dialysis. RN 55(9), 46–53.

Malti, J, & Wellons, D (1989). CAPD: A dialysis breakthrough with its own burdens. RN 51(1), 46–52, 74.

Rydholm, L, & Pauling, L (1991). Contrasting feelings of helplessness in peritoneal and hemodialysis patients. American Nephrology Nurses' Association Journal 18(2), 183–184, 186, 200.

Smith, LJ (1992). Peritoneal dialysis in the critically ill patient. AACN Clinical Issues in Critical Care Nursing 3(3), 558–569.

Twardowski, ZJ, Nolph, KD, & Khanna, R (1990). Peritoneal Dialysis: New Concepts and Applications. New York: Churchill Livingstone.

CHAPTER 17

. . . .
. . . .
. . . .
. . . .
. . . .
. . . .
. . . .
. . . .
. . . .
. . . .

Hemodialysis

Extracorporeal hemodialysis is one of several types of renal replacement therapies for the management of renal failure. It is a technique of filtering the patient's blood outside the body (extracorporeally) by using an "artificial kidney" dialyzing machine. This process eliminates from the blood waste products and fluid that are normally removed by the kidneys. The purposes of hemodialysis are the same as for peritoneal dialysis: to temporarily remove toxic products, both chemical and metabolic, to remove excess fluid, and to correct electrolyte imbalances. Hemodialysis is usually used when the patient's condition cannot be maintained by more conservative medical therapy. The patient's clinical condition is the primary determinant of whether to perform hemodialysis or peritoneal dialysis. Hemodialysis is a faster and more efficient method than peritoneal dialysis for removing urea and other toxic products.

EQUIPMENT

There are several different types of hemodialysis machines; however, the basic principles and components of the machines are essentially the same. Most of the systems include the use of a vascular access device, dialysate solution, dialyzer with cellulose hollow fibers, and a hemodialysis machine (Fig. 17–1). All dialyzers have four basic components: a blood compartment, a dialysate compartment, a semipermeable membrane, and an enclosed structure that supports the membrane.

The physician determines the dialysate concentration to be used. Hemodialysis solutions contain acetate. In contrast, peritoneal dialysis solutions contain lactate. These two substances, acetate and lactate, are the main

233

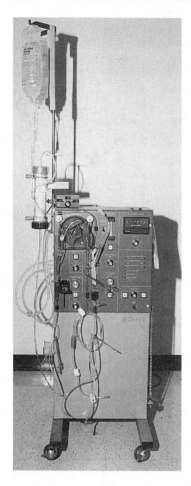

FIGURE 17–1
Century II hemodialysis machine. (Courtesy of St. Francis Regional Medical Center, Wichita, Kansas.)

factors in the conversion of acidemia to mild alkalemia with no serious side effects.

PROCEDURE

In order to perform hemodialysis, access to the patient's bloodstream is necessary. Short-term or temporary access methods for acute conditions include a dual- or triple-lumen central venous catheter (described in Chap. 15) and, less commonly, an arteriovenous (AV) shunt. The creation of an AV shunt is a rather long and tedious surgical procedure. This is one of the reasons why more conservative methods of treatment are preferred over hemodialysis. The type of shunt used varies according to whether the patient has an acute or chronic condition. Generally, the shunt is created with Silastic tubing that can be placed into an artery and a vein (Fig. 17–2). A plastic connector hooks all types of accesses to the machine tubing. They are separated for connection to the artificial kidney and reconnected for the resumption of normal blood flow.

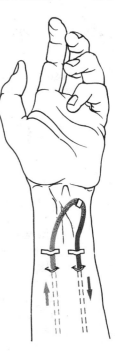

FIGURE 17–2
An arteriovenous shunt in the forearm. One part of the shunt cannula is placed in an artery, the other part in a vein. The ends of the shunt cannula are joined when dialysis is not in progress. (From Ignatavicius, DD, Workman, ML, and Mishler, MA: Medical-Surgical Nursing: A Nursing Process Approach, ed. 2. WB Saunders, Philadelphia, 1995, p 2137, with permission.)

In chronic conditions requiring long-term hemodialysis, the primary AV fistula is the preferred method of shunting. An artery and a vein are grafted together and remain under the skin. A ripening period is required in which the fistulous vein is enlarged. This process is facilitated by applying a tourniquet to the upper arm, allowing backflow of arterial blood into the vein. The pressure from this procedure expands the vein. Another type of shunt is the secondary AV fistula, that is, a Gore-Tex AV graft. Gore-Tex is a synthetic material similar to Teflon. In some cases, a dual-lumen central venous catheter is used for several weeks or longer. Regardless of which type is selected, the patient's blood access is connected to the machine for each procedure by two needle placements or by connection to a blood access catheter.

Once the tubing is connected to the hemodialysis machine, the pump draws blood into the hollow fibers inside the dialyzer, using hydrostatic pressure in the ultrafiltration process. The dialyzer is composed of hollow fibers, which are usually regenerated cellulose or a cuprammonium rayon membrane fiber. These fibers filter toxic materials from the blood into the dialysate. The dialysate solution is on the opposite side of the membrane, where it continuously bathes the hollow fibers while selectively extracting waste products and excess fluids from the blood by diffusion. When the blood leaves the dialyzer, it is returned to the patient through the venous line of the machine into the patient's venous system. The blood takes a circuitous route from the body through the tubing and the dialyzer and back through the body, approximately twice each hour. Hemodialysis requires 3 to 4 hours of dialysis two to three times per week to maintain fluid and electrolyte balance. It can be done in the home setting.

INDICATIONS

Conditions requiring hemodialysis (HD) fall into two general categories: acute and chronic. Acute conditions include overdoses of drugs or poisons, severe hyperkalemia, trauma caused by crushing injuries or shock, hemolysis, and pancreatitis. Patients with an acute condition who are using an artificial kidney require extremely skilled nursing care. Such patients have high mortality rates. In many cases, acute renal failure is reversed, but 30% to 50% of these patients die of other causes, such as injuries or septicemia.

For some patients with *acute* renal failure (ARF), conventional hemodialysis may not be adequate for the fluid removal required. Continuous arteriovenous hemofiltration (CAVH) and continuous arteriovenous hemodialysis and filtration (CAVHD) are examples of other renal replacement therapies for patients with ARF. CAVH is used to remove massive amounts of fluid when the patient is not hemodynamically stable. CAVHD uses a dialysate delivery system to remove nitrogenous or other waste products with the fluid in patients who have cardiac dysfunction or significant hypotension. Both of these procedures are limited to the critical care setting (Chambers, 1995).

Some patients with *chronic* renal disease are maintained by a controlled diet alone and function very well. There are many, however, whose kidney function is so greatly diminished that they must be maintained on an artificial kidney to sustain life.

Criteria must be established for the selection of patients requiring extracorporeal hemodialysis. Selection must be made on an individual basis. Once dialysis is initiated for a patient with a chronic condition, it is usually performed two or three times a week and must be continued for life or until the patient receives a successful kidney transplant. Thus, many factors besides the medical ones must be considered. Some of the major factors that should be evaluated before initiating hemodialysis are listed in Table 17–1.

TABLE 17–1

FACTORS TO CONSIDER BEFORE
INITIATING HEMODIALYSIS

MEDICOPSYCHOLOGICAL FACTORS

Age
Cardiovascular condition
Metabolic conditions
Psychiatric stability

SOCIOREHABILITATIVE FACTORS

Capability of rehabilitation
Motivation to function in family and community

PERSONAL ADJUSTMENT FACTORS

Necessity of role reversal
Possible sexual dysfunction
Increased demands on family
Economic problems

COMPLICATIONS

Complications resulting from hemodialysis are related to either the vascular access or the process itself.

Vascular Access Complications

The most common problems associated with the vascular access include thrombosis or stenosis, infection, ischemia, and aneurysm. Most of these complications are seen in patients with AV fistulas.

Thrombosis (clot formation) in the vascular access is the most frequently occurring complication. Localized clots can be removed surgically or the access can be revised if it is stenosed.

Infection, usually introduced by repeated punctures of the dialysis access, is caused most often by *Staphylococcus aureus*. Strict sterile technique is needed when the patient is placed on the hemodialysis machine.

Local aneurysms (outpouching of the vessels) of a fistula are caused by repeated needle puncture sites. If an aneurysm is present, the access may need revision if it is to function properly.

Localized ischemia can result because the blood from the distal fingers is diverted to the fistula (steal syndrome). The patient complains of cold, numb fingers. The ischemia may progress to gangrene, although this is not common.

Complications of Hemodialysis

Complications of the hemodialysis procedure include those resulting from rapid removal of fluid, electrolytes, and waste products. Hypovolemia and shock can occur, during or after dialysis, as a result of rapid fluid removal. Signs and symptoms of hypovolemia are hypotension, restlessness, dizziness, tachycardia, and anxiety.

One of the most common complications of the hemodialysis procedure is dialysis disequilibrium syndrome during hemodialysis or after hemodialysis is completed. Rapid decreases in extracellular electrolytes and wastes such as blood urea nitrogen (BUN) can result in cerebral edema and increased intracranial pressure, manifested by neurologic changes, such as decreased level of consciousness, nausea, vomiting, and seizures.

Because hemodialysis requires blood access, blood-borne infectious diseases are another potential problem associated with hemodialysis. Hepatitis B is less of a threat today than it was years ago, but patients are still at risk for infection with the hepatitis B virus, especially with the use of multiple blood transfusions. Patients with chronic renal failure are typically anemic because of decreased production of erythropoietin by the kidneys. In recent years, recombinant erythropoietin (EPO) has been available for direct administration to the patient to combat anemia and decrease the number of transfusions required.

Long-term hemodialysis can result in improper regulation of the growth hormone by the hypothalamus, which is significant in the treatment of children and adolescents with chronic renal failure.

As with any mechanical treatment, technical problems can endanger the patient's life. Among the more common technical problems that can develop are blood leakage in the system and machine problems. Pyrogenic reactions caused by bacterial growth have been known to occur. Routine cultures of the equipment should be taken.

Heparinization is vital to keep the blood from clotting, because blood has a normal tendency to clot when it comes in contact with a foreign surface, for example, a dialyzer. Two methods may be used. The systemic method involves direct intravenous injections of heparin or calcium citrate into the patient. The dosage is determined by the clotting time within the dialyzer, which must be carefully monitored. Patients receiving erythropoietin therapy may need a significant increase in the dosage (Norton et al, 1992).

The second anticlotting method is regional. A heparin solution is allowed to flow into the dialyzer. Protamine sulfate solution neutralizes the blood as it is returning to the patient. This method is especially useful when the patient is in danger of hemorrhaging.

Nursing Interventions

The blood access site is the patient's lifeline to the artificial kidney and requires special attention from the nurse. The nurse should observe for warmth and pulsation, and auscultate for bruits (swishing pulsation). When the shunt is connected outside the body, rather than to the AV fistula, the nurse should observe for a stringy appearance of the blood or any suspicion of clotting, and should report these observations to the physician at once. Clamps and a tourniquet should be readily available in case the shunt comes apart or the tubing comes out of the blood vessel.

Problems that directly affect shunts are coagulation defects, vascular problems (usually due to thrombophlebitis or inadequate blood vessels), hemorrhage, inadequate blood flow, blood vessel spasms, improper needle placement, and sepsis.

Before each treatment, the patient's laboratory values are evaluated, vital signs are taken, and weight is recorded. During the procedure, the patient usually reclines and is closely monitored by the nurse. Nursing interventions such as diversional activities and provision of comfort measures can increase the patient's acceptance and compliance.

Postdialysis care includes recording the weight and vital signs, and observation for side effects of the treatment. These manifestations include hypotension, nausea and/or vomiting, malaise, headache, dizziness, and muscle cramping. Extreme hypotension may require rehydration with intravenous fluids, such as normal saline. Because heparinization is required, all invasive procedures should be avoided for 6 to 8 hours after the hemodialysis treatment. The nurse observes for signs of bleeding for at least 1 hour after the dialysis is completed.

Peer counseling, a process using renal patients who are successfully coping with long-term hemodialysis, can assist and educate new chronic renal patients. The primary focus areas for peer counselors are such topics as patient compliance with treatment regimens, especially instructions concerning fluids and nutrition; common reactions to the therapy; and emotional concerns that can lead to depression and self-pity. Although survival rates for patients on hemodialysis are acceptable, the quality of life may be limited.

REFERENCES AND READINGS

Alt, D, Balduf, R, & Thompson, E (1986). When a vascular access site complicates care. RN 49(10), 36–39.

Biggers, PB (1991). Administering epoetin alfa—more RBCs with fewer risks. Nursing91 21(4), 43.

Brinkley, LS, & Whittaker, A (1992). Erythropoietin use in the critical care setting. AACN Clinical Issues in Critical Care Nursing 3(3), 640–650.

Campbell, M (1991). Terminal care of the ESRD patient foregoing life-sustaining dialysis therapy. American Nephrology Nurses' Association Journal 18(2), 202–204.

Chambers, JK (1995). Interventions for clients with chronic and acute renal failure. In DD Ignatavicius, ML Workman, & MA Mishler. Medical-Surgical Nursing: A Nursing Process Approach (2nd ed, pp 2111–2156). Philadelphia: WB Saunders.

Chambers, JK (1993). Renal insufficiency: Implications for care of the medical-surgical patient. MEDSURG Nursing 2(1), 33–40.

Dirkes, S (1994). How to use the new CVVH renal replacement systems. American Journal of Nursing 94(5), 67–73.

Gutch, CF, Stoner, MH, & Corea, AL (1993). Review of Hemodialysis for Nurses and Dialysis Personnel. St. Louis: CV Mosby.

Hampton, JK (1992). Long-term effects of hemodialysis in diabetic patients with end-stage renal disease. American Nephrology Nurses' Association Journal 19(5), 455–456.

Jameson, JD, & Wiegmann, TB (1990). Principles, uses, and complications of hemodialysis. Medical Clinics of North America 74(4), 945–960.

Lawyer, LA, & Velasco, A (1989). Continuous arteriovenous hemodialysis in the ICU. Critical Care Nurse 9(1), 29–41.

Lundin, AP (1991). Recombinant erythropoietin and chronic renal failure. Hospital Practice 26(4), 61–69.

Molzahn, AE (1991). The reported quality of life in selected home hemodialysis patients. American Nephrology Nurses' Association Journal 18(2), 173–180, 194.

Norton, J, et al (1992). Varying heparin requirements in hemodialysis patients receiving erythropoietin. American Nephrology Nurses' Association 19(4), 367–372.

Peschman, P (1992). Acute hemodialysis: Issues in the critically ill. AACN Clinical Issues in Critical Care Nursing 3(3), 545–557.

Price, CA (1991). Continuous renal replacement therapy: The treatment of choice for acute renal failure. American Nephrology Nurses' Association 18(3), 239–244.

Taylor, T (1990). Preventing complications from hemodialysis. Dimensions of Critical Care 9(4), 210–215.

PART THREE **QUESTIONS**

• • • • • • • • • • •

1. What is the most reliable method for checking nasogastric tube placement?
   a. Placing the tube under water
   b. Having a chest/abdominal x-ray examination
   c. Injecting air and auscultating over the stomach
   d. Testing the pH of the aspirate

2. A patient has a nasogastric tube connected to low Gomco suction. What position should the patient be placed in?
   a. Dorsal recumbent
   b. Orthopneic
   c. Sidelying
   d. Semi-Fowler's

3. What acid-base imbalance is likely in a patient with long-term nasogastric suction?
   a. Metabolic alkalosis
   b. Metabolic acidosis
   c. Respiratory alkalosis
   d. Respiratory acidosis

4. How often should a tube-feeding administration set be changed?
   a. Every 8 hours
   b. Every 12 hours
   c. Every 24 hours
   d. Every 48 hours

5. What is the primary purpose of the water that is used to flush a feeding tube?
   a. To keep the tube from clogging
   b. To dilute the enteral formula
   c. To prevent dehydration
   d. To prevent acidosis

6. What nursing action should be done first when infiltration occurs while administering IV fluid?
   a. Slow the infusion.
   b. Stop the infusion.
   c. Call the physician.
   d. Monitor the site.

7. Which of the following nursing interventions is not necessary for the patient receiving total parenteral nutrition in the hospital setting?

a. Take vital signs every 4 hours.
b. Weigh every day.
c. Monitor electrolytes.
d. Monitor blood glucose.

8. What type of solution is commonly used for partial or total parenteral nutrition?
   a. Hypotonic
   b. Mesotonic
   c. Isotonic
   d. Hypertonic

9. What is the primary advantage of continuous ambulatory peritoneal dialysis (CAPD) for the patient?
   a. It is the least expensive method of dialysis.
   b. It requires a vascular access.
   c. It takes a shorter time to perform the treatment.
   d. It allows patient mobility and control.

10. A patient receiving peritoneal dialysis (PD) complains of abdominal pain, nausea, and general malaise. What complication of PD might the patient have?
    a. Internal hemorrhage
    b. Hyperkalemia
    c. Peritonitis
    d. Hepatitis B

11. What is the advantage of peritoneal dialysis over hemodialysis?
    a. Decreased length of treatment
    b. No requirement for vascular access
    c. Less chance of infection
    d. Less fluid and electrolyte imbalances

12. What is the primary transport system by which peritoneal dialysis supplements electrolytes and removes wastes from the body?
    a. Diffusion
    b. Osmosis
    c. Active transport
    d. Filtration

13. Over recent years the incidence of hepatitis B from hemodialysis has decreased. What is the probable explanation for this decline?
    a. The availability of recombinant erythropoietin
    b. The availability of the hepatitis B vaccine for patients
    c. The use of improved vascular access devices
    d. The use of sterile technique

14. A patient's BUN decreased from 94 to 32 following hemodialysis. What assessment finding should the nurse watch for in connection with this rapid change?
    a. Increased urinary output
    b. Muscle cramping
    c. Decreased level of consciousness
    d. Uremic frost

15. A patient who had hemodialysis complains of dizziness. The patient's blood pressure is 84/50 mmHg. What fluid or electrolyte imbalance should the nurse suspect?
   a. Fluid overload
   b. Fluid deficit
   c. Hypernatremia
   d. Hypokalemia

PART FOUR

SELECTED CLINICAL DISORDERS THAT AFFECT FLUID, ELECTROLYTE, AND ACID-BASE BALANCE

CHAPTER 18

. . . .
. . . .
. . . .
. . . .
. . . .
. . . .
. . . .
. . . .
. . . .

Respiratory Disorders

Respiration is one of the primary regulators of fluid and electrolyte homeostasis. Changes in respiration that affect the acid-base balance have already been discussed (see Chap. 3). This chapter will briefly outline the respiratory problems that occur most frequently in the clinical situation.

PHYSIOLOGY

An understanding of the physiology of respiration and the anatomy of the respiratory system is necessary to enable the nurse to cope competently with respiratory problems that affect acid balance.

In normal respiration, air is inhaled through the external respiratory tract and drawn through the bronchial tree to the alveolar level, where the exchange of gases occurs. As oxygen is consumed, carbon dioxide forms and combines with water to create carbonic acid. Carbonic acid has a direct effect on the pH balance of the extracellular fluid. The lungs control the extracellular level of carbonic acid, and thus the pH balance, by exhalation or retention of carbon dioxide (Fig. 18–1).

Respiration can be divided into three interdependent processes: alveolar ventilation, alveolar-capillary diffusion, and perfusion. Alveolar ventilation is the process by which atmospheric gases move into the alveoli. Alveolar capillary diffusion is the passage of alveolar gases into the arterial capillary to be exchanged with venous gases across the thin alveolar-capillary membrane. Perfusion is the circulation of oxygenated blood through the alveolar-capillary membrane, then movement of the blood from the lungs to the tissues, and the subsequent inundation of the body cells with oxygenated blood.

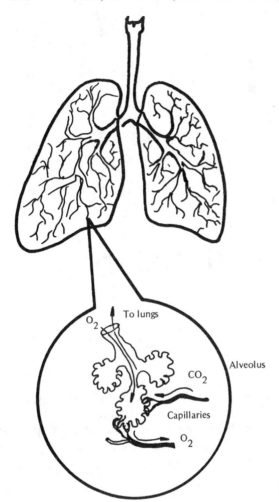

FIGURE 18–1
Cross section of lung with enlargement of alveolus.

One of the factors regulating the exchange of arterial and venous gas is partial pressure. Partial pressure is the force per unit area exerted by a gas in a mixture of gases. The partial pressure of oxygen and carbon dioxide exert varying degrees of pressure on the alveolar-capillary membrane, which allows the oxygen to diffuse into the blood and the carbon dioxide to move back into the alveoli for removal from the body. In diffusion, each gas moves from an area of higher pressure to an area of lower pressure. Partial pressure is an important factor in the respiratory regulation of pH.

The anatomy of the respiratory system influences the efficiency of respiration. The amount of air inhaled and exhaled in quiet breathing averages about 300 to 600 mL and is called the tidal volume. The total volume of air in the lungs depends on body weight and averages about 5 to 8 mL/kg. Any variation in the tidal volume will be reflected by a change in alveolar ventilation and a subsequent change in the entire respiratory process.

The alveoli are the functional units of the respiratory system. Much of the anatomy of the respiratory system is composed of passive airways. These are referred to as physiologic dead space because the air in this space does not participate in the exchange of gases in the passageways. Dead space includes the nasopharynx, the larynx, the trachea, and the main bronchi, and contains approximately 150 mL of air.

The numerous factors involved in the control of respiration include the carbon dioxide concentration, the hydrogen ion concentration (pH), the chemoreceptor system (oxygen deficiency), exercise, arterial pressure, sensory impulses, and speech. This concept of respiratory control is called the multiple factor theory.

CLASSIFICATION

Respiratory diseases, as well as various neuromuscular diseases, interrupt the normal respiratory cycle. These diseases may be classified as chronic or acute and as reversible or irreversible. Among the more common causes of chronic respiratory diseases are smoking, allergies, and chronic infections. Acute respiratory diseases may be caused by infection, tumors, atelectasis, obstruction, and chest trauma. These causes can lead to conditions known as acute adult respiratory syndrome or acute respiratory failure. (Hyaline membrane disease in infants is also an acute respiratory syndrome.) Reversible respiratory diseases are those involving infections, edema, bronchial spasms, and obstruction. Irreversible respiratory diseases are those in which permanent destruction has occurred; these include emphysema, cancerous tumors, and infiltrative diseases.

Assessment of Chronic Respiratory Disease

Chronic respiratory diseases are diseases of the lung characterized by slow pathologic changes and prolonged continuance. Table 18–1 outlines the most common types of chronic lung diseases, their assessment, and treatment. Smoking is probably the greatest factor in chronic respiratory disease, since it is thought to be a major factor in cancer of the lung. Although smoking is not considered a direct cause of emphysema, there is a definite correlation between heavy smoking and emphysema. Air pollution has recently been recognized as another causative factor in respiratory diseases, and it exacerbates the condition of patients with chronic obstructive pulmonary disease (COPD). Chronic obstructive pulmonary disease, most currently referred to as chronic airflow limitation (CAL), can lead to acute decompensation of cardiopulmonary homeostasis. Maintenance of patients being treated for COPD depends primarily on cardiac reserve.

Pulmonary emphysema is the most common of the chronic respiratory diseases or types of COPD (Table 18–1). Emphysema is a destructive, irreversible process in which the total lung capacity and the residual volume are increased. The primary difficulty in emphysema is in exhalation, as shown by the elevated $Paco_2$ reading. Elevated $Paco_2$ readings are clinically identified as carbon dioxide narcosis, evidenced in the late phases by stupor and coma. Diffusion is also a problem in that there is insufficient capillary surface through which blood gases may be exchanged. Fluid and mucus accumulation in the alveoli further inhibits diffusion. The elasticity of the lung is decreased and the alveoli are gradually damaged.

TABLE 18–1
CHRONIC RESPIRATORY DISEASE

Disease	Description	Recognition	Laboratory Findings	Treatment
Bronchitis ("blue bloater")	Inflammatory and/or infectious disease with airway obstruction; person has normal lung volume, is usually a smoker; closely related to emphysema	Productive cough; cyanosis; puffiness; cardiac involvement (cor pulmonale); CO_2 narcosis; chronic fatigue	Increased hemoglobin; increased hematocrit	Antibiotics; intermittent positive pressure breathing (IPPB); expectorants; nebulizers; humidification; stop smoking
Emphysema ("pink puffer")	Primarily exhalation problem; lung capacity increased; residual volume increased; elasticity decreased; alveolar damage; airway obstruction	Respiration labored; barrel-chested; weight loss; cor pulmonale; chronic fatigue; CO_2 narcosis	Pa_{CO_2} increased; globin deficiency in blood	Bronchodilators; IPPB; tranquilizers; control infection; stop smoking; sodium restriction; diuretics
Pulmonary fibrosis	Ventilation and diffusion dysfunction; total lung capacity decreased	Dyspnea; cyanosis; clubbing of fingers; cor pulmonale; hyperventilation; chronic fatigue; CO_2 narcosis	Pa_{O_2} decreased; Pa_{CO_2} decreased (diagnosed by lung tissue biopsy)	Corticosteroids

Emphysema is a chronic condition in which it is easy to precipitate acute respiratory failure. In some cases even the infectious process of a common cold can lead to acute respiratory failure.

Assessment of Acute Respiratory Disease

Table 18–2 lists the most common types of acute respiratory diseases. Pneumonia, the most deadly of the acute infectious respiratory diseases, is one of the most common causes of death in the United States today. Acute pneumonia is an infectious process caused by bacterial and viral pathogens. Pneumococcus appears to be the most frequent bacterial source, followed by the gram-negative bacilli and *Staphylococcus aureus*. The tissues and the air spaces of the lungs become filled with fluid because of the inflammation usually caused by these bacteria. As a result, oxygen diffusion is decreased, resulting in a low arterial oxygen partial pressure. This low oxygen level explains why disorientation and confusion are the earliest manifestations of pneumonia in the elderly. The neurons of the brain require oxygen to function properly. Chest pain and fever are also commonly present in patients with pneumonia.

Acute adult respiratory distress syndrome (ARDS) is a comprehensive term used to describe a condition characterized by sudden dyspnea, forced expiration, tachycardia, and hypotension. Causes most commonly associated with this syndrome include administration of excessive volumes of fluid in the treatment of trauma such as burns, major surgery, sepsis, shock, and multiple blood transfusions. Other causes associated with this syndrome are disseminated intravascular coagulation (DIC), fat embolism, and artificial ventilation. The overloading can lead to radiographic lung changes typical of pulmonary edema.

Because of the problems associated with ARDS, the following precautions should be taken when it is necessary to administer fluids to patients after major surgery or trauma. First, the pulmonary arterial pressure must be monitored. During shock, it is suggested that a 5% albumin solution be used instead of a crystalloid solution. The 5% albumin solution will maintain blood pressure and provide adequate intravascular osmotic pressure, but the precise fluid (sodium, water, colloid) combination required to adequately treat severe respiratory failure has yet to be established.

Severe respiratory failure is a crisis in which the patient is unable to maintain an adequate exchange of gases and must be assisted by intubation and/or ventilation to sustain sufficient oxygenation of the blood and removal of waste products. It can be more specifically defined as a PaO_2 less than 50 mmHg and a $PaCO_2$ more than 50 mmHg. Failure to provide adequate respiratory assistance to the patient will result in a progressive deterioration of the patient's status in the following sequence: hypoxemia, hypercapnia, acidemia, and cardiac arrest.

Surgical patients experiencing respiratory failure may have several combined problems. Infection may be the underlying cause of the respiratory complication. Many patients who have undergone surgery are weak and debilitated, which further hampers normal respiration. The combined problems result in rapid, shallow breathing, thus promoting the development of atelectasis and hypostatic pneumonia.

TABLE 18–2
ACUTE RESPIRATORY DISEASES

Disease	Description	Recognition	Laboratory Tests	Treatment
Bronchial asthma	Acute airway obstruction due to allergic/nervous response; little or no hypoxemia	Respiratory distress; wheezing	CO_2 usually normal; O_2 decreased	Medication (adrenalin): IPPB; expectorants; bronchodilators
Bronchitis	Airway obstruction due to edema secondary to inflammation	Fever; chest pain; dyspnea; cough	White blood cell count (WBC) increased	Antibiotics; IPPB; expectorants; humidification
Pneumonia	Infection by pneumococci, bacterial pathogens, gram-negative bacilli, viruses, fungi, *Staphylococcus aureus*	Fever; productive cough; pulmonary infiltration	WBC increased; PaO_2 decreased; $PaCO_2$ may be increased	Antibiotics; IPPB; expectorants; humidification; hydration to thin secretions
Pulmonary edema	Acute heart failure; hypernatremia and fluid retention due to inability of lymphatics to rapidly remove excess fluid	Edema; weight gain; hypertension; dyspnea	Urine Cl decreased; serum sodium normal; hemoglobin normal or decreased	Diuretics; salt and fluid restriction
Pulmonary emboli	Capillaries plugged with clots	Chest pain; increased blood pressure	Lactic/dehydrogenase (LDH) increased	Anticoagulants
Respiratory depression	Insensitive respiratory center; sometimes accompanies obesity	CO_2 narcosis	PaO_2 decreased; $PaCO_2$ increased	Endotracheal tube; supportive breathing; respiratory stimulants
Spontaneous pneumothorax	Ruptured lung	Blood pressure and pulse decreased; movement of chest wall decreased; cyanosis	X-ray	Release air by inserting catheter into intrapleural space through second intercostal space anteriorly; prophylactic antibiotics
Respiratory anaphylaxis	Allergic reaction due to antigen-antibody response	Acute respiratory distress syndrome	History of dust, pollen, food, or antibiotics such as erythromycin sterate	Ventilation and adrenalin

Children who die unexpectedly are frequently the victims of respiratory tract diseases. Respiratory disease may play a role in sudden infant death syndrome (SIDS).

The most common causes of death in acute respiratory failure are sedation (including sedation from use of aspirin), infection, inhalers (overuse causes dysrhythmia), digitalis preparations, and electrolyte imbalance. Digitalis is often administered for cor pulmonale (right-sided heart failure), which can result from respiratory disease. The lung disease as well as the cardiac complication must be treated.

COLLABORATIVE MANAGEMENT

The goal in management of respiratory problems is ultimately to correct the underlying disease or defect. More immediately, it requires control of the circumstances that threaten the patient until the respiratory system recovers sufficiently to provide adequate ventilation.

Medical Treatment

A Swan-Ganz catheter or other hemodynamic monitoring system, which measures pulmonary arterial pressure, is an important tool for the physician. It enables the physician to evaluate the adequacy of the oxygen delivery system and the effects of fluid therapy. In addition, cardiac output should be adequate to maintain normal cerebral and coronary perfusion.

Intubation and ventilation of the patient become necessary when the pH drops to 7.25 and tidal volume is decreased. Ventilation is not adequate when respirations are rapid and shallow. The rapid, shallow breathing causes hypercapnia and acidemia.

In patients with acute respiratory failure, there may be twice as much dead space as normal; therefore, the ventilator setting should provide twice the normal ventilation per minute. The patient is removed from the ventilator by a weaning process. An immediate increase in oxygen consumption (by as much as 40%) occurs when patients in severe respiratory failure are weaned from ventilatory assistance. Patients who can be weaned slowly are comatose patients, those with adequate but unstable cardiac output, and those with weak but recovering ventilatory mechanics. Hyperalimentation regimens may also need to be taken into consideration in weaning the patient from the ventilator, since parenteral nutrition regimens increase the production of carbon dioxide ($Paco_2$).

Endotracheal suctioning assists in clearing fluids and mucus from the tissue, thereby increasing aeration.

Nursing Interventions

For patients with respiratory disease the nurse should:
- Evaluate ventilation.
- Keep sedatives to a minimum.
- Maintain low-flow oxygen (1 to 2 L) in patients with chronic pulmonary diseases.

- Increase alveolar ventilation to remove carbon dioxide using proper respiratory therapies; for example, deep breathing, pursed lip breathing, and physical therapy.
- Be aware that, in chronic obstructive pulmonary disease, acute respiratory failure can be precipitated by infection, congestive heart failure, drugs, atelectasis, or bronchitis.
- Be alert to a decrease in aeration, an increase in pulse rate, restlessness, sweating, sensorium changes, somnolence, and increased breathing effort.
- Keep the airway free of secretions by encouraging coughing.

REFERENCES AND READINGS

Finesilver, C (1992). Perfecting the art: Respiratory assessment. RN 55(2), 22–29.

Fraser, R, & Pare, JA (1991). Diagnosis of Disease of the Chest. Philadelphia: WB Saunders.

Hunter, FC, & Mitchell, S (1993). Managing ARDS. RN 56(7), 52–58.

Kersten, LD (1989). Comprehensive Respiratory Nursing: A Decision Making Approach. Philadelphia: WB Saunders.

Mason, SG (1992). When a ventilator patient is going home. RN 55(10), 60–64.

Noll, ML (ed) (1990). Respiratory care in adults. AACN Clinical Issues in Critical Care Nursing 1(2), 237–326.

Stiesmeyer, JK (1992). Care of the elderly mechanically ventilated patient: Preserving the fragile environment. AACN Clinical Issues in Critical Care Nursing 3(1), 129–136.

CHAPTER 19

. . . .
. . . .
. . . .
. . . .
. . . .
. . . .
. . . .
. . . .
. . . .

Congestive Heart Failure

Congestive heart failure (CHF) is a common type of heart (pump) failure that occurs when the heart is no longer able to maintain cardiac output sufficient to meet tissue demands.

CLASSIFICATION OF DISORDERS

Congestive heart failure is classified in several ways: right or left, backward or forward, organic or functional, acute or chronic, low output or high output, and compensated or decompensated.

A common way to differentiate types of heart failure is by referring to the side of the heart that failed—right or left. Right-sided heart failure results in an overload of the venous system. The signs and symptoms of right failure, then, are systemic, such as edema and hypertension. Left-sided heart failure results in an overload in the pulmonary system. The signs and symptoms associated with left failure are pulmonary in nature, such as crackles and dyspnea. Although the pump may initially fail on one side, both sides are usually affected as the disease progresses.

Another way to classify heart failure is by referring to it as backward or forward. Backward failure occurs when the ventricle cannot maintain an adequate workload. It compensates, in general, by hypertrophy of the left ventricle. Forward failure occurs when the kidneys cannot adequately excrete the excess fluid.

Organic and functional classifications distinguish between physical changes and process changes. Organic relates directly to changes in the heart. The main contributory conditions to organic heart disease are arteriosclerosis (hardening of the arteries), atherosclerosis (fat deposits, usually at bifurcations of vessels), hypertension, and valvular disorders, most commonly caused by rheumatic fever secondary to streptococcal infections. Functional conditions involve nonheart tissue as the primary cause. Functional causes include such factors as infection and thyrotoxicosis.

Cardiac disorders also may be classified by the timing of onset, either acute or chronic. For example, a myocardial infarction is an acute situation. Congestive heart failure is usually a chronic state. However, acute and chronic situations frequently compound each other. An acute myocardial infarction may be accompanied by congestive heart failure even though congestive heart failure is more commonly a chronic condition occurring as a result of chronic hypertension or myocardial changes over the years.

Lastly, cardiac failure may be classified as compensated or decompensated. These terms are very helpful in planning the treatment and nursing care regimes because the operational definitions relate to quality of circulation. Compensated states exist when the patient can still maintain normal circulation, but develops cardiac hypertrophy and tachycardia, and has some restrictions with vigorous activity. Decompensation is a term used when poor circulation results, leading to an inefficient system. The results are inadequate circulation to the peripheral tissue and compromised or sacrificed cellular metabolism jeopardizing both proper nutrition to the cells and/or the removal of waste products. In decompensated states, activities have to be curtailed because of a hypoxic state that continues even with drug and oxygen therapy. Decompensated states are characterized by key fluid and electrolyte features: edema with decreased renal perfusion, and electrolyte and acid-base imbalances. For this reason, congestive heart failure is intimately involved with fluid, electrolyte, and renal problems.

RISK FACTORS

Hereditary factors that contribute to the development of cardiovascular disease are generally related to biologic factors of age (elderly persons), gender (men), race (African American), and family history.

Other factors over which an individual has more control include sedentary lifestyle, dietary intake high in lipids, history of smoking, obesity, and a lifestyle that produces tension (Table 19–1). Nicotine causes the release of catecholamines, which stimulate the autonomic nervous system, with a resultant increase in the diastolic blood pressure. Lifestyle and Type A personality are also correlated with arteriosclerotic heart disease. The Type A person manifests competitiveness, excessive ambition, aggressiveness, and a sense of time urgency. These activities also cause stimulation of catecholamines with sympathetic nervous system response. Diabetes mellitus is also a factor.

The physiologic ramifications of these risk factors include elevated serum lipid levels, increased cholesterol levels, and impaired glucose tolerance. Hypertension accelerates the formation of heart disease; elevations of both diastolic and systolic pressure predict an even greater risk. The hypertensive state is further exacerbated in clients who smoke.

TABLE 19–1

RISK FACTORS ASSOCIATED WITH
CARDIOVASCULAR DISEASE

UNCONTROLLABLE RISK FACTORS

Age (elderly persons)
Gender (men)
Ethnic/racial background (African American, Native American)
Genetic predisposition
Inability to cope with stress

CONTROLLABLE RISK FACTORS

Sedentary lifestyle
Smoking
High-fat and/or high-cholesterol diet
High serum cholesterol and/or lipids
Diabetes mellitus
Atherosclerosis
Hypertension
Obesity

CAUSATIVE FACTORS

Congestive heart failure is the clinical condition of increased right and/or left arterial pressures. The most common cause of heart failure is chronic hypertension leading to decreased ventricular contractility. The increased peripheral vascular resistance over the years results in an enlargement of the left ventricle to overcome the resistance. Thus, congestive heart failure secondary to left ventricular contractility causes increased atrial pressure, increased left ventricular end-diastolic pressure, and decreased stroke volume, all of which result in decreased cardiac output.

Other common causes are myocardial infarction and valvular lesions. Less common causes are uncontrolled tachycardia, such as in atrial fibrillation, atrial flutter, and paroxysmal atrial tachycardia (PAT). Other causes, which are not common but may occur, include rheumatic carditis, congenital heart lesions, ventricular aneurysm, anemia, and thyrotoxicosis.

Coronary artery disease and, consequently, myocardial infarction continues to be a major disease in modern society. Other noncoronary artery disease causes of cardiomyopathy are classified as infective, metabolic, infiltrative, toxic, and nonspecific. Examples are given below:

Infective: Viral (e.g., coxsackievirus A and B, AIDS), parasitic, and bacterial diseases.

Metabolic: Diabetes mellitus; anorexia nervosa; thyroid disorders, especially thyrotoxicosis; and Cushing's disease. Vitamin and mineral deficiencies, such as scurvy and thiamine deficiency, may also create metabolic disorders resulting in cardiomyopathy (see Chap. 20).

Infiltrative: Neoplastic disease, sarcoidosis, glycogen disorders, and hemochromatosis

Toxic: Alcohol, lead, cocaine, steroids, and selected antibiotics.

Nonspecific: Sickle cell anemia, neuromuscular and postpartum disorders.

PATHOPHYSIOLOGY

Congestive heart failure occurs when the ventricles, particularly the left ventricle, become weakened, causing a reduction in the contractile force that empties the left ventricle during systole. Thus, the amount of cardiac output and the stroke volume are decreased. A reduction in systolic emptying occurs concurrently with increased diastolic filling as a compensatory mechanism. Both of these generally cause an elevation in left ventricular volume and pressure. Over a period of time, as the myocardium attempts to compensate, it becomes hypertrophied and heart enlargement can be seen on a chest x-ray.

In congestive heart failure, the heart is unable to pump with sufficient force to adequately perfuse the body's tissues. For proper metabolism, the heart must pump adequate oxygen and nutrients to all body tissues via the systemic circulation. Oxygen is particularly crucial for the metabolism of glucose and fats to carbon dioxide and water. The stage at which the heart is no longer able to function adequately is called compensatory decompensation or frank decompensation.

Congestive heart failure is associated with three pathophysiologic processes: inadequate ventricular filling, increased workload of the heart muscles, and deterioration of the myocardium caused by a decreased blood supply.

Inadequate Ventricular Filling

Inadequate ventricular filling may be caused by obstructions, such as mitral stenosis secondary to rheumatic heart disease. In this disease process, inflammation eventually causes scarring and stenosing of the opening.

Another possible cause of inadequate ventricular filling is a pressure buildup outside of the ventricles that decreases the area in which the ventricles can expand. This type of extracardiac pressure can be found in dysfunctions such as chronic pericarditis or cardiac tamponade in which excessive fluids gather around the heart.

A third cause of inadequate filling is dysrhythmia. An example is tachycardia, in which the increased heart rate does not allow sufficient time for the ventricle to fill adequately, thereby decreasing cardiac output. Cardiac output, which is the amount of blood pumped from the right or left ventricle per minute, is equal to the heart rate times the stroke volume. Stroke volume is the amount of blood ejected by the left ventricle with each beat. With a pulse rate of 70, the average cardiac output is 4 L/min. Increased cardiac output is achieved with a more effective pump, improved venous return, and a normal pulse rate.

Increased Workload

The second major pathologic factor is an increased workload for the heart muscle. This may be caused by increased resistance to the peripheral vascular system because of narrowed vessels (Fig. 19–1). With vasoconstriction the heart must pump harder to overcome the increased resistance. In hypertension, the peripheral vessels may have become narrowed because of arte-

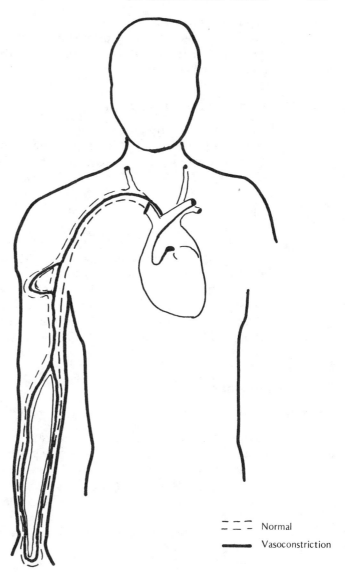

FIGURE 19–1
Vasoconstriction. The *broken line* indicates a normal vascular bed. The *solid line* shows the reduction of the vascular bed due to vaso-constriction.

- - - Normal
━━━ Vasoconstriction

riosclerosis (hardening of the arteries), thereby heightening the peripheral resistance. Peripheral hypertension specifically magnifies the work of the left heart. Pulmonary hypertension makes it harder for the right heart to pump blood into the lungs via the pulmonary artery. Pulmonary hyperten-sion may be the result of pulmonary or vascular diseases. Systemic hyper-tension results in elevation of both systolic and diastolic pressures, and is operationally defined as a pressure of 140/90 mmHg or greater.

Another cause of an increased workload for the heart is anemia. The lack of red blood cells and the resulting lack of hemoglobin decreases the amount of oxygen carried to the cells; therefore the heart must work harder to circulate the blood more rapidly. Any disease that increases the body's need for oxygen and nutrients will increase the workload of the heart.

Decreased amounts of fluid in the body also cause the heart to work harder and faster to carry oxygen and nutrients to the cells by increasing the rate of circulation. If the patient has a failing heart, this added strain exacerbates the failure.

Decreased Functional Blood Supply

The third pathophysiologic process in heart failure is a deterioration of the myocardium caused by a decreased functional blood supply. The heart, like every other muscle in the body, must receive oxygen in order to perform its task. An inadequate blood supply will cause failure of its pumping action. An inadequate blood supply may be caused by coronary ischemia secondary to plugging of the coronary vessels by fatty plaques or clots, with consequent death of tissue (myocardial infarction). Tissue death results in inadequate myocardial pump action because the dead or scarred muscles have lost their contractibility. This can create a circular, compounding problem because the irregular beats and rhythms will lessen the normal filling time of the ventricles, further decreasing the cardiac output, thereby decreasing even further the amount of blood going to the coronary vessels. Muscles in an anaerobic cycle produce lactic acid, inducing metabolic acidosis. In other words, metabolic acidosis is caused by inadequate tissue perfusion.

Specific Disorders

Atherosclerosis

The fatty plaques of atherosclerosis are accumulations of lipid and fibrous tissue and cause a narrowing of vessels. Atherosclerotic disease primarily results from deposits of fatty plaques on the inner lining (tunica intima) of blood vessels. This deposition of plaque generally occurs at the bifurcation of major arteries. Blood flow to the heart then becomes compromised and the blood encounters increased resistance to flow as the lumen narrows. If the disease process is allowed to continue, vascular changes occur and impair the ability of the diseased vessel to dilate. The myocardium may develop a lesion as a result of the decreased oxygen supply and increased demand.

Arteriosclerosis

Arteriosclerotic heart disease usually results from calcium deposits in the tunica media. These accumulate over a period of time and give rise to the condition commonly known as hardening of the arteries (arteriosclerosis).

Angina

Both atherosclerosis and arteriosclerosis can eventually lead to ischemia from decreased blood flow, which can lead in turn to angina pectoris. When attacks of angina occur, the client generally complains of pain in the chest or jaw, or down the left arm.

Angina is precipitated by activities, such as exercise, that increase the need for oxygen. It can be diagnosed earlier by stress electrocardiograms.

Angina can generally be relieved with a few minutes of rest or administration of nitroglycerin. If the attacks last longer than 30 to 45 minutes, death of heart tissue can occur, with irreversible cellular damage.

Myocardial Infarction

Actual damage to the heart tissue can result when a myocardial infarction occurs as the result of a major blockage or obstruction. Such damage can result in a necrotic or infarcted area of the myocardium, usually in the left ventricle, and is commonly known as a heart attack.

An infarction can also occur in the anteroseptal, anterolateral, and inferolateral sections of the myocardium. Infarctions involving over 40% of the myocardium are generally associated with cardiogenic shock (see Chap. 24).

The prognosis following an acute myocardial infarction is linked to the extent of the myocardial necrosis, as well as to the presence of saline edema and subsequent pulmonary edema. A classification system has been developed by Killip to categorize patients into four hemodynamic subsets according to physical assessments. This classification is important to the determination of both short-term care and long-term care as well as prognosis (Ignatavicius et al., 1995). The criteria that characterize Killup classes are listed below:

Class	Criteria
Class I	No signs of left ventricular failure
Class II	S_3 gallop and/or pulmonary congestion limited to basal lung segment
Class III	Acute pulmonary edema*
Class IV	Shock syndrome*

*Note that acute pulmonary edema is an acute fluid excess state; shock syndrome is cardiogenic shock state.

Mechanisms of Cardiac Failure

The mechanisms of cardiac failure are very complex and at present are not completely understood. The clinical syndrome of congestive heart failure presents as systemic hypoperfusion and tissue hypoxia (forward failure); and, as pulmonary and venous congestion (backward failure).

There are two basic theoretical models: the forward theory and the backward theory. The forward theory describes a condition in which the heart cannot pump out the amount of blood needed by the peripheral vascular system. There is a decrease in cardiac output, progressing into the low output syndrome seen in cardiogenic shock. Low cardiac output results in ischemia in all the periphery. Because one fourth of the cardiac output perfuses the kidneys, the kidneys are affected, producing a low renal output. The body's response to decreased cardiac output is to decrease circulation to the kidneys.

The backward theory describes a situation in which the heart is receiving more blood than it can pump out. The blood, therefore, backs away from the heart like water behind a dam. The blood can be trapped at two places. The increased pressure is generally reflected backward, and right-sided congestive heart failure results in systemic edema. This edema is manifest in enlarged neck veins, liver, abdomen, periphery, arms, and legs. In left-sided heart failure, backward pressure is reflected from the left atrium to pulmonary veins and leads to pulmonary congestion. As the pulmonary congestion increases, fluid will extrude into the interstitial spaces of the lungs, causing pulmonary edema. Pulmonary edema is the result of increased hydrostatic pressure in the pulmonary capillaries, commonly known as wedge pressure. Pulmonary edema can be demonstrated by rales and rhonchi in dependent areas of the chest and by fluid in the chest, which can be visualized on x-ray examination. Congestive heart failure is usually the result of both forward and backward mechanisms.

COMPENSATORY MECHANISMS

In congestive heart failure, all the organs of the body endeavor to counteract the effects of the failing pump. Some of the compensatory mechanisms become self-defeating. Figure 19–2 illustrates these self-defeating effects.

Cardiac Response

During congestive heart failure, the decreased pumping action leads to incomplete emptying of the ventricles; therefore less blood circulates through the lungs and less oxygenation takes place. The incomplete evacuation of

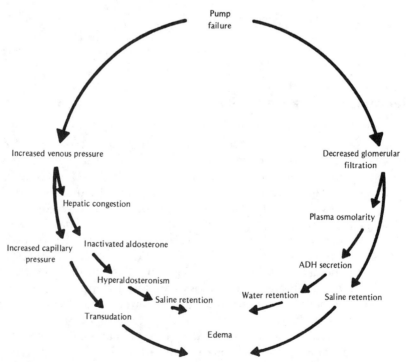

FIGURE 19–2
Schematic representation of the self-defeating effects of the compensatory mechanisms.

the ventricles and decreased arterial blood volume cause increased venous blood volume leading to pulmonary edema. While the inadequate emptying of the left heart is producing pulmonary edema, the back pressure of the right heart is causing systemic edema. Pulmonary edema may be recognized by signs of anoxia, such as paroxysmal nocturnal dyspnea (PND) and orthopnea, and also by early signs of anoxia to the brain and heart. Systemic edema can be recognized by engorged neck veins, liver congestion, and pedal and presacral edema. Edema in the periphery and abdomen is caused by an elevated hydrostatic pressure in the veins secondary to the increased venous blood volume. The elevated hydrostatic pressure pushes fluid out of the veins into the interstitial spaces. The cardiac compensatory response is to increase output by increasing the rate of contraction of the heart muscle. This is ultimately a self-defeating reaction because extremely rapid heart contractions do not allow sufficient time for the ventricles to fill adequately.

Adrenocortical Response

The anterior pituitary interprets decreased arterial blood pressure as a stress situation. The body cells indicate a need for more fluid (saline) to increase the arterial blood supply to the head. Through the negative feedback mechanism and corticotropin-releasing factor, the anterior pituitary is stimulated to release adrenocorticotropic hormone (ACTH), which in turn causes the adrenal cortex to secrete various steroids: glucocorticoids, mineralocorticoids, and androgens. The mineralocorticoids especially cause the retention of sodium and chloride and thus water. Retention of extracellular fluid actually compounds congestive heart failure by increasing the already excessive blood volume. It also adds all the problems inherent in an extracellular fluid excess.

An additional complication caused by the retention of sodium and chloride may be the loss of potassium through the urine.

Neural Response

The neural response, like the adrenocortical response, is triggered by the lowered arterial blood pressure. The lowered arterial blood pressure decreases the perfusion of the brain, stimulating the posterior pituitary to release antidiuretic hormone (ADH). ADH causes the distal tubules of the kidney to retain sodium and water in an attempt to increase the arterial blood pressure. This compensatory response is also self-defeating, because it increases the burden on the failing heart. This response also causes an increase in potassium elimination.

Renal Response

Decreased cardiac output and peripheral vasoconstriction decrease renal perfusion. In response to the stress hormones (glucocorticoids and mineralocorticoids), the kidney tubules retain sodium, chloride, and water as shown in Figure 19–3. The kidneys also release renin, a parahormone, to raise the blood pressure. Erythropoietin, another parahormone of the kid-

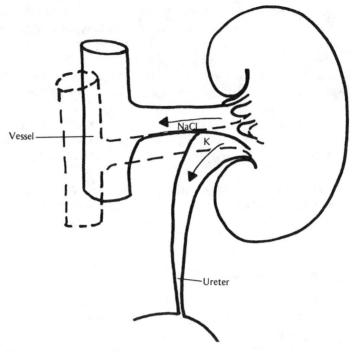

FIGURE 19-3
Schematic representation of electrolyte movement. Because of a stress response, sodium and chloride are reabsorbed into the bloodstream, while potassium is excreted.

neys, may be secreted, as part of the stress response. It increases the release of red blood cells from the bone marrow to carry more oxygen to the peripheral tissues.

Hepatic Response

The damming up of the venous blood flow as a result of peripheral vasoconstriction increases the amount of blood in the liver and causes congestion. Liver congestion accompanies right heart failure. Normally, the liver breaks down excess hormones (steroids, ADH); however, with congestion, this function fails and the excessive hormone levels cause the undesirable effects already listed when discussing the adrenal and renal responses.

Peripheral Vascular Response

The decreased cardiac output and arterial blood pressure eventually progress to cardiogenic shock. The sympathetic nervous system initiates the stress response by releasing adrenalin and noradrenalin, causing vasoconstriction of all major peripheral vessels, except the coronary vessels. Vasoconstriction decreases the amount of oxygen available to the body cells to carry on metabolism (aerobic metabolism). Without oxygen, the carbohydrates are catabolized only to the lactic acid stage and lactic acidosis ensues. Vasoconstriction increases arterial resistance and forces the heart to pump harder, thus aggravating the pump failure. Vasoconstriction also causes increased venous return to the heart, adding to the already existing overload. Again, this mechanism is actually self-defeating.

A further complication of the peripheral vascular response is caused by the increased level of adrenalin, which in addition to stimulating vasoconstriction, also directly stimulates the heart. Because the heart cannot respond to this stimulus with effective beats, the increased adrenalin may result in further dysrhythmias.

Respiratory Response

The compensatory action of the respiratory system responds to two stimuli: adrenalin in the blood and lactic acidosis. Adrenalin, which is produced by the sympathetic nervous system and the adrenal medulla, causes the bronchioles of the lungs to dilate and increases the rate of breathing. In lactic acidosis, which is the result of an inadequate amount of oxygen to metabolize glucose, the bronchioles dilate and the rate of breathing increases in an attempt to increase the amount of available oxygen to alleviate the lactic acidosis and to convert glucose into the energy so desperately needed by the body.

FLUID, ELECTROLYTE, AND ACID-BASE IMBALANCES

Congestive heart failure can result in a number of fluid, electrolyte, and acid-base imbalances, including fluid overload, hypokalemia, and metabolic acidosis.

Fluid Overload

Several factors combine to create the extracellular fluid excess found in congestive heart failure. As the heart fails, renal perfusion is decreased, causing the kidney tubules to retain sodium, chloride, and water, as well as nitrogenous waste products. The stimulated stress response results in the production of steroids (especially mineralocorticoids), which also cause saline retention.

Right heart failure, which frequently causes backward pressure, further aggravates the situation by increasing the venous blood pressure. As venous pressure builds up, it elevates the hydrostatic pressure in the plasma, which pushes blood into the interstitial spaces.

These combined factors of extracellular fluid excess can lead to abdominal distention from liver congestion (right heart failure), ascites (third spacing), and pulmonary edema (left heart failure).

The edema, whether it be central or peripheral, results from the change in the proportion of saline fluid, that is, movement from the plasma to the interstitial fluid spaces. Treatment, then, is directed to improvement of cardiac and renal output. Additional attention must be directed to the hypertensive state, to lowering blood cholesterol, and attention to the lipoproteins. If the patient is concomitantly anemic, the edematous state is worsened as is the oxygenation state. The hematocrit and hemoglobin are reduced because of the extracellular fluid excess.

Hypokalemia

Hypokalemia, like extracellular fluid excess, is caused by several interrelated factors. When the kidneys are stimulated by the steroids secreted in response to stress, they retain sodium and chloride and excrete potassium. Vomiting and diarrhea, which also accompany the stress response, further deplete the body's potassium. A potassium deficit is compounded by the administration of potent diuretics and digitalis preparations. The deficit can become so acute that it results in digitalis intoxication.

A potassium deficit may exist as a result of the conditions listed above, and yet the serum potassium reading may be elevated because of the presence of lactic acidosis. As the acidosis is corrected, hydrogen ions move out of the cell and potassium ions move back into the cell, magnifying the extracellular potassium deficit.

If congestive heart failure becomes more severe and oliguria develops, the low potassium level will rapidly increase. In this scenario, hyperkalemia is the ultimate result of renal failure and metabolic acidosis.

Metabolic Acidosis

A chain of events leads to this imbalance: decreased cardiac output causes decreased renal perfusion; this results in the retention of nitrogenous waste products, which increases the retention of hydrogen ions. Hypoxia causes anaerobic metabolism of carbohydrates, which increases the production of lactic acid and results in metabolic acidosis. Metabolic acidosis is more frequently seen with a cardiopulmonary arrest.

ASSESSMENT

Heart failure can be either chronic or acute. *Chronic* heart failure occurs when injury or decompensation takes place slowly. The symptoms develop over a period of days, weeks, or months. The cardinal signs are fluid congestion in the peripheral tissues and hypoxia caused by the diminished blood flow to the tissues as a result of pump failure. The most common *early* symptoms are dyspnea and fatigue.

In *acute* heart failure, there is a sudden loss of effective cardiac contractions and thus a reduced cardiac output. Acute heart failure results in an engorgement of the lungs that presents as acute pulmonary edema. When fluid enters the lungs, the situation becomes a medical emergency. Diagnosis is confirmed with radiographic evidence of cardiomegaly and pulmonary vascular redistribution.

Physical Assessment

Physical findings vary depending on whether the left or right side of the heart is affected.

Left Heart Failure

Both ventricles may fail simultaneously or independently; however, left ventricular failure is more common. Left heart failure is predominantly a backward failure in which the blood "dams" back into the lungs, causing acute pulmonary edema. Pulmonary edema is a severe sign of fluid excess manifested in lung tissue. The extra blood in the pulmonary capillaries may escape into the alveoli, causing the patient to drown in his or her own fluids. Fluid in the main bronchi causes congestion and produces mucus and a productive cough, possibly with blood-tinged sputum. The blood-tinged mucus is caused by the rupture of distended capillaries.

Cheyne-Stokes respirations may occur. These deep, irregular respirations are followed by intermittent periods of apnea. On examination of the apical pulse, the point of maximal impulse (PMI) is displaced to the left and downward, as the heart dilates and hypertrophies. If pleural effusion is also present, the nurse notes dullness on percussion. Moist crackles can be heard as air flows through the fluid in the alveoli.

The most common symptom is dyspnea even at rest, including orthopnea. The patient may awaken at night with a type of dyspnea known as acute paroxysmal nocturnal dyspnea, a sign of acute pulmonary edema. Dyspnea is an action initiated by the dilated pulmonary vessels. This symptom is frequently coupled with a cough producing white, frothy mucus. The reduced size of the respiratory air sacs causes inadequate oxygenation of the blood, further aggravating the problem.

Right Heart Failure

Right heart failure is usually a backward failure causing systemic venous congestion. Right heart failure manifests in the periphery, with edema occurring systemically. This is noted by jugular vein distention, peripheral edema, and ascites. The spleen and liver enlargement cause tenderness in the area on palpation. Peripheral edema is usually of the pitting type. When the liver is compressed, a hepatic jugular reflex may be noted.

Although it is important to distinguish between the types of heart failure, it is more critical for the nurse to recognize the degree and the rate of congestive heart failure than to be able to distinguish which side of the heart is failing. Table 19–2 gives signs and symptoms elicited by altered function.

Diagnostic Assessment

The narrowing of the lumen and the increase in oxygen demand (greater than the diseased vessels can supply to the myocardium) may cause a localized heart ischemia. This generally occurs in the lower part of the left ventricle and can be verified by a stress electrocardiogram, showing inverted T waves and ST segment depression.

Laboratory tests of cardiac, renal, and hepatic functions are needed to evaluate a patient with congestive heart failure. Analyzed together, these test results, listed in Table 19–3, indicate the degree or confirm the presence of congestive heart failure or related heart problems.

Other major diagnostic tests may include radiography, echocardiography, and cardiac catheterization. Associated laboratory abnormalities include elevation of blood urea nitrogen (BUN) and creatinine levels, hyponatremia (from fluid overload), hypohemoglobinemia (from fluid overload),

TABLE 19-2
SIGNS AND SYMPTOMS INDICATIVE OF ALTERED FUNCTIONS

Signs and Symptoms	Altered Functions
Dyspnea, shortness of breath with or without exertion, paroxysmal nocturnal dyspnea, orthopnea, Cheyne-Stokes respiration	Pulmonary congestion with fluid from left-sided failure; pulmonary capillary pressure increases, lessening available space to diffuse gases, leading to hypoxemia, hypercapnia, and hemoptysis; increased pulmonary capillary pressure pushes blood into alveoli leading to foamy, blood-tinged sputum; increased pulmonary congestion heightens work of the right ventricle leading to failure
Apprehension, irritability, restlessness, sensorium decreased to stupor, coma, cough reflex	Dyspnea leads to less oxygen to central nervous system; nerve tissue unable to function without oxygen and nutrients; increased fluid in lungs stimulates cough reflex
Systemic edema, pitting edema, liver congestion, ascites, sacral edema when recumbent, neck vein distention, increased central venous pressure	Retention of fluid in interstitial spaces secondary to increased hydrostatic pressure in capillaries, resulting from increased venous pressure due to retention of saline; capillary pressure high and tissue pressure low in veins draining upper and lower parts of body
Anorexia, nausea, vomiting, diarrhea, or constipation	Decreased blood flow to GI system in *stress* response; vagal reflexes initiated by cardiac changes also stimulate GI nerves; reflex vomiting possibly due to hypoxia of brain tissue; may be decreased or increased peristalsis in stress
Fever, diaphoresis	Sympathetic nervous system (*stress* response) stimulated to increase heart rate to augment cardiac output; speeds metabolism
Palpitation, tachycardia, gallop rhythm, pulsus alternans	Change in heart rhythm related to hypoxia of the myocardium
Oliguria	Decreased cardiac output leads to inadequate renal perfusion; retention of nitrogenous waste products and hydrogen ions results in acidosis

and elevated liver enzymes, such as aspartate transferase and alanine transferase (right heart failure). Increased right-sided pressure is measured by central venous pressure (CVP) or hemodynamic monitoring.

COLLABORATIVE MANAGEMENT

The major goals in the treatment of cardiac failure are to increase the contractility of the heart muscles (the pump action) and to decrease the peripheral demands on the heart for oxygen and nutrients until the pump action is restored.

TABLE 19–3
RESULTS OF LABORATORY TESTS BY FUNCTION

Function	Laboratory Results
Cardiac	Evaluate electrical impulses (ECG) for dysrhythmias; x-ray for size and position of heart and great vessels; cardiac output decreased; circulatory time increased; venous pressure increased (CVP normal = 6–12 cm of water); CPK increased with damage to the myocardium
Renal	BUN increased with prerenal failure or renal failure from decreased perfusion; creatinine increases after BUN; urine osmolarity increased; high specific gravity; serum osmolarity decreased; proteinuria; serum Na decreased (ICF excess); serum hemoglobin and hematocrit decreased (ECF excess)
Hepatic	AST, ALT lactic dehydrogenase (LDH), and alkaline phosphatase increased due to decreased liver flow; liver hypoxia; slight increase of bilirubin; hypoxia results in failure to synthesize albumin, increasing the edematous state because of decreasing colloid osmotic pressure in blood; some prolongation of prothombin time due to failure to synthesize; hypoglycemia; inadequate storage of glycogen in liver and increased catabolism of glucose to lactic acid with hypoxia; serum CO_2 decreased, pH decreased with lactic metabolic acidosis

CVP, central venous pressure; CPK, creatinine phosphokinase; BUN, blood urea nitrogen; ICF, intracellular fluid; ECF, extracellular fluid; AST, aspartate transferase; ALT, alanine transferase.

Medical Treatment

The primary treatment for congestive heart failure consists of the "three Ds": digitalis, diuretics, and diet. Cardiotonic drugs (glycosides) are the most frequently used means of improving the pump action, with digitalis still being the primary mode of therapy. Rest and limited activity during the acute phase decrease the workload of the heart. Aminophylline decreases bronchospasm, increases cardiac output, and promotes diuresis. Morphine lessens anxiety and pain; it also decreases dyspnea because it lessens the myocardial demand for oxygen. Oxygen by nasal cannula is usually needed. Diuretics, especially furosemide (Lasix), administered by intravenous (IV) push or orally, are used to decrease the amount of fluid circulating through the heart and enhance renal output. Anticoagulants may be used to decrease clot formation.

If acute pulmonary edema occurs as a result of left heart failure, the patient is usually placed on a ventilator as well. Some hospitals still advocate the use of rotating tourniquets to decrease blood return to the heart. In this procedure, tourniquets (blood pressure cuffs) are placed on three of four extremities at the same time. Every 15 minutes, the tourniquets are rotated to include one new extremity. Distal circulation must be checked carefully to avoid ischemia.

If cardiogenic shock ensues, primarily after a myocardial infarction, treatment may incorporate the use of vasopressors. Forward failure, which is cardiogenic shock, is the least understood aspect of the shock syndromes. This type of shock results in inadequate tissue perfusion. Mortality, even with treatment, is 80% or more. Pacing may be required if a bradycardia or heart block develops with the hypoxia.

Key treatment of heart failure following a myocardial infarction includes reducing the fluid load, usually with diuretics. Digitalization is appropriate with cardiomegaly and S_3 gallop existing, the latter indicative of a fluid load. Afterload can be reduced with angiotensin-converting enzyme (ACE) inhibitors. ACE inhibitors are used to decrease the levels of angiotensin II (for the hypertension) and aldosterone (to lower fluid afterload). Kidney function must be monitored with use of these drugs because they can trigger hyperkalemia, especially in clients with diabetes, renal disease and the concomitant congestive heart failure. Also, note that ACE inhibitors should not be given concurrently with potassium-sparing diuretics because this would enhance hyperkalemia.

Dietary therapy is also important in treating congestive heart failure (CHF). Dietary treatment for CHF specifically includes a low-sodium diet and attention to hypokalemic states secondary to poor nutrition, digitalis, or diuretics (Poleman & Peckenpaugh, 1991).

Salt and foods high in sodium increase the propensity to accumulate fluid (i.e., develop extracellular volume excess), especially when renal perfusion is compromised. Thus, a sodium-restricted diet is recommended to reduce pulmonary and peripheral edema, both of which have negative consequences. Pulmonary edema decreases the oxygenation of all tissues and increases the risk of lactic acidosis. Peripheral (systemic) edema decreases the cellular metabolic rate and leads to tissue breakdown.

A sodium-restricted diet is a normal, adequate diet that has been modified by decreasing the sodium content so that daily intake is within the range of 250 to 2000 mg per day. The usual intake in the United States is much too high, ranging from 3000 to 7000 mg per day. The diet regimes are classified as having strict (500 mg or less), moderate (1000 mg), and mild (2000 to 4000 mg) restriction. Salt (sodium chloride) is nearly half sodium by weight (about 40%), and a teaspoon of salt contains 2100 mg of sodium. Thus, in congestive heart failure, the use of table salt should be entirely eliminated; in addition the intake of foods with high sodium content should be decreased. Persons are encouraged to gradually adapt to the lessened salt intake by using other materials to enhance food appeal, for example, spices, herbs, lemon juice, or vinegar.

Sodium restriction improves the effectiveness of diuretic therapy, but it also tends to increase the loss of potassium. Potassium can be replaced parenterally after evaluating the pH state and assessing serum potassium levels. Oral replacement of potassium relies primarily on fruits and vegetables as dietary sources. Fruits, such as bananas and oranges, are good sources. Potatoes, legumes, and fresh vegetables are appropriate. (See Chap. 4 for a discussion of hypokalemia.)

Nursing Interventions

Table 19–4 lists the nursing responsibilities associated with the care of a patient with congestive heart failure, and provides the reasons for their implementation. The treatment plan must be followed closely and adapted as necessary. It is especially crucial to note subtle changes in edema, respira-

TABLE 19–4
NURSING RESPONSIBILITIES AND THEIR RATIONALE

Responsibility	Rationale
Monitor cardiac patterns	Identifies rhythm changes secondary to hypoxia
Start keep vein open (KVO) intravenous fluid according to institutional policy*	May be needed for emergency drugs
Administer oxygen as directed	Lessens hypoxia and metabolic lactic acidosis
Provide emotional support	Decreases tension, thus decreases oxygen need; avoids vicious cycle of tension and hypoxia
Observe and record vital signs and report changes	Indicate fluid overload; cardiogenic shock; dysrhythmias
Some sources do not recommend taking temperatures rectally even if oxygen is being administered; axillary temperature may be acceptable.	Thermometer may stimulate the vagus nerve in the rectal sphincter causing or aggravating dysrhythmias.
Have emergency drugs readily available according to *institutional policy*	Administer drugs for crisis
Lidocaine bolus and/or drip	Used commonly for dysrhythmias
Atropine	Used commonly for bradycardia
Sodium bicarbonate	Used commonly for metabolic acidosis due to cardiac arrest
Administer drugs as ordered; observe for side effects of diuretics and digitalis (indirect diuretic)	Diuretics may cause potassium deficit
Elevate head of bed slightly unless patient is in shock	Decreases venous return; lowers abdominal pressure; increases lung space, i.e., ventilation; facilitates respiration
Suction orally as needed	Clears airway to aid ventilation and prevent hypoxia
Note and chart the appearance of skin for moistness, dryness, coolness	Changes may indicate further stress response
Observe for signs of increased edema; obtain daily weight	Increases may be indicative of further fluid retention
Chart fluid intake and output	Assesses adequate perfusion of kidneys
Restrict fluids as ordered	Decreases fluid volume
Restrict sodium in diet as ordered	Decreases fluid volume caused by saline excess
Assure rest according to the physician's orders. Guidelines (unless specified otherwise):	Lessens the workload and the burning of nutrients in decreased oxygenated state
Do *not* allow patient to pull self up in bed, strain on bedpan or commode, cross legs, or massage limbs	
Do actively move the patient's arms and legs gently several times daily and assist patient to turn	Maintains muscle strength; discourages formation of peripheral emboli
Rotate tourniquets	Helps relieve skin breakdown secondary to anoxia
Initiate rehabilitative teaching relative to circumstances	Assists patient and family to cope with change in body image and/or lifestyle
Identify possible social agency referrals pertinent to patient/family	Uses agencies that have services available to aid the patient and family in time of need

*These are suggested procedures that vary with the institution; standing orders are recommended.

tion, consciousness, apprehension, urine output, and circulatory status. Complications such as peripheral emboli must be prevented. The major treatment objectives for the patient with congestive heart failure are always to reduce the workload of the heart and to increase circulation to the periphery to keep the body properly oxygenated.

REFERENCES AND READINGS

Braunwald, E (1992). Heart Disease: A Textbook of Cardiovascular Medicine. Philadelphia: WB Saunders.

Brown, KK (1993). Boosting the failing heart with inotropic drugs. Nursing93 23(4), 34–43.

Galvao, M (1990). Role of angiotensin-converting enzyme inhibitors in congestive heart failure. Heart and Lung 5, 505–511.

Hartshorn, J, Lanborn, M, & Noll, ML (1993). Introduction to Critical Care Nursing. Philadelphia: WB Saunders.

Ignatavicius, DD, Workman, ML, & Mishler, MA (1995). Medical-Surgical Nursing: A Nursing Process Approach, ed 2. Philadelphia: WB Saunders.

Kelly, DP, & Fry, ET (1993). Heart Failure. In M Woodley & A Whelan (eds), Manual of Medical Therapeutics (p 105) Boston: Little, Brown.

Letterer, RA, et al (1992). Learning to live with congestive heart failure. Nursing92 22(5), 34–41.

Poleman, CM, & Peckenpaugh, NJ (1991). Nutrition: Essentials and Diet Therapy (6th ed). Philadelphia: WB Saunders.

Ramsey, PG, & Larson, EB (1993). Medical Therapeutics. Philadelphia: WB Saunders.

Schwertz, DW, & Piano, MR (1990). New inotropic drugs for treatment of heart failure. Cardiovascular Nursing 26(2), 7–12.

Smith, DF, & Bumann, R (1993). Assessing and treating decreased cardiac output. MEDSURG Nursing 2, 351–357.

Yakabowich, M (1992). What you should know about administering nitrates. Nursing92 22(9), 53–55.

CHAPTER 20

· · · · · · · ·
· · · · · · · ·
· · · · · · · ·
· · · · · · · ·
· · · · · · · ·
· · · · · · · ·
· · · · · · · ·
· · · · · · · ·
· · · · · · · ·
· · · · · · · ·

Endocrine Disorders

THE ENDOCRINE SYSTEM

The endocrine glands manufacture chemicals called hormones, which are secreted into the bloodstream in low concentrations and which cause marked metabolic and biochemical effects on various tissues. As a group, hormones are of three types: proteins, amines, and steroids.

The endocrine system responds to the neurologic system in maintaining body metabolism. The neurologic system receives information from internal as well as external stimuli; this information is transmitted to the hypothalamus.

The hypothalamus is the chief intermediary between the neurologic and endocrine systems and is responsible for the production of chemicals known as corticotropin-releasing factors (CRF). These corticotropin-releasing factors stimulate the pituitary gland (hypophysis) to produce specific hormones known as master gland hormones, and these, in turn, affect target glands throughout the body.

The pituitary gland, referred to as the master gland, is located in the sella turcica and is composed of anterior and posterior lobes. The target glands are the thyroid, parathyroids, adrenals, islands of Langerhans of the pancreas, and the gonads. There are other structures that have an endocrine function, such as the gastrointestinal mucosa and the placenta. Other organs that are specifically influenced by the endocrine system are the liver, kidneys, heart, and thymus, but all body cells are influenced by endocrine hormones.

Control

Hormone production is usually controlled by a negative feedback mechanism in which a specific plasma hormone level is assessed by the pituitary gland. The pituitary gland responds by secreting a specific master hormone. The master hormone, released into the bloodstream, stimulates a target gland to produce the target hormone. An example of this mechanism is the release of thyroid-stimulating hormone (TSH) by the pituitary gland as the amount of thyroxin (T_4) is decreased. When TSH increases, this causes the thyroid to increase production of T_4. As the amount of T_4 decreases, the pituitary gland secretes more TSH. Not all negative feedback mechanisms are directly related to hormone-to-hormone action.

The feedback mechanism may also be a chemical-to-hormone action. An example of this is the calcium and parathyroid hormone relationship. As the calcium level drops in the blood, the parathyroid glands will produce parathyroid hormone, which in turn increases the level of calcium in the blood.

Functions

The principal hormones from the pituitary gland are designated as anterior and posterior pituitary hormones. The anterior pituitary hormones, based on the portion of the gland producing them, are luteinizing (LH), adrenocorticotropic (ACTH), somatotropic (SH), thyroid-stimulating (TSH), luteotropic (LTH), and follicle-stimulating (FSH). Luteotropic hormone is also known as prolactin. The posterior pituitary produces two basic hormones, the antidiuretic hormone, also known as vasopressin, and oxytocin.

Target tissues affected by these pituitary hormones are thyroid, parathyroid, pancreas, adrenal, kidney, and uterus, and vessels and other cells of the body.

The endocrine system is primarily responsible for maintaining and regulating vital functions including:

1. Response to stress and injury. This is controlled primarily by adrenocorticotropic hormone which targets the adrenal gland.
2. Growth and development, which are primarily accomplished by somatotropic hormone, which affects all cells of the body. Thyroid-stimulating hormone affects the thyroid gland, which stimulates general body metabolism and growth.
3. Reproduction, which is controlled by follicle-stimulating hormone, luteinizing hormone, and luteotropic hormone in the female, and interstitial cell hormone in the male. Oxytocin stimulates smooth muscle of uterus in childbirth.
4. Ionic homeostasis, which involves fluid and electrolyte balance.
5. Energy metabolism. Thyroid hormones affect energy metabolism in general, while insulin and glucagon play roles in specific metabolic processes.

Without a normally functioning endocrine system, one would fail to grow, reach maturity, or reproduce. The hypothalamic-pituitary axis is chiefly responsible for growth, maturity, and fertility. The hypothalamic-pituitary adrenal axis is primarily responsible for the maintenance of ionic homeostasis and the compensatory stress mechanism. This axis maintains a normal internal environment in which all cells are in constant contact with the internal environment through the balance of sodium, potassium, water, and acid base. Adrenocorticotropic hormone, mineralocorticoids, and anti-diuretic hormones are the specific hormones responsible for these functions. The concentrations of these fluids and electrolytes are also controlled by endocrine functions that are based upon a biochemical reaction between the cells and neuroactivation of muscle cells. The parathyroids are responsible for regulating calcium and phosphate metabolism. The endocrine system is also the principal regulator of energy metabolism, primarily through the thyroid hormones. The thyroid-stimulating hormone stimulates the thyroid gland to produce thyroxin (T_4) and triiodothyronine (T_3). In addition to thyroid control of metabolism, nutrients are also made available to all cells through the integrated action of gastrointestinal and pancreatic hormones.

SELECTED DISORDERS

Metabolic diseases are essentially of two types: excess and deficit. Excess is usually the result of an overabundant secretion of the hormones, which causes specific pathologic effects in tissue. A deficit can usually decrease the normal metabolic responses of the hormones.

Adrenal Disorders

The adrenal glands are divided into two parts: the cortex and the medulla. The adrenal cortex produces three groups of hormones: glucocorticoids, mineralocorticoids, and sex hormones (androgens and estrogens). All of these hormones are steroids. The glucocorticoids (e.g., cortisol) are responsible primarily for converting starches, fats, and proteins into glucose, in that order. The mineralocorticoids (e.g., aldosterone) are responsible for helping to maintain blood pressure by retaining sodium, chloride, and water. They also cause a loss of potassium. The mineralocorticoids act on the distal tubules of the kidneys. The androgenic hormones are not significant unless an abnormality exists. Normally, all three groups of adrenal hormones are secreted in response to stress or an alteration in function.

All the steroid hormones are stimulated by adrenocorticotropic hormone (ACTH), which is secreted by the anterior pituitary.

Pathophysiology

Overproduction or underproduction of hormones by the adrenal glands will result in electrolyte and pH imbalances. The following overview classifies adrenal problems as being related to either underproduction (hypoadrenalism) or overproduction (hyperadrenalism). Conditions involving underproduction of hormones include acute failure with adrenal crisis, as in the Waterhouse-Friderichsen syndrome (meningococcal infection); chronic primary deficiency (Addison's disease); chronic secondary deficiency (pituitary hypofunction); and hypoaldosteronism (lack of aldosterone). Conditions involving overproduction of hormones or their metabolites include Cushing's syndrome and hyperaldosteronism.

Adrenal Crisis

Adrenal crisis is usually a complication of a chronic deficiency and results when the adrenal glands are unable to cope with an increased stress situation superimposed on the chronic deficiency. Increased stress can arise from infection, colds, drugs, trauma, hemorrhage, or surgery. Any kind of stress requires increased secretion of hormones. In hypofunction, sufficient mineralocorticoids and glucocorticoids are not available to be secreted, even with increased stimulation. Crisis may also be induced by adrenalectomy or by a too-rapid withdrawal from prolonged treatment with steroids.

Assessment

The more common signs and symptoms include a slow development of weakness, progressing to vascular insufficiency. Increased stress from a meningococcal infection may cause a severe hyperthermia to progress to hypothermia, coma, and shock. This type of crisis develops as a result of a decrease in blood volume caused by failure to retain sodium, chloride, and water. Thus, it actually leads to an extracellular volume deficit.

Laboratory tests show a reduced plasma cortisol level (cortisone) and a reduced aldosterone level because less hormones are produced. The urine should also be checked for the presence of electrolytes. The ratio of sodium to potassium in the urine is reduced during hypoadrenalism.

Collaborative Management

Treatment includes intravenous (IV) administration of 100 mg of the hormones cortisol phosphate or succinate, repeated in doses of 50 mg every 6 hours. Oral steroid replacement may follow IV replacement. Treatment of shock, electrolyte deficits, and hypoglycemia may also be necessary. These problems may be counteracted by glucose, saline, and vasopressors, respectively. If infection is the cause of the adrenal crisis, it must also be treated.

Chronic Adrenal Insufficiency (Addison's Disease)

Chronic adrenal deficiency, or Addison's disease, is a disease of the adrenal cortex that results in a deficiency in the amount of corticosteroids produced over an extended period.

Causative Factors

Chronic adrenal deficiency, by strict definition, includes diseases of hypofunction restricted to the spontaneous and slower secretion of adrenal hormones. Hypofunction may be the result of destruction of the cortex by cancer metastasis, tuberculosis, fungus, infection, or atrophy of unknown origin. Chronic deficiency does not usually manifest itself until 90% of the cortex is gone. A superimposed stress will produce temporary symptoms because of the increased need for the scantily available hormones. The disease is more common in adults and occurs predominantly in women. A possible hereditary tendency has been suggested.

Assessment

The cardinal symptoms of Addison's disease are increased pigmentation of the skin, physical and mental asthenia (easy fatigability), hypotension, and gastrointestinal disorders.

Increased pigmentation of the skin is caused by excess adrenocorticotropic hormone (ACTH) and melanocyte-stimulating hormone (MSH). Darkening of freckles and milk-white patches of skin surrounded by normal pigmentation are apparent. The skin may first have a dingy, smoky appearance that eventually progresses to a dark amber or bronze. Pigmentation changes are most marked in regions exposed to light and friction. The skin acquires the consistency of leather, and body hair usually decreases. These skin changes are seen only in chronic, not acute, deficiency. Table 20–1 categorizes by body system the basic signs and symptoms of hypoadrenalism.

Carbohydrate metabolism is affected by chronic adrenal insufficiency and results in a decreased amount of glycogen stored in the liver. The patient has fasting hypoglycemia, occasionally associated with diabetes mellitus.

Laboratory tests show decreased sodium and chloride levels in the blood, with an increased blood potassium level. When deprived of salt in the diet, patients with Addison's disease lose sodium and chloride in the urine, because the steroids needed for retention are not being adequately produced. Thus, the amount of sodium and chloride in the urine is increased, and the urine potassium is decreased.

Metabolic acidosis is the pH imbalance associated with an increased blood potassium level. There is a correlation between increased BUN and a decreased extracellular fluid volume. Fasting hypoglycemia may be present because of the decreased glucocorticoids. Urinary studies may show decreased amounts of 17-ketosteroids, corticosteroids, and aldosterone.

When the production of steroids is increased, the number of eosinophils, a type of white blood cell, decreases. When ACTH is given to a

TABLE 20–1

SIGNS AND SYMPTOMS OF CHRONIC PRIMARY ADRENAL INSUFFICIENCY (HYPOADRENALISM)

System	Signs and Symptoms
Integumentary	Increased pigmentation due to excess ACTH and melanocyte-stimulating hormone (MSH)
Central nervous	Personality disturbances such as listlessness, apathy, depression, irritability, and negativism, possibly progressing to psychosis; other possible symptoms include neuralgia, headache, vertigo, tinnitus, insomnia
Cardiovascular	Systolic/diastolic hypotension due to decrease in blood volume; small heart; orthostatic hypotension; circulatory collapse in crisis
Gastrointestinal	Anorexia; nausea, vomiting, diarrhea, occasional steatorrhea; weight loss; sensitivity to salt, and sweet, bitter, and sour tastes
Reproductive	Decreased libido; menses usually normal

person with Addison's disease, the usual drop in the eosinophil count will not occur because the adrenal cortex cannot produce the hormones normally released in response to ACTH stimulation. Because patients with Addison's disease have delayed excretion of water loads, large volumes of water may be given as a diuretic test of adrenal function. The amounts of steroids, especially plasma cortisol, in the serum are decreased.

Collaborative Management

Treatment includes 25 to 50 mg per day of steroids (prednisone). When stress is increased, the daily dose of steroids must be increased. Diets high in sodium (up to 10 g daily) may be needed.

Secondary chronic adrenocortical deficiency is related most commonly to the administration of steroids over long periods of time for the treatment of other illnesses, such as arthritis or bronchitis. When steroids are administered for an extended period, the adrenal cortex atrophies. Thus, when stress is encountered, the adrenals cannot produce enough hormones. For this reason, the steroid dosage must be increased in stress situations.

Nursing interventions include the following:

- Identify patients who are taking steroids; report these to the physician.
- Observe for negative effects of steroid administration, such as electrolyte imbalances.
- Observe for responses to stress and for symptoms that show an increased need for hormones (such as symptoms of hypovolemia and/or hypoglycemia proceeding to shock).
- Assist in regulating a slow withdrawal from steroid drugs.
- Observe for and report signs and symptoms that indicate ulcer formation (steroids increase the production of hydrochloric acid).
- Perform meticulous skin care.
- Observe for symptoms of increasing loss of electrolytes.
- Administer intravenous steroids carefully and oral steroids with milk or antacid.

Adrenal Hyperfunction (Cushing's Syndrome)

Adrenal hyperfunction, or Cushing's syndrome, is the overproduction of hydrocortisone or other steroids.

Causative Factors

Causes of Cushing's syndrome fall into two major categories. In the first, the cause is found directly in the adrenal glands. Such problems as adenomas and carcinomas of the adrenal glands are examples of direct adrenal malfunction. Hyperplasia of the adrenals, also a direct malfunction, is caused by overactivity of the hypothalamus, which in turn increases ACTH production. In the second category, the cause is the overproduction of ACTH despite an increased level of plasma cortisol. In other words, the negative feedback mechanism is not working. This is particularly evident in pituitary adenomas or nonendocrine tumors of the ACTH-secreting pituitary.

Assessment

The major symptoms of adrenal hyperfunction are related to the increased amount of glucocorticoids and mineralocorticoids in the blood. Changes in the skin are frequently seen. Loss of elasticity results in fragility of the blood vessels and the development of purplish streaks on the skin. Patients bruise easily, with purple to black ecchymoses. The face is usually flushed and oily. Patients with Cushing's syndrome usually develop excessive body and facial hair (hirsutism) and a thinning of scalp hair. This is caused by an excess of androgens. Another effect of excessive androgens in women is acne. Table 20–2 lists by body system the basic signs and symptoms of hyperadrenalism.

The serum hormonal level of steroids is increased. A urinalysis will show increased levels of 17-hydroxycorticoids and 17-ketosteroids.

If, following administration of ACTH, a serum hormonal test indicates increased hormonal production, hyperfunction exists. Aldosterone levels may be normal or elevated. Androgens are increased; estrogens may also be increased.

Collaborative Management

In some cases, the patient with adrenal hyperfunction undergoes spontaneous remission; however, this is rare. If the problem is caused by a tumor, surgery to remove the tumor is indicated. In cases of hyperplasia, a bilateral adrenalectomy may be necessary. The adrenals may be irradiated to decrease the amount of adrenal tissue producing hormones. Suppressive drugs may also be used. With any of these treatments, hormone production may drop too low.

Nursing interventions include the following:

- Monitor and report signs and symptoms suggestive of extracellular volume excess (hypertension, edema, breathing difficulty).
- Report symptoms suggestive of hyperglycemia (lethargy, nausea, vomiting, diarrhea, polyuria, excessive thirst).

TABLE 20–2
SIGNS AND SYMPTOMS OF OVERPRODUCTION OF ADRENAL HORMONES

System	Signs and Symptoms
Central nervous	Emotional changes including depression, anxiety, irritability, apathy
Cardiovascular	Hypertension; enlarged heart progressing to congestive heart failure and terminating in an edematous state (sodium retention and potassium loss)
Gastrointestinal	Ulcers due to increased pepsin and hydrochloric acid; hyperglycemia
Reproductive	Loss of libido; impotence; absence of menses
Musculoskeletal	Heavy trunk, thin extremities, moon face, buffalo hump (kyphosis of the spine), fat pads under cheeks, bulging eyeballs, eventual development of stretch streaks on hips and shoulders

- Remove excess oils from skin.
- Initiate ulcer prevention measures by the use of drugs or milk products according to physician's directions.
- Provide emotional support, recognizing personality changes.
- Observe closely for shock after adrenalectomy (the sympathetic nervous system will replace the adrenaline lost by the adrenalectomy).

Thyroid Disorders

Thyroid disorders are usually of two types: hypersecretion and hyposecretion. Hypersecretion [known as hyperthyroidism (Grave's disease)] is usually caused by excessive amounts of thyroxin (T_4), although one can have excessive amounts of both T_4 and T_3 (triiodothyronine). Occurrences are caused by thyroiditis or thyroidism; hypersecretion; and autoimmune diseases. Signs of these disorders are usually associated with increased metabolic rates: tachycardia, tachypnea, excessive weight loss, inability to maintain normal weight even when eating heavily, nervousness, irritability, and headaches. A certain percentage of patients with hyperthyroidism also exhibit exophthalmos, in which the globe of the eyeball protrudes.

Treatment is generally resection of the gland or use of drugs, such as Lugol's solution, to decrease the secretions of thyroxin. On occasion radiation is used.

Hypothyroidism (hyposecretion of thyroxin) generally results in slowed metabolic processes exhibited by lethargy, fatigue, mental disturbances (usually depression), and anxiety. Also noted are hair loss, dry skin, brittle nails, and menstrual irregularities, including amenorrhea. Decrease or loss of libido can also be present. Blood pressure is generally reduced, and pulse and respirations are slowed. A general appearance of neurasthenia characterizes hypothyroidism.

The disorder can be caused from decreased production of hormone or hyposecretion of the hormone resulting from pressure from a tumor, such as adenocarcinoma. In geographic areas of iodine deficiency or when the diet is deficient in iodine, goiters may be the cause of hypothyroidism.

Symptomatology includes retention of fluid, a slight tendency toward edema, particularly pedal edema, and loss of potassium, possibly aggravating complaints of fatigue, excessive sleepiness, and withdrawal from social environment.

Treatment is usually replacement of thyroid hormones on a daily basis.

Parathyroid Disorders

Symptoms of hyperparathyroidism are primarily caused by hypercalcemia, which will result in neurologic changes. Parathyroid adenomas cause primary hyperparathyroidism with increased serum calcium levels. These adenomas are often asymptomatic; but some adenomas, with calcium levels up to 13 or 14 mg/dL or more, may cause polyuria, polydipsia, and mental symptoms such as lethargy and mental obtundation. Increased serum calcium is a symptomatic condition that needs to be treated with parathyroidectomy, if it is caused by hyperparathyroidism, or in an acute situation with normal saline and furosemide (Lasix), which cause a calcium diuresis. In some cases, mithramycin is also used. Lowering of the calcium level tends to improve the mental symptoms.

Hypoparathyroidism usually follows surgery for removal of the parathyroid glands intentionally or as a complication of a thyroidectomy, usually for removal of a malignant thyroid gland. The manifestations related to hypocalcemia are tetany, carpopedal spasms, dysreflexia, hyperreflexia, and possible seizures.

The treatment is to administer 1000 to 3000 mg of elemental calcium a day, as well as 50,000 to 100,000 units of vitamin D. For acute conditions, calcium may need to be given intravenously, followed by oral administration. There are also a few cases of idiopathic hypoparathyroidism that are treated with large doses of calcium and vitamin D and may have newer types of vitamin D such as 1,25-dihydroxyvitamin D_3 or dihydrotachysterol (DHT, Hytakerol).

Diabetes Mellitus

Disorders of the pancreas are primarily related to diabetes mellitus, which is a genetically determined disorder of metabolism. The etiology is manifested by insulin loss from deficient or inadequate production as well as loss of carbohydrate tolerance. Diabetes is a disorder of protein, fat, and sugar metabolism and affects the entire body response. From 10% to 50% of otherwise healthy aging adults have glucose intolerance.

Diabetes mellitus is classified as type I (insulin-dependent) or type II (non-insulin-dependent). Patients with type I diabetes are typically children, adolescents, or young adults. The beta cells of their pancreas are unable to produce insulin. Type II diabetes is usually diagnosed in later years, often following a stressful episode. Patients typically present with the classic "three Ps" of polydipsia (excessive thirst), polyphagia (excessive hunger), and polyuria (excessive urination). The beta cells are able to produce insulin but in insufficient amounts to meet the body's needs.

Diabetic ketoacidosis (DKA), in which there is a very high blood glucose level, is a complication of type I diabetes that is potentially fatal. If the condition is not treated in a timely fashion death can occur from fluid depletion, shock, metabolic acidosis, and electrolyte imbalance. Even after hospital admission potentially fatal problems can develop as a result of inappropriate fluid administration, potassium deficiency, cerebral edema, or hypoglycemia. Postadmission causes of death are avoidable with a knowledgeable and skillful medical team and appropriate therapy and monitoring. With a skilled medical team, remarkable improvement can be expected within a few hours.

Hyperglycemic, hyperosmolar nonketotic syndrome (or coma) (HHNS or HHNC) is a medical emergency that can occur in type II diabetics when their blood sugar greatly increases. Dehydration becomes so severe that the mortality rate is extremely high, especially in elderly persons. Table 20–3 differentiates the assessment findings of DKA and HHNS.

TABLE 20–3

DIFFERENCES BETWEEN DIABETIC KETOACIDOSIS AND HYPERGLYCEMIC HYPEROSMOLAR NONKETOTIC SYNDROME

	Diabetic Ketoacidosis	Hyperglycemic Hyperosmolar Nonketotic Syndrome
ONSET	Sudden	Gradual
PRECIPITATING FACTORS	Infection Other stressors Stopping insulin administration	Infection Other stressors Poor fluid intake
MANIFESTATIONS	Ketosis: Kussmaul respirations (with a "fruity" or sweet smell), nausea, abdominal pain Dehydration or electrolyte loss; polyuria; polydipsia; weight loss; dry skin; sunken, soft eyeballs; lethargy; coma	Altered CNS function with neurologic symptoms Dehydration/electrolyte loss (same as for DKA)
MONITORING VARIABLES	ECG: Hyperkalemia: peaked T waves, widened QRS complex, prolonged PR interval, flattened or absent P wave Hypokalemia: depressed ST segment, flat or inverted T waves, increased ventricular dysrhythmias	ECG: Hypokalemia (same as for DKA) Hemodynamic measurements: CVP: > 3 mmHg below baseline PADP and PAWP: > 4 mmHg below baseline
LABORATORY FINDINGS		
Serum glucose	> 300 mg/dL	> 800 mg/dL
Osmolality	Variable	> 350 mOsm/L
Serum ketones	Positive, large amount	Negative
Serum pH	< 7.3	> 7.3
Serum HCO_3^-	< 15 mEq/L	> 20 mEq/L
Serum Na^+	< 137 mEq/L	Elevated, low, or normal
Serum K^+	Normal, rises with acidosis, falls with hydration	Normal or low
BUN	> 20 mg/dL	> 20 mg/dL
Creatinine	> 1.5 mg/dL	> 1.5 mg/dL

CNS, central nervous system; CVP, central venous pressure; PADP, pulmonary artery diastolic pressure; PAWP, pulmonary artery wedge pressure.

Source: Ignatavicius, DD, Workman, ML, and Mishler, MA: Medical-Surgical Nursing: A Nursing Process Approach, ed 2. Philadelphia, WB Saunders, 1995, p 1862. Reprinted with permission.

Assessment

Classic manifestations of diabetes include polydipsia, polyphagia, and polyuria. These symptoms result from fluid volume deficit as well as the possible losses or redistribution of potassium. As the hyperglycemia increases because of lack of sufficient insulin, the body begins to utilize fats as an energy source. With increased fat metabolism, ketones are produced, leading to metabolic acidosis.

TABLE 20-4
COMMON MACROVASCULAR AND MICROVASCULAR
COMPLICATIONS OF DIABETES MELLITUS

Macrovascular Complications	Microvascular Complications
Cardiovascular disease	Diabetic neuropathy
• Myocardial infarction • Hypertension	• Peripheral neuropathy: decreased sensation, infection, gangrene, amputation • Autonomic neuropathy: cardiovascular (orthostatic hypotension), gastrointestinal (malabsorption), urinary (retention)
Peripheral vascular disease (intermittent claudication) Cerebrovascular disease (stroke)	Ocular complications: blindness, retinal detachment, macular degeneration, glaucoma Diabetic nephropathy (renal failure)

In many patients, diabetes is diagnosed only when vascular complications have already begun. Table 20-4 lists common macro- and microvascular complications of diabetes mellitus.

Collaborative Management

A complete discussion of the various treatment options for diabetes is beyond the scope of this book. However, because hyperglycemic emergencies (DKA and HHNS) and hypoglycemia result in marked fluid, electrolyte, and acid-base imbalances, these complications are described.

Hyperglycemic Emergencies

Treatment of diabetic ketoacidosis (DKA) and hyperglycemic, hyperosmolar nonketotic coma relies on administration of intravenous fluids and insulin. It is divided into several phases.

INITIAL PHASE • Fluid depletion and electrolyte imbalances are the most important immediate concerns in the patient with diabetic ketoacidosis (DKA) and must be treated vigorously. Rehydration is important for maintaining tissue perfusion. The administration of intravenous fluids without insulin will lower the blood glucose level so that insulin administration is not the most immediate concern. Because fluid depletion and shock are the most immediate considerations, the hydrating fluid is chosen accordingly. The initial effort is to restore and maintain intravascular volume and renal blood flow. In the presence of severe fluid depletion and shock, albumin, plasma, dextran, or blood may be used as a plasma expander; however, these substances are rarely necessary. The most commonly used initial hydrating fluids are lactated Ringer's solution and normal saline. Either can be used and both are effective. Normal saline is often preferred because lactic acidosis is easily confused with DKA. Lactic acidosis may occur in older diabetic patients. Although the condition is rare, the use of additional lactate, such as lactated Ringer's solution, can be fatal. In most cases, normal saline is safe and effective. The initial dose is usually 20 mL/kg over the first hour. If renal blood flow is not then established, the administration of normal saline solution may be repeated. If renal output is not established with the second administration, renal failure should be suspected.

SECOND PHASE • Once the initial hydrating solution has been administered and the shock condition corrected, several options can be pursued. In conventional insulin therapy with subcutaneous insulin administration, fluid therapy usually consists of a multielectrolyte intravenous solution such as one-half normal saline with potassium added. With the new methods of intravenous insulin delivery, normal saline with added potassium is continued until the blood glucose level is about 250 mg/dL and is then followed by a multielectrolyte solution with added glucose.

Continuous infusion of low-dose intravenous insulin is the typical method of treating diabetic ketoacidosis. With this method, as soon as adequate renal blood flow (urine output) is established, potassium, as potassium phosphate, is added to the intravenous fluids. The amount of potassium is determined by the serum electrolyte measurements. In addition, bicarbonate is added as needed. During this phase of management, normal saline is administered in amounts calculated to meet maintenance requirements (100 mL/kg in children; 50 mL/kg in adults) plus replacement of the remaining deficit over the next 24 hours.

THIRD PHASE • When the blood glucose level falls to about 200 to 250 mg/dL, the fluids are changed to 5% dextrose in a multielectrolyte solution, most commonly one-half normal saline with appropriate potassium added. Maintenance rates, plus continuity of deficit replacement, are continued until hydration is complete and the patient is able to tolerate oral fluids.

FOURTH PHASE • As soon as the patient is able to retain oral fluids, fluids containing carbohydrates (juices and commercially produced and sweetened soft drinks, such as Kool-Aid) should be offered. The intravenous rate is decreased as oral intake is increased. When the patient is able to meet daily requirements orally, the intravenous fluid is discontinued and a regular diabetic diet begun.

SPECIAL CONSIDERATIONS • Potassium depletion can be quite severe in the diabetic patient with diabetic ketoacidosis (DKA), especially the newly diagnosed diabetic who has been in a catabolic state for some time. The serum potassium measurements may not reflect the extent of the deficit. Replacement must be vigorous or potentially fatal cardiac dysrhythmias can develop. The patient should be placed on a cardiac monitor and serum potassium levels should be closely monitored. Potassium can be replaced using potassium chloride or potassium phosphate. The latter is preferred because the combination of sodium chloride solution and potassium chloride can lead to a chloride excess.

Another consideration is that the body may suffer from phosphate depletion. Potassium phosphate is given in a concentration of 40 mEq/L of fluid or occasionally as high as 80 mEq/L, with the dosage calculated according to the serum potassium concentration and the cardiac monitor. Replacement should be according to the calculated deficit plus maintenance according to standard formula. The concentration depends upon fluid flow rates and the calculated need.

Chromium supplementation will assist in decreasing the fasting glucose level in chromium-responsive individuals. Additionally, insulin responsiveness changes with the aging process and degree of obesity. The significance of the changes associated with aging are unknown, since some studies are reporting increases in insulin responsiveness and some are reporting decreases. The significance of the study is that it indicates that chromium levels should be noted.

Bicarbonate deficits are presently being treated less vigorously than in the past. New data indicate that correction of peripheral bicarbonates (serum bicarbonate levels) may enhance acidosis in the central nervous system. Bicarbonate is given according to the standard formula calculated from the base excess (deficit). Sodium bicarbonate is added to the intravenous fluids when the arterial pH is below 7.1 or when the serum bicarbonate level is below 7 mEq/L.

Patients with diabetic ketoacidosis should be placed on a cardiac monitor in an intensive care unit. T waves should be closely monitored to assess potassium levels (see Chap. 4).

Urinary output should be measured hourly, and the intravenous fluid rate should be adjusted to maintain adequate output. Blood pressure and pulse should be monitored frequently to check the hydration status.

Blood glucose should be monitored hourly. Laboratory measurement of blood glucose and electrolytes every 2 to 4 hours may be supplemented by blood glucose determination by fingerstick in the interim hours.

The amount of fluid to be administered is usually calculated on the basis of 2500 to 3000 mL of fluid per square meter of body surface area every 24 hours, with the exact amount depending on the state of fluid depletion. Calculation of the amount of sodium required for body maintenance and to replace deficits is often done by subtracting the actual serum sodium level in milliequivalents per liter from 140 mEq/L and multiplying by the sodium space of roughly 60% of the body weight in adults and 75% of the body weight in children.

Hypoglycemia

Hypoglycemia is a lowered blood glucose level, usually caused by an excess of insulin. Hypoglycemia may occur during treatment of diabetic ketoacidosis, but it is an avoidable complication if patients are carefully monitored. In diabetic ketoacidosis, blood glucose levels as well as pulse, sweating, and elevations of blood pressure should be carefully monitored. The unconscious patient with diabetic ketoacidosis (DKA) can go from a hyperglycemic to a hypoglycemic coma without regaining consciousness.

There are several stages of hypoglycemia, depending on the severity or rate of fall of the blood glucose level. The mildest symptoms are those referable to a sympathetic response. Release of epinephrine is a potent counterregulatory mechanism to protect the brain from hypoglycemia. Epinephrine release or sympathetic response is triggered more by the rate of fall of blood glucose than by the absolute blood glucose level. Sympathetic symptoms can then be triggered at fairly high blood glucose levels if the blood glucose level is falling rapidly. For this reason, it is preferable to drop the blood glucose levels slowly during treatment of DKA. Sympathetic symptoms are shakiness, sweating, hunger, dilated pupils, rapidly bounding pulse, and symptoms that could be reproduced by administering epinephrine. These symptoms are uncomfortable but not harmful and serve only to indicate a falling blood glucose level that needs to be corrected.

As blood glucose falls below normal (40 mg/dL), brain symptoms begin to occur and evidence of glucose deprivation to the brain becomes manifest. Brain symptoms are inappropriate behavior (laughing, crying, and so forth), memory loss, inappropriate responses, inability to think and perform normal functions, and finally, loss of consciousness and convulsions.

As hypoglycemia becomes more pronounced, more of the brain becomes involved, and glucose is metabolized at a faster rate in relation to the brain involved, that is, cerebral hemispheres and cerebellum parts. The last part of the brain to be affected is the medulla oblongata, which has the lowest metabolic rate among the various parts of the brain. Physiologic effects are a fast, strong pulse; pale, cold, clammy skin; weakness; dilated pupils; and lowered temperature. The condition can progress to decreased respiratory rate, bradycardia, and coma. Seizures are possible, as the decrease in metabolic rate caused by the lack of glucose may also influence the electrical activity of the brain. There may be residual neurologic involvement with blood sugar levels of less than 20 mg/dL.

In a mild state of hypoglycemia brought about by epinephrine or sympathetic response, food is the treatment of choice if the patient is able to eat. Orange juice is a good choice; milk is an excellent food because it contains a simple sugar, lactose, for short-term action and protein for sustained action. An intake of 40 to 80 calories is usually sufficient. For a moderate reaction, especially when the individual is noticeably symptomatic, simple sugar (20 to 40 calories) should be given, to be followed in 10 to 15 minutes by a small amount of food. For a severe reaction, especially if there is danger of impaired swallowing, intravenous administration of 50% glucose is the treatment of choice. The usual dose is 10 to 20 mL of 50% glucose. At home, glucagon (1 mg for children over 3 years of age and adults; and 0.5 mg for children 3 years of age or younger) should be given intramuscularly, to be followed with a simple sugar when the person responds. Food should then be administered when the nausea, if any, subsides.

If seizures from hypoglycemia occur, the immediate danger is an obstructed airway and the immediate treatment is airway maintenance. Electroencephalogram (EEG) readings taken after a seizure from hypoglycemia will usually be abnormal but will return to normal in 4 to 6 weeks. The person often experiences a 6- to 12-hour postictal state, which is the brain's mechanism to assist in the healing process. The patient should be allowed to sleep but should be periodically awakened to take nourishment. There may even be temporary paralysis during the postictal state, but this condition should not lead to alarm because patients usually recover from these states.

REFERENCES AND READINGS

Casey, C (1992). Diabetes complications that can cripple. RN 55(8), 36–42.
Deakins, DA (1994). Teaching elderly patients about diabetes mellitus. American Journal of Nursing 94(4), 38–42.
Fain, JA (ed) (1993). Diabetes. Nursing Clinics of North America 28(1), 1–120.
Fleckman, A (1993). Diabetic ketoacidosis. Endocrinology and Metabolism Clinics of North America 22(2), 181–208.

Halloran, TH (1990). Nursing responsibilities in endocrine emergencies. Critical Care Nursing Quarterly 13(3), 74–81.

Handerhan, B (1992). Recognizing adrenal crisis . . . How to respond to severe steroid withdrawal. Nursing92 22(4), 33.

Lammon, CA, & Hart, G (1993). Recognizing thyroid crisis. Nursing93 23(4), 33.

Macheca, MKK (1993). Diabetic hypoglycemia: How to keep the threat at bay. American Journal of Nursing 93(4), 26–30.

McMorrow, ME (1992). The elderly and thyrotoxicosis. AACN Clinical Issues in Critical Care 3(1), 114–119.

Mulcahy, K (1992). Hypoglycemic emergencies. AACN Clinical Issues in Critical Care 3(2), 361–369.

Peterson, A, & Drass, J (1993). How to keep adrenal insufficiency in check. American Journal of Nursing 93(10), 36–39.

Sauve, DO, & Kessler, CA (1992). Hyperglycemic emergencies. AACN Clinical Issues in Critical Care 3(2), 350–360.

Walworth, J (1990). Parathyroidectomy: Maintaining calcium homeostasis. Today's OR Nurse 12(4), 20–24, 31–33.

Wilson, BA (1994). What nurses don't know about managing NIDDM. MEDSURG Nursing 3(2), 152–154.

CHAPTER 21

· · · · · · · · ·
· · · · · · · · ·
· · · · · · · · ·
· · · · · · · · ·
· · · · · · · · ·
· · · · · · · · ·
· · · · · · · · ·
· · · · · · · · ·

Acute Renal Failure

Acute renal failure (ARF) is a condition in which renal function rapidly becomes impaired or ceases. In most cases, the earliest sign of impending renal failure is a decrease in the urinary output. Nitrogenous waste products accumulate rapidly in the blood and the patient develops azotemia. This trend can be reversed in many instances by prompt and adequate treatment.

RENAL PHYSIOLOGY

The functioning and health of the entire renal system must be evaluated in treating renal problems. The kidneys are retroperitoneal organs whose major functions are maintaining the electrolyte composition, acid-base balance, and fluid volume in the body and excreting products of metabolism. The nephron is the functional unit of the kidneys (Fig. 21–1). It contains the glomerulus (located inside Bowman's capsule), the proximal convoluted tubules, the loop of Henle, and the distal convoluted tubules.

About one fourth of the blood pumped by the heart passes through the kidneys. The adult kidneys filter approximately 160 to 170 L of blood every 24 hours, but this volume varies with the size of the individual. The kidneys extract 1 to 1.5 L of fluid from the blood and concentrate the fluid into urine. The normal kidneys regulate the volume and composition of the blood by filtration, selective reabsorption, and secretion.

Filtration of the plasma takes place in the glomerulus, where the fluid and dissolved substances are filtered through the capillary walls, the basement membrane, and the visceral layer of Bowman's capsule. Glomerular

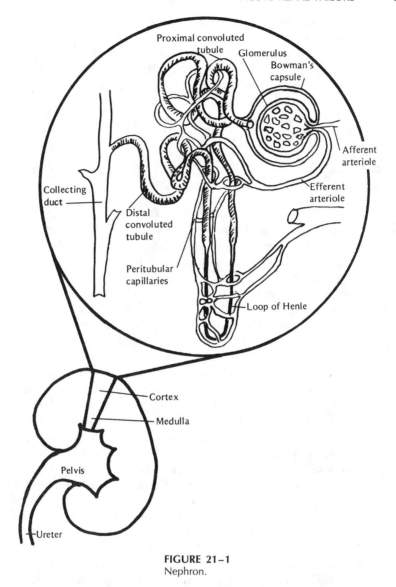

FIGURE 21–1
Nephron.

filtrate, the product of this process, is a plasmalike substance that lacks protein and cells.

Selective reabsorption into the capillaries occurs as the glomerular filtrate passes through the renal proximal tubules. The proximal tubules reabsorb essential substances such as potassium, sodium, glucose, calcium, magnesium, and amino acids. The proximal tubules are also involved in nutrition. Glucose, proteins, vitamins, amino acids, and acetoacetate ions are almost completely reabsorbed. Acid-base balance is partially regulated by the proximal tubules via the reabsorption or filtration of bicarbonate, according to the body's needs.

Secretion takes place in the distal tubules. Hydrogen ions, potassium, and uric acid are secreted into the tubular fluid. Urea, a by-product of nitrogen metabolism, and creatinine, an end-product of muscle metabolism, are examples of nitrogenous waste products excreted as a result of kidney function.

The kidneys also produce two substances that have important roles in the body—renin and erythropoietin. As discussed in Chapter 1, renin helps maintain blood pressure through the renin-angiotensin-aldosterone mechanism. Erythropoietin stimulates the bone marrow to produce red blood cells. If kidneys do not function properly, red blood cell production decreases and anemia results.

PATHOPHYSIOLOGY

When disease interrupts kidney function, the filtrate is reabsorbed out of the tubules into the blood, taking water, electrolytes, and the nonvolatile products of metabolism with it. If one kidney ceases to function, the other kidney can adequately compensate by increasing its function. The functioning kidney undergoes cell proliferation and increases its weight, sometimes by as much as 50%. The enlarged functioning kidney can maintain up to 80% of the function of two normal kidneys, but continuation of the disease process can lead to renal failure.

An interruption in the flow of urine can originate anywhere within the renal structure, from the proximal tubule to the urethra. The discussion of renal failure can be divided into three categories: prerenal, postrenal, and intrarenal (Fig. 21–2). Prerenal failure is caused by events that take place before the blood reaches the kidneys and involves the subsequent reaction of normal kidneys to these stimuli. Postrenal failure may result from an obstruction in the system that leads away from the kidney. The third type, intrarenal failure, may be caused by primary parenchymal damage, or it may be a sequela of pre- or postrenal failure.

In any discussion of renal failure, it should be noted that deficiencies of urinary concentration in infants and young children will require health management different from that in adults. Two primary factors must be taken into consideration. The infant cannot increase or decrease his or her own intake of fluid based on the thirst response. The infant's or young child's minimum daily requirement for water is, proportional to size, two or three times greater than an adult's requirement because of the immature kidneys' inability to concentrate urine.

PRERENAL FAILURE

Prerenal failure occurs anywhere in the body system before the blood reaches the kidneys. Most often, the problem results in inadequate blood flow to the kidneys, which diminishes their function.

Causative Factors

Prerenal failure is caused by inadequate blood perfusion of the kidneys that results in oliguria and kidney dysfunction. This type of acute renal failure can be reversed if the systemic problem is corrected before renal cell

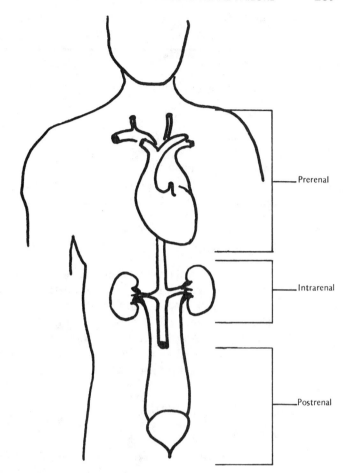

FIGURE 21-2
Types of renal failure.

damage occurs. Table 21–1 lists the causes and altered functions of prerenal failure.

Assessment

Physical Assessment

The nurse should be alert to a decrease in urinary output and should be aware of conditions that predispose the patient to oliguria. The most common conditions that cause oliguria include intracellular and extracellular fluid depletion, diabetic acidosis, shock, and cardiac failure. Other visible signs or symptoms of prerenal failure are rare.

Problems associated with prerenal failure in newborn children are perinatal asphyxia or septic shock. Occasionally, infants with complex congenital heart disease experience prerenal failure with oliguria following cardiac catheterization. Another cause of decreased renal blood flow is the use of large amounts of hyperosmotic dye in cardiac catheterization. In most cases, the prerenal failure resolves once the underlying cause is corrected.

TABLE 21-1
CAUSES AND ALTERED FUNCTIONS OF PRERENAL FAILURE

Cause	Altered Function
Hypovolemia/hypotension	Decreased fluid volume reduces renal blood flow and glomerular filtration rate
Decreased cardiac output	Kidneys not receiving the normal cardiac output (normally 25% of the body's blood)
Vascular failure, vasogenic shock, neurogenic shock	Interference with vascular elasticity collapse or pooling of the blood (vascular dilation)
Hepatorenal syndrome	Diseased liver causes body to falsely sense a volume deficit, thus retaining fluids; liver becomes grossly endematous

Diagnostic Assessment

Blood urea nitrogen (BUN) and creatinine levels are elevated, with a ratio of 20:1 or greater. Specific gravity is usually greater than 1.020, because the kidneys are trying to concentrate the urine. The urine sodium level will be less than 20 mEq/L. It may drop much lower than 20 mEq/L if the patient has a saline deficit because the kidneys retain sodium in the serum as a result of hypoperfusion. Urine osmolality is above 450 mOsm/kg and is usually greater than plasma osmolality.

Collaborative Management

Medical Treatment

The treatment of prerenal failure is aimed at restoring adequate circulation to the kidneys and correcting the underlying cause.

Mannitol, an osmotic diuretic, may be administered intravenously to determine whether the condition is prerenal or intrarenal failure. Mannitol is both a diagnostic aid and a temporary treatment. If the condition is prerenal, there should be an adequate increase in urine output within 3 hours. If a fluid deficit is thought to be the cause of the prerenal failure, 1500 mL of 5% dextrose in half normal saline solution, lactated Ringer's solution, or normal saline solution may be given intravenously. Colloids such as albumin may also be used. If fluid deficit is the cause, the urine flow will increase.

Nursing Interventions

The nurse is typically the first member of the healthcare team to become aware of a decrease in the urinary output. At the first suspicion of urinary decrease, the nurse should measure urinary output hourly with a urometer. The physician should be notified immediately if oliguria exists or is developing. The nurse's prompt report to the physician and immediate execution of the physician's orders may prevent the development of acute renal failure.

When an oliguria condition exists, the nurse should assume the following responsibilities:

- Observe for changes in vital signs.
- Observe for fluid overload.
- Observe for extracellular fluid excess.
- Decrease the patient's potassium intake.
- Maintain accurate intake and output records.

POSTRENAL FAILURE

Postrenal failure involves obstructions and malfunctions that occur below the kidney. Structurally, this involves the ureters, bladder, and urethra.

Causative Factors

The urinary tract (postrenal) can become involved in renal failure when disease or obstruction causes retention of urine and consequent fluid and electrolyte imbalance. Urine passes from the kidney through the ureters into the bladder and is then excreted through the urethra. Inflammation caused by infection, scarring, calculi, and tumors is a frequent cause of obstruction.

The short length of the female urethra, approximately 1½ inches in length, provides easy access to organisms that can cause bladder infections. These infections sometimes result in scar tissue, obstruction, and pyelonephritis.

The male urethra is approximately 9 inches long. It provides a channel for excreting urine as well as for passing spermatic fluid. The prostate gland encircles the urethra and the neck of the bladder. Prostate enlargement and tumors can cause urethral obstruction.

Fluid and electrolyte problems can ensue when the urine is retained because of obstruction and other impediments to urination. Table 21–2 identifies the causes and altered functions of postrenal failure.

TABLE 21–2
CAUSES AND ALTERED FUNCTIONS OF POSTRENAL FAILURE

Cause	Altered Function
Urethral obstruction (prostatism, tumor, pressure from uterine tumor)	Complete obstruction causes anuria; partial obstruction usually causes anuria alternating with polyuria
Ureteral obstruction (calculi, tumors, trauma)	Complete obstruction causes anuria; partial obstruction usually causes anuria alternating with polyuria; edema following instrumentation
Sulfa or urate precipitation	Blockage due to precipitation or crystal formation
Anticholinergic drugs and ganglionic blocking agents	Acute urinary retention possible because nerves not adequately stimulated; however, this is unusual.

Assessment

Physical Assessment

With postrenal failure, the patient becomes anuric (less than 10 mL urine output in 24 hours). When only an intermittent obstruction exists, anuria alternates with polyuria. Diuresis will follow the release of the obstruction. Postobstructive diuresis is primarily caused by decreased reabsorption of solutes from the proximal tubules and the loops of Henle. The amount of fluid replacement should be adequate to maintain the patient, but not equal to the amount of diuretic loss. Replacing total excessive diuretic losses will prolong the diuresis.

Diagnostic Assessment

Urinalysis may reveal scanty sediment, occasional white and red blood cells, and hyaline or finely granular casts.

X-rays may be taken to assess kidney size and to determine if renal stones or obstructions are present. A renal ultrasound can detect stones, tumors, or other structural abnormality.

Collaborative Management

The treatment for postrenal failure is to relieve the obstruction. Nursing care is similar to that for other types of renal failure. The nurse should:

- Report the occurrence of oliguria or anuria to the physician immediately.
- Report pain to the physician because this can aid in diagnosis.
- Check for fluid and electrolyte depletion during diuresis.
- Maintain accurate intake and output records.

INTRARENAL FAILURE

Intrarenal failure involves damage to the functional portion of the kidney.

Causative Factors

The most common causes of intrarenal failure are acute tubular necrosis (ATN), acute glomerulonephritis, and pyelonephritis. Table 21–3 identifies the causes and altered functions of intrarenal failure. Angiography with insufficient hydration can result in acute renal failure; however, angiography with sufficient hydration is not considered a significant hazard. Acute renal failure can also be secondary to interstitial lupus nephritis (glomerulonephritis).

Assessment

Physical Assessment

Decreased urinary output or oliguria exists with elevated BUN and serum creatinine. As this azotemic condition develops, so do central nervous system symptoms.

TABLE 21–3

CAUSES AND ALTERED FUNCTIONS OF INTRARENAL FAILURE

Cause	Altered Function
Acute tubular necrosis Ischemia (circulatory shock)	Shock decreases renal perfusion causing circulatory insufficiency; resultant hypoxia causes tissue death
Drugs, poisons, metals, e.g., glycols, pesticides, carbon tetrachloride, mercuric ion, contrast media (Fanconi's syndrome)	Nephrotoxic substances cause tissue damage and scarring
Hemolysis	Mismatched blood transfusion causes agglutination of red blood cells which obstruct tubules
Hypercalcemic crisis	Excess calcium crystallizes in tubules, causing obstruction; condition varies according to pH of urine
Trauma	Burns and crushing injuries result in extracellular deficit causing insufficient tissue perfusion (circulatory shock); myoglobin causes sludging in tubules
Leukemia	Hypokalemia due to excessive urinary potassium excretion caused by renal tubular defects
Acute glomerulonephritis After streptococcal infection	Antigen-antibody reaction in glomerulus causes inflammation and decreased glomerular blood flow
Vascular Renal arterial thrombosis	Thrombi in renal vessels cause occlusion
Acute cortical necrosis Pregnancy (abruptio placentae)	Intravascular coagulation possibly due to amniotic fluid in the plasma; anuria without obstruction
Sepsis (especially in pregnant women and patients on steroid therapy)	White blood cells fail to handle toxins as they would under normal conditions
Papillary necrosis Urinary tract infection	Papillae slough off; particles obstruct the tubules
Intrarenal precipitation Urates and sulfonamides	Precipitation and crystal formation cause obstruction
Multiple myeloma	Precipitation of myeloma protein into kidney
Pyelonephritis	Infection of kidney causes scarring of tubular system; concentrating defect due to initial injury to renal medulla—primary defect unknown
Lupus erythematosus (a collagen disease)	Damaged glomeruli and small renal vessels

Intracellular and extracellular fluid depletion, diabetic acidosis, shock, and cardiac failure are predisposing conditions to acute renal failure.

There are few visible signs. The most significant indicators are changes in urinary output and laboratory tests.

Diagnostic Assessment

Laboratory tests can be very significant in diagnosing acute renal failure. Tests of the glomerular filtration rate (GFR) will determine the extent of decreased renal function. Other common tests are blood urea nitrogen (BUN) and serum creatinine, 24-hour endogenous creatinine clearance, and 24-hour protein excretion tests. The BUN and serum creatinine levels are elevated. When oliguria is present, the BUN reading will rise approximately 20 mg/dL or more per day. A creatinine clearance test will often show a marked decrease before BUN and serum creatinine levels show significant elevation.

The specific gravity of the urine is less than 1.015. Urine osmolality is below 450 mOsm/kg and is less than or near plasma osmolality. The urine sodium level is above 40 mEq/L, because the kidney is not able to reabsorb sodium.

The ratio of urine urea nitrogen to BUN and of urine creatinine concentration to serum creatinine concentration is less than 10 to 1. In acute renal failure, the serum concentration of phosphate is elevated because phosphate is normally handled by the kidneys. As a result, the serum calcium level is decreased.

In acute tubular necrosis, the urinalysis shows the presence of renal tubular cells, tubular cell casts, coarse granular casts, red blood cells, hemoglobin, and red blood cell casts. In acute glomerulonephritis, urinalysis shows hematuria, proteinuria, and the presence of red blood cell and hemoglobin casts.

Collaborative Management

Medical Treatment

Acute renal failure is a condition of parenchymal damage that is temporarily irreversible. Treatment is designed to maintain the patient until the acute lesion is healed and normal renal function resumes. Some sources recommend that acute tubular necrosis be treated aggressively even when there are other complications.

In acute renal failure, the oliguric phase usually lasts 10 days to 3 weeks. Fluid intake should be carefully regulated during this period. Fluid allowance is calculated on the basis of 800 mL for insensible loss (800 mL per day), minus the amount acquired from catabolism (approximately 400 mL). Thus the fluid allowance is about 400 mL per 24-hour period, which should result in a weight loss of 250 g per day. The amount of fluid allowed for insensible loss may vary according to the amount lost through diaphoresis and the gastrointestinal tract. The restriction of fluid is essential to prevent a fluid excess and the consequent possibility of congestive heart failure and pulmonary edema. Increases in extracellular fluid are further reduced by eliminating sodium from the diet. Sodium is usually not administered during the oliguric phase, unless a deficit exists or excessive losses occur through diaphoresis or the gastrointestinal tract.

The diet should meet caloric needs through the consumption of carbohydrates in order to reduce endogenous catabolism. Because it is difficult to provide sufficient calories, a 20% to 50% glucose solution is often administered in a large central vein, such as the subclavian, to supply 100 to 200 g of sugar daily.

Potassium is not administered because acute renal failure increases the likelihood of hyperkalemia, which may result in cardiac arrest. Although there is no potassium intake, tissue catabolism increases the serum potassium level. Acute and severe hyperkalemia can be effectively treated by intravenous infusion of sodium bicarbonate, calcium gluconate, or glucose and insulin. Kayexalate, a cation-exchange ion, given orally or rectally, is also effective but acts more slowly. In order to limit the amount of urea in the serum, protein intake is severely limited or omitted from the diet. The addition to the diet of proteins containing essential amino acids helps to minimize negative nitrogen balance.

Renal acidosis will develop because the kidneys are unable to excrete the 50 to 70 mEq of nonvolatile acids produced daily. Peritoneal dialysis or hemodialysis may be used to alleviate the uremia if other methods are inadequate.

Blood transfusions may be necessary in acute renal failure due to the inability of the kidney to produce erythropoietin, which facilitates the manufacture of red blood cells. Whole fresh blood should be used because cell breakdown causes a higher potassium content in stored blood. Anemia may develop in 10 to 14 days following the onset of acute renal failure.

Anabolic steroids, such as testosterone propionate in oil, may be administered to decrease protein catabolism.

One of the primary causes of death in acute renal failure is the patient's susceptibility to infection; for this reason, indwelling catheters are seldom inserted. Because many antibiotics are excreted through the kidneys, the amount administered should be reduced during acute renal failure. Some antibiotics, though not all, can cause nephrotoxicity in acute renal failure. Penicillins are nontoxic and can be administered in normal dosages. Tetracyclines should be avoided because their action enhances catabolism and causes the BUN level to rise. Methenamine mandelate (Mandelamine), cephaloridine (Keflin), and nalidexic acid (NegGram) should not be used. Kanamycin sulfate (Kantrex), polymyxin (Aerosporin), and gentamycin sulfate (Garamycin) are usually given only every 2 to 3 days. Keflin should be reduced. Antibiotic levels in the blood must be determined; antibiotics may have to be reduced or discontinued because of nephrotoxicity. The length of time for which the antibiotic is retained in the blood can be determined by using the following formula:

Serum creatinine level $\times$ 4 = Number of hours the antibiotic is retained

Thus, if the serum creatinine level is 6 mg/dL, the antibiotic will be retained in the blood for 24 hours (6 $\times$ 4 = 24 hours). All patients who receive nephrotoxic antibiotics for over 1 week should be monitored with a 24-hour creatinine clearance test.

The oliguric phase may be followed by a diuretic phase, in which the patient experiences excessive water losses and accompanying changes in electrolyte balance. Recovery of the glomerular filtration rate before recovery of tubular function can increase the length and the severity of the diuresis associated with the diuretic phase of acute renal failure. The fluid replacement should be sufficient to maintain the patient but not equal to the amount of the diuretic loss. Replacing total excessive diuretic losses will prolong the diuresis.

The maintenance of fluid in infants and children, as in adults, revolves around accurate recording and replacement of fluid loss related to both sensible and insensible losses. Endogenous fluid resulting from metabolism should be subtracted from water intake. Endogenous fluid can be calculated at the rate of approximately 1 mL/kg of body weight per day.

Two problems associated with supportive therapy are hyperkalemia and fluid overload. Potassium should not be administered, and dietary intake of potassium should be kept to a minimum. Rectal or oral administration of Kayexalate (1 g/kg of body weight) usually corrects a mild hyperkalemia. More severe hyperkalemia may require dialysis.

Sodium should not be administered unless there is a urinary sodium loss and no expansion of fluid volume.

Fluid problems are compounded in infants with acute renal failure secondary to perinatal asphyxia or to hypoxia associated with pulmonary disease. The water limitation required to prevent fluid excess is often exceeded by the amount of fluid necessary to administer the parenteral medications necessary for cardiorespiratory support. Excess fluid balance is the rule, not the exception, in such cases. If cardiac or pulmonary functions are threatened, peritoneal dialysis should be initiated.

If the patient with acute renal failure does not respond to conservative medical treatment, renal replacement therapy may be required. These invasive therapies include hemodialysis, peritoneal dialysis, and continuous renal replacement therapy (CRRT). In many hospitals, CRRT is used most often for acute renal failure than the other types (Price, 1992).

Like the conventional hemodialysis described in Chapter 17, CRRT is an extracorporeal (out of the body) circulation of whole blood from an artery to a vein through a very porous hemofilter. It can be used for a period of 12 to 24 hours and works by the processes of ultrafiltration, clearance, and diffusion. Types of CRRT include:

- Continuous arteriovenous hemofiltration (CAVH)
- Continuous arteriovenous hemodialysis (CAVHD)
- Continuous arteriovenous ultrafiltration (CAVU), slow continuous ultrafiltration (SCUF)

Nursing Interventions

In addition to the responsibilities already described and those discussed in the dialysis chapters, the nurse should:

- Maintain accurate intake-output records.
- Accurately measure body weight.
- Practice medical asepsis.
- Encourage coughing and deep breathing.
- Assist with mouth care: teeth should be brushed and an antiseptic mouthwash used at least every 2 hours.
- Assist with emotional/mental needs.

The prolonged stress of acute renal failure may lead to psychotic behavior; this possibility usually increases when the patient is in protective isolation. Helping the patient to cope with confusion and restlessness presents an additional nursing care challenge. Heavy sedation must be avoided because many sedatives and tranquilizers are excreted through the kidneys. A pleasant atmosphere and access to radio, television, and current news may help the patient remain oriented to reality during this stressful situation.

REFERENCES AND READINGS

Baer, CL (1990). Acute renal failure. Nursing90 20(6), 34–40.

Baer, CL, & Lancaster, LE (1992). Acute renal failure. Critical Care Quarterly 14(4), 1–21.

Chambers, JK (1993). Renal insufficiency: Implications for care of the medical-surgical patient. MEDSURG Nursing 2(1), 33–40.

Douglas, S (1992). Acute tubular necrosis: Diagnosis, treatment, and nursing implications. AACN Clinical Issues in Critical Care Nursing 3(3), 688–697.

Finn, WF (1990). Diagnosis and management of acute tubular necrosis. Medical Clinics of North America 74(4), 873–891.

Innerarity, SA (1992). Hyperkalemic emergencies. Critical Care Quarterly 14(4), 32–39.

King, BA (1994). Detecting acute renal failure. Nursing94 24(3), 34–40.

Price, CA (1992). An update on continuous renal replacement therapies. AACN Clinical Issues in Critical Care Nursing 3(3), 597–604.

Stark, JL (1992). Acute tubular necrosis: Differences between oliguria and nonoliguria. Dimensions of Critical Care 9(4), 210–215.

Toto, KH (1992). Acute renal failure: A question of location. American Journal of Nursing 92(11), 44–57.

CHAPTER 22

. . . .
. . . .
. . . .
. . . .
. . . .
. . . .
. . . .
. . . .
. . . .
. . . .
. . . .

Chronic Renal Failure

Chronic renal failure (CRF) is often asymptomatic until it is manifested by complications. The condition progresses gradually and causes varying degrees of irreversible damage to the kidneys, making them unable to filter all the waste products from the blood, especially protein by-products.

PATHOPHYSIOLOGY

The basic functions of the renal system are discussed in Chapter 21. To manage patients with chronic renal failure, it is necessary to have an awareness of body chemical interactions that affect this disease. The primary defect in CRF is in the intrarenal segment of the renal system. The disease usually damages the nephrons, decreasing or destroying the glomerular or tubular function. Most commonly, the damage is caused by one of four factors: antigen-antibody deposits in the glomerular basement membrane (lupus erythematosus, poststreptococcal glomerulonephritis); thickening or occlusion of arterioles caused by renal vascular disease (diabetic nephropathy, nephrosclerosis); obstructive substances (calcium or uric acid stones); and scarring caused by necrosis and infection (pyelonephritis). The frequency with which thickening and occlusion of the arterioles are caused by hypertension has resulted in an increased concern over the roles of renin and angiotensin.

Renin

Renin is an enzyme produced by the kidney when the arterial pressure decreases or the extracellular volume (ECV) decreases. Renin is released by the juxtaglomerular cells, primarily in reaction to a reduced glomerular filtration rate, regardless of the cause. After its release, renin interacts with the vasopressor angiotensin to form angiotensin II. This conversion takes place in the lungs, although the kidneys and other organs may also be conversion sites.

Angiotensin

Angiotensin II has two immediate direct effects on arterial blood pressure. It is a powerful vasoconstrictor that contracts the peripheral arterioles, thereby increasing total peripheral resistance. A small amount of angiotensin (one ten-millionth of a gram or less) can increase arterial pressure by as much as 10 to 20 mmHg. This powerful agent is inactivated by angiotensinase enzymes 1 or 2 minutes after its formation.

Aldosterone

Angiotensin II increases the total amount of body sodium by stimulating the adrenal cortex to secrete aldosterone. The result of this action is increased extracellular fluid (secondary to sodium retention) and increased cardiac output. This is an interactive process that also involves the potassium regulation of aldosterone secretion.

Because there is a feedback sequence among renin, angiotensin, and aldosterone, angiotensin II also has an indirect effect on the production of renin through regulation of the baroreceptor mechanism, the retention of sodium, and the stimulation of the arteriole sympathetic nerves of the juxtaglomerular apparatus.

These three mechanisms may function independently or interdependently. Although the function of renin in the maintenance of blood pressure is considered a normal homeostatic response, its relationship to normal blood pressure and hypertension is for the most part unknown.

CAUSATIVE FACTORS

Although the primary defect for chronic renal failure is intrarenal, the initial cause may result from a combination and interaction of several processes. Three common causes of chronic renal failure are glomerulonephritis, pyelonephritis, and benign nephrosclerosis.

Glomerulonephritis is a process in which the glomerular membrane becomes progressively thicker because of precipitation of antigen-antibody complex followed by glomerular inflammation. The function of the nephrons is gradually destroyed as the glomeruli are replaced by fibrous tissue.

Pyelonephritis affects primarily the ability of the kidneys to concentrate urine. An infectious or inflammatory process invades the renal parenchyma and destroys the renal tubules, glomeruli, and other structures.

Nephrosclerosis occurs when vascular lesions cause renal ischemia. The most common causes of nephrosclerosis are (1) arteriosclerosis of the smaller renal vessels; (2) atherosclerosis of the renal or large arteries; and (3) fibromuscular hyperplasia, as in the occlusion of large arteries. The kidneys become smaller and develop a nodular surface as kidney tissue is replaced by areas of fibrous tissue. This process, commonly seen in old age, causes a progressive decrease of renal blood flow and renal plasma clearance, with consequent reduced urine output. Table 22–1 lists other causes of and altered functions in chronic renal insufficiency.

TABLE 22–1
CAUSES AND ALTERED FUNCTIONS
OF CHRONIC RENAL INSUFFICIENCY

Cause	Altered Function
Glomerulonephritis	
Proliferative	Diffuse increased cellularity of glomeruli; lesions due to antigen-antibody reaction; inflammation and dysfunction of glomerulus; most commonly seen in post-streptococcal glomerulonephritis; frequently a recoverable lesion
Membranous	Thickening of the basement membrane; usually idiopathic (also seen in diabetic renal disease); leads to heavy loss of protein; nephrotic state
Focal	Nondiffuse scattered lesions of proliferative, membranous, or membranous-proliferative type; frequently benign; most common symptom is hematuria
Pyelonephritis	Kidney damage attributed to chronic infection with microorganisms beginning in the renal pelvis and proceeding to renal parenchyma; urinary stasis
Nephrosclerosis	Hardening of renal tissue resulting in narrowing or occlusion of vessels: arteriosclerosis related to aging process, which is accelerated in uncontrolled hypertension; atherosclerosis
Other	
Acute renal failure	Incomplete recovery can lead to chronic renal failure
Lupus erythematosus	A collagen-vascular disease; no set pattern of events; renal involvement most often focal but can cause proliferative or membranous glomerulonephritis
Nephrocalcinosis	High concentration of calcium in the blood causes calcium deposits with subsequent renal damage; seen in many diseases; such as hyperparathyroidism and sarcoidosis
Papillary necrosis	Obstruction and sloughing of the papillae may be related to diabetes, infection, analgesic drug abuse, urinary obstruction
Polycystic kidney disease	Congenital hereditary disease causes formation of cysts whose enlargement causes pressure and atrophy of functioning glomeruli
Renal vascular disease	Vascular lesions leading to renal ischemia and renal tissue death
Sickle cell anemia	Hereditary S-shaped or sickle-shaped hemoglobin cells cause sloughing into and plugging of the glomeruli; significant to African Americans
Uric acid nephropathy	Uric acid causes blockage of tubules, e.g., gout

COMPLICATIONS

Although chronic renal failure (CRF) is a disease of the kidneys, almost every area of the body becomes involved. Most of the complications are the result of the accumulation of waste products, such as urea and creatinine, or imbalances of fluid, electrolytes, or acidity. The acid-base imbalance is typically metabolic acidosis.

Table 22–2 lists and explains the common complications of CRF. Two of the complications, anemia and osteodystrophy, need further discussion.

Severe anemia usually occurs as the chronic renal failure becomes more critical. It is suspected that the damaged kidneys are no longer capable of producing erythropoietin. Erythropoietin stimulates red blood cell production in the bone marrow. A decline in erythropoietin results in a reduction in the number of red blood cells, which is manifested by anemia. The level of red blood cells in the body is also affected by other factors such as the high concentration of urea, hydrogen ions, and potassium.

The kidneys affect the ability of the bones to absorb calcium. Vitamin D plays a vital role in the absorption of calcium. The vitamin is processed initially by the liver; however, it must also pass through a chemical process in

TABLE 22–2
COMPLICATIONS OF CHRONIC RENAL FAILURE

System	Complication
Neural	Decreased mental alertness; impaired vision and hearing; glossitis; parotitis; tingling; numbness (blood supply to nerves decreased); seizures (high BUN toxic to central nervous system); sensorium changes (ICF excess); hyperreflexia; asterixis
Respiratory	Infection (pneumonia, pleurisy); fluid retention (pulmonary edema)
Cardiovascular	Congestive heart failure; uremic pericarditis; hypertension (ECF increased); dysrhythmias (elevated potassium in terminal stages)
Gastrointestinal	Nausea; vomiting; hematemesis; paralytic ileus; melena
Reproductive	Decreased ovulation or spermatogenesis; vaginal hemorrhage; impotence
Skeletal	Osteodystrophy due to secondary hypertrophy of parathyroid glands; this causes increased vascular and extraskeletal calcifications; hypertrophy of parathyroid glands occurs because the body is attempting, without success, to keep the serum calcium and phosphate normal
Integumentary	Itching due to microscopic calcium and phosphate deposits in the skin caused by overactive parathyroid glands
Hematologic	Anemia due to three factors: decreased erythropoietin (a hormone produced by the kidneys that stimulates the bone marrow); shorter life span of red cells due to uremic factors of hemolysis in the blood; oozing blood through the bowels

ICF, intracellular fluid; ECF, extracellular fluid.

the kidneys in order to complete the conversion of vitamin D_3 into 1,25-dihydroxycholecalciferol, a chemical necessary for the bones to absorb calcium. Any disturbance in the conversion process will result in a weakened, vulnerable skeletal structure. In addition, serum phosphate levels increase in renal failure, resulting in a reciprocal decrease in calcium (see Chap. 7).

ASSESSMENT

Chronic renal disease is usually well advanced before it is recognized. Its signs and symptoms are also indicative of other disease processes.

Physical Assessment

Symptoms include loss of appetite, headache, nausea, vomiting, and itching. Hyperventilation may be present because the lungs attempt to "blow off" excess carbon dioxide as compensation for metabolic acidosis. Polyuria accompanied by nocturia may occur early in the disease, but is followed by oliguria or anuria in end-stage renal disease (ESRD).

Other recognizable signs include a sallow complexion, listlessness, edema, peripheral neuropathy (footdrop, loss of ankle jerk), and elevated blood pressure.

The patient may develop renal metabolic acidosis (Chap. 3) as a result of the inability of the renal tubules to reabsorb sufficient bicarbonate, which effects a net acid excretion. Table 22–3 lists the key features of chronic renal failure, which are assessed by the nurse and other health team members.

Diagnostic Assessment

Laboratory findings that are commonly seen in patients with chronic renal failure include:

- Increased serum creatinine (waste product)
- Increased blood urea nitrogen (BUN) (waste product)
- Decreased or normal serum sodium
- Increased serum potassium
- Increased serum phosphorus (phosphate)
- Decreased serum calcium
- Increased serum magnesium

In addition to these serum lab findings, a urinalysis reveals a fixed specific gravity of 1.010 to 1.015, indicating an inability to concentrate and dilute urine. An elevated protein level of 1+ to 4+ in the urine suggests damage to the filtering system of the glomerulus. The urine may also contain red blood cells, bacteria, and casts. White blood cells and bacteria are present when a urinary tract infection occurs. Creatinine clearance, assessed from a 24-hour urine specimen, is decreased as the kidneys are unable to remove this waste product from the blood. Normal creatinine clearance is about 100 to 130 mL/min.

Other tests may indicate normochromic anemia, metabolic acidosis, and an abnormal glucose tolerance test. As the kidneys lose function, they allow glucose to "spill" into the urine.

TABLE 22–3
EFFECTS OF RENAL FAILURE ON ELECTROLYTE BALANCE

Electrolyte	Effects of Renal Failure	Problems	Treatments
Potassium	Retained with oliguria	Hyperkalemia Cardiac dysrhythmias Asystole	Kayexalate, oral or rectal Regular IV insulin with 5% to 50% dextrose IV calcium gluconate Dialysis
Sodium	Retained	Dilutional hyponatremia Fluid volume excess Hypertension Congestive heart failure Pulmonary edema	Diuretics, until kidneys no longer responsive Dialysis Fluid restriction Sodium restriction
Phosphate	Retained	Hyperphosphatemia Metastatic calcium phosphate deposits Renal bone disease	Phosphate-binding agents Limit phosphorus intake Vitamin D analogues

Calcium	Decreased
	  gastrointestinal
	  absorption
	Binds to phosphate
		Bone demineralization
		Pathologic fractures
			Replace vitamin D
			Calcium supplements

Hydrogen	Retained	Binds with bicarbonate
		  for excretion through
		  respiratory
		  compensation
			Dialysis
			Bicarbonate supplements

Bicarbonate	Depleted	Used for blood buffering
		  to prevent metabolic
		  acidemia
			Dialysis
			Bicarbonate supplements

Magnesium	Retained	Potential for
		  hypermagnesemia
			Avoid magnesium-containing
			  antacids and laxatives

Source: Ignatavicius, DD, Workman, ML, and Mishler, MA: Medical-Surgical Nursing: A Nursing Process Approach, ed 2. Philadelphia, WB Saunders, 1995, p 2114. Reprinted with permission.

In the terminal stages of ESRD, the potassium level may increase and cause elevated T waves in the electrocardiogram (ECG). Radiographic findings may show structural abnormalities of the kidneys, depending on the cause of the renal disease. Chapter 4 describes hyperkalemia in detail.

COLLABORATIVE MANAGEMENT

The goal in the treatment of chronic renal failure is to assist the patient to maintain optimal renal function. The patient generally responds well if he or she adheres to the treatment plan.

Medical Treatment

Chronic renal failure (CRF) is usually controlled conservatively with proper diet and fluid restriction. A low-protein and high-caloric diet is used most frequently. Sodium is usually restricted because about 80% of patients with chronic renal failure retain salt. The remaining 20% lose salt. If the urine sodium level is below 20 mEq/L, the patient is usually considered a salt retainer. If the urine sodium level is over 20 mEq/L and if extracellular fluid depletion results in a postural blood pressure decrease while on a salt-restricted diet, the patient is a salt loser. Treatment focuses primarily on maintaining optimal levels of protein, sodium, potassium, and water. Levels of each nutrient are individually adjusted according to the progression of the illness, type of treatment being used, and the patient's response to treatment.

Infections that occur with renal failure are treated with antibiotics; however, because of the compromised clearing of nitrogenous and other toxic materials related to the disease process, doses must be adjusted on the basis of glomerular filtration rates. As the function of the kidneys declines, drugs clear more slowly and can build up past therapeutic, if not to toxic, levels. For example, if gentamicin is ordered, the dosage may be adjusted by lengthening the interval between doses.

Increased serum phosphorus is treated with a phosphate-binding antacid that does not contain magnesium, for example, Amphojel or Alu-Caps.

In recent years, recombinant erythropoietin (EPO, Epogen) has been available to combat the chronic anemia of renal failure. The drug dosage is based on the patient's hemoglobin level and hematocrit; the average dose is 75 to 100 units/kg three times a week (Gahart, 1994).

In cases that do not respond to these conservative treatments, the more complex procedures of peritoneal dialysis (Chap. 16), hemodialysis (Chap. 17), or kidney transplants must be used.

Nursing Interventions

When a patient with chronic renal disease is hospitalized, it is usually for diagnostic purposes, evaluation, complications, or treatment of unrelated diseases. Regardless of the reason, the nurse plays an important part in helping to maintain the patient's health and manage symptoms of disease. It is particularly important for the nurse to be aware of the patient's emotional status. Expression of negative feelings should be reported to a psychiatrist or psychologist. For example, discussions of dreams have been known to be helpful in identifying a patient's area of concern. In many cases, it is advisable for a psychiatrist or psychologist to be available to help the patient make the adjustment to his or her chronic disease.

The nurse should:

- Maintain accurate intake-output records.
- Take accurate daily weight measurements.
- Implement diet restrictions.
- Calculate fluids as ordered.
- Be alert to changes in vital signs (elevated blood pressure, extracellular fluid excess, postural blood pressure changes).
- Report abnormal electrolyte readings immediately.
- Take seizure precautions.
- Provide emotional support.
- Educate the patient and family about the disease.

- Be aware that a crisis situation could occur if any of the following problems develop: elevated blood pressure, renal metabolic acidosis, hyperventilation, elevated potassium level resulting in ECG changes, oliguria, decreased calcium, elevated blood urea nitrogen (BUN) and creatinine levels, or extracellular fluid excess.

For patients receiving renal replacement therapy (dialysis), nursing interventions are described in Chapters 16 and 17.

REFERENCES AND READINGS

Berg, J (1990). Assessing for pericarditis in the end-stage renal disease patient. Dimensions of Critical Care 9(5), 266–271.

Biggers, PB (1991). Administering epoetin alfa—more RBCs with fewer risks. Nursing91 21(4), 43.

Calkins, ME (1993). Ethical issues in the elderly ESRD patient. American Nephrology Nurses' Association Journal 20(5), 569–571.

Compher, C, Mullen, JL, & Barker, CF (1991). Nutritional support in renal failure. Surgical Clinics of North America 71(3), 597–608.

Gahart, BL (1994). Intravenous Medications. St. Louis: Mosby.

Harum, P (1993). Vitamin, mineral, and hormone interaction in renal bone disease. Journal of Renal Nutrition 3(1), 30–35.

Miller, SB (1992). Dosage adjustments of drugs in renal failure. In Woodley, M, & Whelan, A (eds). Manual of Medical Therapeutics (pp 529–533). Boston: Little, Brown.

Prowant, BF, et al (1991). Nephrology nursing care plan and patient education for the patient receiving Epogen. American Nephrology Nurses' Association Journal 18(2), 188–194.

CHAPTER 23

. . . .
. . . .
. . . .
. . . .
. . . .
. . . .
. . . .
. . . .
. . . .
. . . .

Shock

Shock may have physical or psychological causes; both result in a discrepancy between the circulating blood volume and the size of the vascular bed. Figure 23–1 illustrates this effect. Shock is a pathologic problem rather than a disease.

CLASSIFICATION

Shock has been classified by the site of origin of the underlying problem or disease, for example, hypovolemic shock, cardiogenic shock, vasogenic shock, and septic shock. These terms are still used clinically, but are not the most current classifications.

The current classification system (Singleton & Workman, 1995) includes the following types of shock, which are based on the specific functional impairment manifested:

- Hypovolemic shock
- Cardiogenic shock
- Distributive shock
- Obstructive shock

Hypovolemic Shock

As the name implies, hypovolemic shock is caused by body fluid deficit, most often from hemorrhage and dehydration. Hemorrhage can result from trauma, surgery, gastric or duodenal ulcers, or diseases in which clotting is inadequate, such as cirrhosis, cancer, and hemophilia. Dehydration can re-

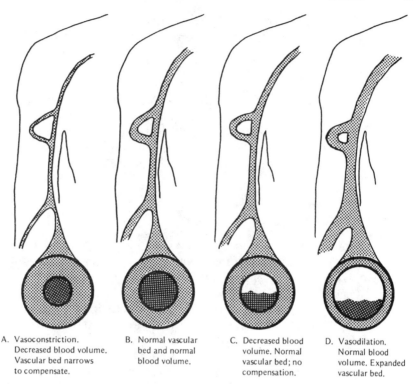

A. Vasoconstriction.
Decreased blood volume.
Vascular bed narrows
to compensate.

B. Normal vascular
bed and normal
blood volume.

C. Decreased blood
volume. Normal
vascular bed; no
compensation.

D. Vasodilation.
Normal blood
volume. Expanded
vascular bed.

FIGURE 23–1
Discrepancy between amount of blood volume and size of circulating bed.

sult from gastrointestinal (GI) losses, inadequate fluid intake, diuresis, and hyperglycemia. Chapter 2 discusses dehydration in detail.

Cardiogenic Shock

Cardiogenic shock is caused by direct heart or pump failure. This complication typically results from severe myocardial infarction, cardiomyopathies, ventricular dysrhythmias, or cardiac arrest.

Distributive Shock

Distributive shock is caused by decreased vascular tone or volume. It can be further divided into neurally induced distributive shock and chemically induced distributive shock. Common conditions that can cause neurally induced distributive shock include spinal cord injury, neurologic damage, severe pain, and psychologic stress.

Chemically induced distributive shock is caused by anaphylaxis (type I hypersensitivity reaction) and sepsis. Sepsis leading to shock occurs when pathogens, usually bacteria, invade the bloodstream. Both gram-negative bacteria, such as *Echerichia coli* and *Pseudomonas aeruginosa*, and gram-positive bacteria, such as *Staphylococcus aureus*, can cause sepsis and sepsis-induced distributive shock. Some patients develop disseminated intravascular coagulation (DIC) as well. This clotting disorder results when toxins are released into the bloodstream and affect the normal clotting cascade.

Obstructive Shock

Obstructive shock, like cardiogenic shock, results from heart failure. However, in cardiogenic shock the heart itself is damaged. In obstructive shock, other external conditions affect the ability of the heart to pump effectively. Some of the causes include cardiac tamponade, pulmonary embolism, constrictive pericarditis, and aortic aneurysm (Houston, 1990).

PATHOPHYSIOLOGY

The body depends on three main factors to maintain the correct proportion of blood to the size of the vascular bed: blood volume, cardiac output, and vascular tone. A significant variation in any of these three hemodynamic factors may result in shock.

The average-sized man (75 kg) has a total blood volume of 4.5 to 5.5 L. Plasma is approximately 90% water and 10% solutes, which are either crystalloid or colloid. The solutes may be classified according to their ionizing ability as electrolytes or nonelectrolytes. When the volume of plasma changes, the type of filtrate reaching the cells for tissue metabolism also changes. Blood plasma, however, is only one of the body's three fluids. The others are the interstitial fluid and the intracellular fluid. A loss of any of the body's fluids may precipitate shock, because they all ultimately affect the volume of blood circulating through the body.

The primary function of the circulatory system is transportation. The cells will not receive sufficient oxygen and nutrients if the blood volume is inadequate. The result of this deficiency is cellular hypoxia.

In shock, the circulatory system becomes unable to meet the demands for moving nutrients and oxygen into the cells and also unable to remove waste products. Shock is essentially defined as a discrepancy between the circulating blood volume, cardiac output, and the peripheral vascular bed, in which the dynamic interchange between these systems is interrupted. If the quantity of blood does not change, the amount of blood that circulates to the tissues results from an interaction between cardiac output and peripheral tone. The term peripheral tone includes both vasoconstriction and vasodilation. Blood pressure is determined by the force with which the heart is pushing the blood (indicated by the systolic reading) and the resistance or back pressure the heart meets (indicated by the diastolic reading). The resistance of the vessels is largely determined by their size. The more the vessels are constricted, the higher the resistance and the harder the heart must pump to circulate the blood through the vessels. Thus shock is not defined merely as a low blood pressure. In fact, the blood pressure can remain unchanged, while the amount of the blood circulating to the tissues varies. Figure 23–2 demonstrates a comparison of these three factors. As the figure illustrates, the amount of urinary output is a more reliable guide to tissue perfusion than is blood pressure.

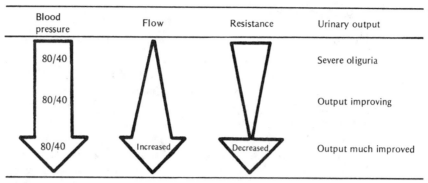

Blood pressure	Flow	Resistance	Urinary output
80/40			Severe oliguria
80/40			Output improving
80/40	Increased	Decreased	Output much improved

FIGURE 23–2
Hemodynamics in shock. Although blood pressure remains constant, as the cardiac output is increased and the peripheral vascular resistance is decreased, kidney perfusion and urinary output improve. This indicates an improved total body tissue perfusion.

The circulating blood volume may be reduced from actual hemorrhage or from plasma loss (third spacing or fluid shift out of the bloodstream) that is typically seen in patients with bowel obstruction, cirrhosis with ascites, and burns. (Also see Chap. 2 for discussion of third spacing under Fluid Deficit). The peripheral vascular bed will have increased resistance because of pressure (particularly diastolic pressure) that becomes increased because of the release of catecholamines. Catecholamines (epinephrine and norepinephrine) have a vasoconstrictive effect on peripheral arterioles and therefore on the oxygenation of peripheral tissue; however, catecholamines increase the rate and force of myocardial contraction, causing vasodilation in the brain, myocardium, and skeletal muscle as a protective mechanism. The perfusion of the brain, heart, and muscles as a compensatory stress mechanism is generally at the expense of organs that are not necessary for immediate survival, particularly the vessels of the gastrointestinal tract, the kidneys, and the liver.

In general, shock can cause the following significant changes:

- Cellular anoxia or hypoxia
- Metabolism of glucose without oxygen, leading to anaerobic metabolism of carbohydrates and a resultant buildup of lactic acid in cells
- Metabolic acidosis resulting in a decreased serum pH
- Loss of tone in precapillary arterioles, leading to decreased oxygenation of the periphery because of capillary bed engorgement and increased resistance
- Decrease in circulating blood volume caused by direct fluid loss (e.g., hematogenic shock), or decreased cardiac output (e.g., cardiogenic shock)
- Impairment of kidney function following prerenal failure with oliguria, caused by decreased perfusion of the kidney as a result of extracellular fluid loss

Prerenal failure that is not corrected can progress to intrarenal failure and death.

CLINICAL MANIFESTATIONS

The major manifestations of shock from decreased tissue perfusion are the following:

- The level of consciousness is usually changed by anxiety caused by hyperactivity of the sympathetic nervous system from increased secretion of catecholamines.
- Anxiety may progress to agitation, lethargy, restlessness, and confusion caused by hypoxia of the brain; coma and death may occur.
- Increased catecholamine secretions may cause tachycardia and tachypnea. The tachycardia is an attempt to increase the circulating blood volume to vital organs. The compensatory tachypnea is an attempt to increase cellular oxygenation. It is generally ineffectual because the respirations are too rapid and shallow.
- Catecholamine secretion may cause the skin to become cool, cold, or clammy. This is generally called *cold shock*. In conditions of septic shock, *warm* or *hot shock* can occur.
- Diastolic blood pressure initially increases caused by catecholamine secretion. Subsequently there is a fall, first in diastolic pressure, then in systolic pressure due to extracellular fluid deficit.
- Oliguria provides the best indication of compromised tissue perfusion. This condition is due to decreased perfusion of the kidneys.

The kidneys' obligatory urine flow is 20 mL/h. A urine output of 30 to 50 mL/h can indicate an impending oliguria and should be monitored. Urine output of 20 mL or less per hour indicates oliguria and a critical situation.

The decreased tissue perfusion and oxygenation of cells, the failure of internal respiratory gas exchange, and the buildup of waste products can all result in lactic metabolic acidosis. If not corrected, this situation leads to cell death (see Table 23–1).

CHARACTERISTICS OF TYPES OF SHOCK

Hypovolemic shock is the most common type of shock seen in the clinical setting, and is therefore discussed in the most detail.

Hypovolemic Shock

Hemorrhage and gastrointestinal bleeding are typical causes of hypovolemic shock; however, vomiting, diarrhea, burns, peritonitis, and trauma are also common causes of fluid loss. Hypovolemic shock can be caused by hemorrhage involving the loss of whole blood or the loss of only the plasma portion of the blood. The latter leads to hemoconcentration.

Shock caused by hemorrhage is related to the decreased volume of circulating blood. Hemorrhage is the loss of blood from the bloodstream caused by a disruption in the continuity of one or more blood vessels (arterial, venous, or capillary). Possible causes include physical or chemical injury to the body. More specific examples are the rupture of or injury to a blood vessel from the slipping of a suture; erosion of a vessel by drainage tubes, tumor, or infection; interference with the clotting mechanism, as in hemophilia; or fragile capillaries, as in vitamin C deficiency. Bleeding may also occur from a body orifice, as in gastrointestinal bleeding, or from an incision site. Blood may collect under subcutaneous tissue as a hematoma, or in a body cavity.

Hemorrhage may be classified as gradual or sudden, internal or external, and by type of vessel. The types of vascular bleeding are further categorized as arterial (bright red and spurting), venous (dark red and continuous), and capillary (either bright red or dark red and oozing). Gradual, internal, and capillary bleeding are the most difficult hemorrhages to detect from early shock symptoms.

When a patient of average weight (75 kg) has lost over 500 mL of whole blood, the effect of this loss becomes evident from certain specific body changes. All of these effects of blood loss reinforce the hypoxia and further increase tissue damage. The decreased capacity of the body to deliver sufficient blood and oxygen to the cells is related to the decreased volume and the decreased number of red blood cells. Part of the problem in hypoxia is that the body's defense mechanisms automatically eliminate general tissue perfusion and instead concentrate on keeping the blood flow steady to the more vital organs.

A decreased blood volume excites three initial responses. The first, vasoconstriction, lessens the size of the vascular bed to maintain systemic blood pressure. This change can be indicated by a slight rise of the diastolic pressure in early shock. The second, the constriction of arterioles, increases peripheral resistance to help maintain systemic blood pressure. The third, the transfer of fluid from the interstitial fluid to the blood, is stimulated by the adrenocortical hormones. This transfer of fluid increases the blood volume, again with the objective of increasing the systemic blood pressure.

The body tries to control local bleeding in several ways. A reflex contraction (vascular spasm) of the blood vessel occurs. Contraction of the lumen of the vessel is based on the physical principle that the smaller the opening, the less blood can be evacuated. The muscles also contract to aid in reducing the size of the lumen of the vessel. Muscle contraction further aids the body by immobilizing the injured part. The inside layer of the vessel curls inward to increase occlusion of the vessel.

The signs and symptoms of hemorrhage vary according to the site of bleeding and the corresponding amount of blood lost. If blood is trapped in a small cavity such as the cranial or spinal cavity, the symptoms appear sooner than when the blood is lost to the exterior or a larger cavity. The warning signs and symptoms of cranial or spinal hemorrhage are the same as those of increased intracranial pressure due to a tumor: headache, vertigo, loss of consciousness, convulsions, slower respiration and pulse, vomiting, widening pulse pressure, and unequal pupillary reaction to light. The pulse and respiration rates decrease only in cranial and spinal hemorrhage because of pressure on the cardiac and respiratory control centers.

Signs of blood escaping without pressure within a cavity are largely related to hypovolemia and hypoxia. These signs are reduced blood pressure (reduced volume), nausea and vomiting (stress response), increased pulse and respiration rates (result of a stimulation of stress mechanisms), and re-

TABLE 23–1
STAGES OF SHOCK AND THEIR CHARACTERISTICS

Stages of Shock	Blood Pressure	Fluid Vol. Deficit	Reversible/ Irreversible	Recognition	Potential Lab. Test Results*	Compensatory Mechanisms
Onset	Down 10%	500 mL	Rev.	Visible signs/symptoms rare; rise of diastolic blood pressure	None	Stress mechanism causing vasoconstriction
Slight	Down 20%	1000 mL	Rev.	Diastolic blood pressure still increased; pulse rate increased; respirations increased; skin cool, pale, dry; basal metabolic rate reduced; listlessness, apathy, weakness, dizziness proceeding toward unconsciousness; decreasing urinary output	Red blood cells normal, decreased, or increased; white blood cells increased; plasma protein decreased; BUN increased; blood sugar increased	Stress mechanism releasing ACTH; posterior pituitary releases ADH to hold water
Moderate	Down 35%	1800 mL	Rev.	Fast pulse but weak and thready or irregular; respiration rapid and more shallow; skin pale, cool, clammy; sensorium dulled; dilating pupils; dim vision; oliguria	White blood cells increased; plasma protein decreased; BUN increased; blood sugar increased; hemoglobin and hematocrit may be increased, normal, or decreased, depending on cause of shock; red blood cells decreased in hemorrhage	Continuation of stress response; red bone marrow stimulated to release mature (and later immature) red blood cells into the system to carry oxygen; liver and spleen release extra red blood cells; kidneys release RPS to increase arteriolar constriction; renin acts on

| Severe | Down 50% | 2500 mL | Possibly irreversible | Diastolic blood pressure decreased; veins collapsing; pulse rapid, extremely thready and irregular, barely perceptible; dysrhythmias appear on cardiac monitor; respiration rapid and more shallow; unconsciousness; peripheral skin ashen gray, cold, clammy; frank cyanosis; kidney failure (oliguria) | Serum sodium decreased; serum CO_2 decreased (acidosis); white blood cells increased; plasma protein decreased; BUN increased; blood sugar increased; hemoglobin and hematocrit may be increased, normal, or decreased, depending on cause of shock; red blood cells decreased in hemorrhage | plasma protein to form angiotensin to further increase vasoconstriction; erythropoietin released by the kidneys to stimulate production of red blood cells; ACTH released to retain sodium and thus water to increase blood volume

Stress mechanisms are failing; signs and symptoms point to cellular death |

*All of these test results may not be seen in every individual. This is an inclusive guide.
ACTH, adrenocorticotropin hormone; ADH, antidiuretic hormone; RPS, renal pressor substance.

duced body temperature (with vasoconstriction). Other signs and symptoms include apprehension, restlessness, and extreme weakness progressing to unconsciousness; thirst; paleness progressing to pallor and eventually cyanosis of the lips, conjunctiva, and skin (a late sign); spots before the eyes; and ringing in the ears.

Cardiogenic Shock

In cardiogenic shock the heart fails to pump enough blood into the systemic circulation to maintain adequate tissue perfusion. This inability of the heart may be caused by congestive heart failure (pump failure) or irregular, fast beating that does not allow enough time for the ventricle to fill or have enough strength to pump adequately (atrial flutter, tachycardia, fibrillation). Alternatively, myocardial infarction may have lessened the amount of healthy muscular tissue needed for adequate contraction. Cardiogenic shock is a normovolemic condition, in which the patient needs correction of the pump failure rather than replacement of fluid.

Distributive Shock

Neurally induced distributive shock is caused by peripheral vascular dilation. There is no reduction in blood volume but rather a relocation of a large proportion of the blood. The increased size of the vascular bed may be caused by nerve stimulation or a nerve block. Simple fainting from severe psychic stimulation is an example of neurogenic shock. The peripheral blood vessels dilate, leading to pooling of the blood in the periphery. This shunts blood away from the brain and the person faints. When the person is prone, blood moves back into systemic circulation because of the force of gravity, and the condition self-corrects.

Other causative factors include spinal anesthesia shock, insulin shock (insufficient glucose to brain cells for effective nervous system functioning), postural hypotension, drugs (such as Thorazine), and deep general anesthesia (depressing the vasomotor center in the medulla).

Another type of distributive shock is chemically induced. For example, in anaphylactic shock, a foreign protein enters the body and initiates the antigen-antibody response, resulting in the release of histamine. Histamine is a strong peripheral vasodilator and bronchial constrictor. In response to its action, blood is shunted from the bloodstream into the interstitial tissues. Once the blood is in the periphery, symptoms of hypovolemic shock follow. This is not hypovolemic shock however; it is normovolemic shock. Administration of large amounts of fluid is not only unnecessary, it is dangerous. Administration of adrenaline (epinephrine) causes peripheral vasoconstriction and bronchodilation. Antihistamine drugs like diphenhydramine hydrochloride (Benadryl) may also be given to lessen the antigen-antibody reaction.

Septic shock is another type of distributive shock in which bacteria release toxins that can act directly on the vessels to cause vasodilation. Gram-positive organisms release an exotoxin; gram-negative organisms release an endotoxin. Gram-negative shock is usually more profound and difficult to treat than gram-positive shock. Antibiotics are an important part of the treatment.

Obstructive Shock

Obstructive shock is caused by indirect heart failure. The best treatment of this type of shock is to adequately treat the underlying cause before heart failure occurs. If shock does result, the treatment is supportive and includes identification and treatment of the underlying condition, such as cardiac tamponade.

STAGES OF SHOCK

Shock can be staged in several ways. One method is to stage it by severity as seen in Table 23–1. Another method is to describe the progression of untreated shock. These steps include the initial stage, nonprogressive stage, progressive stage, and refractory stage. The physiological events that occur in each stage are outlined in Table 23–2.

ASSESSMENT

Time is crucial in the recognition and treatment of shock. Shock is reversible if identified early and treated effectively. As cellular functions decrease, however, shock may become irreversible. To ensure optimal treatment, it is important that the nurse recognize signs and symptoms that indicate an improving as well as a deepening state of shock and then accurately and promptly report these signs to the physician. Shock is *preventable* and *reversible*; however, it is a continuum on which there is a point of no return (Fig. 23–3).

Hematologic Assessment

In conditions of shock, it is extremely important to monitor chemical changes closely and adjust the treatment as the assessment indicates.

The laboratory measurement of hematocrit usually shows an elevation when there has been a plasma loss. The test is a measurement of red blood cells; therefore, the ratio of red blood cells to the plasma is higher following plasma loss. If there has been an equivalent loss of red blood cells and plasma, the hematocrit laboratory test may not show an early change and may give normal readings for the first few hours.

The decline in the hematocrit results from three specific factors: (1) large amounts of whole blood loss; (2) transcapillary pulling of fluids back into the vasculature from tissues; and (3) the time required to rebuild the supply of mature red blood cells. Movement of immature red blood cells can be inferred from an increased reticulocyte count.

Factors Exacerbating Shock

It is important that a person in shock become neither chilled nor too warm. Either extreme will hamper the effectiveness of vasoconstriction. What happens if hot water bottles and too many blankets are placed on the patient in shock? The heat causes peripheral vasodilation, drawing blood away from

TABLE 23–2
PHYSIOLOGIC EVENTS DURING SHOCK

Stage of Shock	Physiologic Event
Initial stage	Decrease in baseline mean arterial pressure (MAP) of 5–10 mmHg Increased sympathetic stimulation • Mild vasoconstriction • Increase in heart rate

Nonprogressive stage	Decrease in MAP of 10–15 mmHg from the client's baseline value
	Continued sympathetic stimulation
	
	• Moderate vasoconstriction
	• Increased heart rate
	• Decreased pulse pressure
	
	Chemical compensation
	
	• Renin, aldosterone, and antidiuretic hormone secretion
	  • Increased vasoconstriction
	  • Decreased urinary output
	  • Stimulation of the thirst reflex
	
	Some anaerobic metabolism in nonvital organs
	
	• Mild acidosis
	• Mild hyperkalemia

Progressive stage	Decrease in MAP of >20 mmHg from the client's baseline value Anoxia of nonvital organs Hypoxia of vital organs Overall metabolism is anaerobic. • Moderate acidosis • Moderate hyperkalemia • Tissue ischemia
Refractory stage	Severe tissue hypoxia with ischemia and necrosis Release of myocardial depressant factor from pancreas Buildup of toxic metabolites

Source: Ignatavicius, DD, Workman, ML, and Mishler, MA: Medical-Surgical Nursing: A Nursing Process Approach, ed 2. Philadelphia, WB Saunders, 1995, p 970. Reprinted with permission.

Reversible shock   Inadequate → Cellular → Acidosis → Compensatory → Cellular   Irreversible shock
tissue   hypoxia   mechanisms   death
perfusion   fail

FIGURE 23–3
Continuum of shock.

the systemic circulation. In addition, the hot water bottles are extremely dangerous, because a person in shock often loses peripheral neural sensations because of the decreased oxygenation of the skin. Thus, he or she may be burned without feeling it. If the patient is shivering, it is best to apply extra blankets, but never hot water bottles. Shivering is an important sign because it indicates that the patient is a step beyond just being cold, and the body is attempting to warm itself. Shivering must be controlled, because it increases the body's metabolic rate. An increased rate of metabolism requires more oxygen, placing further demands on an already limited supply.

Severe pain may initiate or increase the severity of shock, because it may cause extreme vasoconstriction through increased stimulation of the sympathetic nervous system. Vasoconstriction, if not excessive, helps maintain and increase the circulating blood volume; however, severe vasoconstriction leads to further tissue hypoxia, reinforcing the necessary evil of shock. Thus, adequate relief of pain in the shock patient is important. The route of administration for pain-relieving drugs must be carefully selected. Intravenous infusion is the preferred method. If analgesics are administered intramuscularly or subcutaneously and the peripheral circulation is poor, very little of the drug may be absorbed. If, after several doses have been administered with little absorption, the shock state suddenly improves, all the drugs may be absorbed at the same time when the peripheral circulation increases. For example, if the patient has been given several doses of morphine, an improved shock state and restored circulation might result in an overdose of morphine, causing profound respiratory depression.

Anoxia (lack of oxygen) caused by reasons other than shock enhances a shock state. Anoxic anoxia, also called arterial anoxia, is caused by respiratory obstruction, reduced oxygen supply in the air, or inadequate lung tissue across which sufficient oxygen may be diffused into the bloodstream. Because cellular oxygen in the victim of shock is already inadequate because of inadequate blood circulation or an inadequate blood volume, a diminished supply of oxygen to the blood further aggravates the shock state. This is why it is extremely important to maintain an unobstructed airway and sufficient space for lung expansion in the patient in shock. Currently, the use of the Trendelenburg position, previously thought to be the best position for a patient in shock, is being questioned because it forces the abdominal viscera up against the thoracic cavity, decreasing the space available for adequate lung expansion. An alternative position is a supine position in which only the feet are elevated. Adequate availability of oxygen is also important in a first-aid situation, when it may be necessary to encourage a crowd to move away from the injured person to ensure access to the available oxygen.

Anemic anoxia is a deficiency of hemoglobin in which there is a reduced number of red blood cells available to carry oxygen from the respiratory system, across the alveoli, to each cell of the body. Administration of oxygen to a patient in shock may be of limited benefit if there is not sufficient hemoglobin to carry the extra oxygen to the cells. Some of the oxygen, though, dissolves in the plasma.

Circulatory anoxia results from poor circulation of blood. Local anemia caused by a decrease in arterial flow is called ischemia. Increased ischemia of tissues eventually leads to infarction, or death of tissue, from insufficient oxygen and nutrients. A decrease in venous flow is called stagnation. Stagnation leads to an accumulation of waste products (metabolic acidosis). Circulatory anoxia may be caused by a weakened heart muscle or extremely viscous blood.

Histotoxic or metabolic anoxia occurs when the cells do not have the ability to use oxygen. An example of this is carbon monoxide poisoning.

Changes in the patient's position can increase the severity of shock. Quick position changes distort the tone of the peripheral blood vessels and may decrease the amount of blood returning to systemic circulation. For this reason, it is extremely important that a patient in shock be moved very evenly and slowly. It is best not to move a patient with shock at all until the condition improves.

The presence of other diseases or disabling conditions in a patient may increase the difficulty of overcoming shock. For example, there is a decreased resistance to shock with age, hypoproteinemia, anemia, and malnutrition. It is interesting to note that women tolerate shock with or without blood loss better than men.

COLLABORATIVE MANAGEMENT

The principles of treatment can be embodied in the following mnemonic: VIP-PS.

V-ventilation
I-inhalation
P-perfusion
P-pharmacology
S-surgery

Medical Treatment

Ventilation is important because hypoxia is the common denominator underlying all types of shock. The maintenance of external respiration is critical, to provide both a sufficient intake of oxygen to the lungs and an adequate amount of oxygen for internal diffusion into the cells (internal respiration). Metabolic acidosis can develop when there is not enough oxygen in the body to completely metabolize carbohydrates and fats to carbon dioxide and water. The presence of metabolic acidosis in the body causes decreased circulation, poor energy production, fatigue, and the failure of normal compensatory mechanisms to respond in a stress situation.

Inhalation can be ensured by administering oxygen at acceptable rates and positioning the patient in a modified Trendelenburg position (elevation of the legs with a slight elevation of the head). This change in position effects an internal transfusion. Positive pressure ventilators can also be used to introduce oxygen into the alveoli under pressure, and thus to increase diffusion of oxygen across the pulmonary capillary membranes. In emergency situations without available life support systems, of course, cardiopulmonary resuscitation (CPR) would be indicated.

The success of adequate ventilation can be measured by the arterial partial pressure of oxygen and carbon dioxide (Pao_2, $Paco_2$), and by oxygen saturation levels of arterial blood (Sao_2). Levels for Pao_2 indicate the amount of oxygen in a closed system in the vessels: $Paco_2$ relates to the amount of waste product removal by the lungs; and Sao_2 saturation shows the effectiveness of oxygen uptake by hemoglobin, which as oxyhemoglobin ensures the internal respiration of the cells. It may become necessary to provide additional ventilatory assistance by tracheostomy, endotracheal intubation, and/or pulmonary ventilation.

Intravenous fluids and electrolytes are administered to increase intravascular volume and thus increase tissue perfusion of fluids and nutrients to cells, as well as move waste products out of the body. The type of infusion selected is determined by the cause of the shock.

Specific types of fluid loss in shock conditions and types of fluid replacement are as follows: Whole blood loss is replaced with whole blood. In plasma loss (from burns), 5% glucose in dextrose in water and plasma are possible fluids of choice. When the cause is extensive gastrointestinal losses, gastrointestinal solution replacement is used. In conditions of vasogenic (septic) shock, it is crucial to give antibiotics in addition to fluid replacement.

Cardiogenic shock does not require fluid replacement; however, it is important to maintain a patent intravenous line for the administration of drugs, and to administer oxygen. The major treatment in cardiogenic shock is to improve myocardial contractions, usually with inotropic medications such as digitalis preparations. Diuretics are administered if there is fluid overload. If the patient does not respond to drug therapy, an intra-aortic balloon pump (IABP) may be inserted to facilitate left ventricular emptying. The balloon is inflated during diastole and deflated during systole.

Fluid volume replacements are usually the following:

1. Blood is usually given when hemorrhage has occurred and there is a loss of whole blood. It should be administered with a blood pump if the situation is critical. It is also desirable to warm the blood to body temperature because cold blood can cause dysrhythmias as well as cardiac slowing. If a potassium excess is of concern, fresh blood should be used and the use of uncitrated blood may also be relevant.

2. Balanced solutions may be used instead of blood until typing and crossmatching can be done. The most common replacement is 5% dextrose in lactated Ringer's solution. This solution will provide dextrose to decrease protein catabolism, as well as provide physiologic levels of sodium, chloride, calcium, and potassium. Also, lactate will be coverted to bicarbonate and decrease the acidosis.

3. Another type of solution that may be used is hypertonic plasma, volume expander solutions such as dextran, or plasma. These volume expander solutions are particularly indicated in burns and are selected to replace fluids without causing electrolyte or sodium overload. In white bleeding, in which there has been a plasma loss, saline (usually 0.9%) with bicarbonate added may be chosen to correct metabolic acidosis.

4. Protein solutions may be used following the loss of protein. Concentrated albumin solutions may be given. For example, Amigen may be

used to improve the colloid osmotic pressure in the vascular bed. It is hoped that, with the increased protein, water will be drawn from the cells into the vascular bed to increase vascular volume and decrease the shock. Some saline will be pulled from the interstitial space. When fluids are being replaced, it is important to monitor for blood pressure changes, particularly for the desired increase in the systolic pressure. It is also important to assess the central venous pressure (CVP) (on the hospital unit) or evaluate hemodynamic monitoring (in the critical care unit) to ensure a lessening of the hypovolemia. One must also be alert for an overly rapid infusion of fluid, which could cause a pulmonary overload.

Hemodynamic monitoring provides information about blood volume, vascular capacity, pump effectiveness, and tissue perfusion. A triple- or quadruple-lumen pulmonary catheter is centrally inserted to measure right atrial pressures (RAP), indirect left atrial pressures (LAP), or pulmonary artery wedge pressure (PAWP).

Drugs most commonly used for shock include:

1. Vasoconstrictor drugs, such as dopamine hydrochloride (Intropin), epinephrine (Adrenalin), and norepinephrine (Levophed). Levophed administration must be watched closely to avoid infiltration into the peripheral vascular tissues, because this could result in tissue necrosis.
2. Cardiac drugs that enhance contractility and myocardial perfusion, such as atropine sulfate, dobutamine (Dobutrex), amrinone (Inocor), and sodium nitroprusside (Nitropress) for cardiogenic shock. All of these drugs need to be monitored closely. Equipment for defibrillation should be available.
3. Antibiotics for septic-induced distributive shock. It is very important to obtain blood cultures to identify the organism before starting treatment. Most antibiotics have major adverse effects. For example, aminoglycosides, such as gentamycin, can cause kidney dysfunction (especially in elderly persons), so monitoring the blood urea nitrogen (BUN) and serum creatinine levels is important. A more accurate method of monitoring these drugs is by peak and trough blood levels.
4. Steroids for shock conditions that are not responding well to treatment. Steroids tend to decrease peripheral resistance, that is, decrease peripheral vasoconstriction. Steroids will cause an increase in tissue perfusion, and with the increased venous pressure, the flow of fluid back to the heart will be increased and the heartbeat strengthened. When administering steroids intravenously, it is important to be alert for gastrointestinal system complications, such as complaints of pain, stress, ulcer, and gastrointestinal GI bleeding. Prolonged use of steroids may also cause a predisposition to infection.
5. Mannitol, an osmotic diuretic, particularly if there has been an episode of prerenal failure and oliguria. Mannitol increases renal blood flow and perfusion to the kidneys and can help prevent postshock renal failure.

Surgical procedures may be necessary. For example:

- Surgical management of vascular obstructions caused by clots caused by decreased circulation from the shock condition

- Pericardiocentesis to relieve cardiac tamponade
- Incisional drainage of abscesses, following an acute crisis, to prevent further infection

Nursing Interventions

Early discovery of shock through skilled nursing assessment can help prevent irreversible cellular damage and the increased, prolonged difficulties that can follow the failure of stress compensatory mechanisms. Many of the nursing responsibilities regarding shock have been included in the foregoing discussions of a particular pathophysiology or treatment. Thus, parts of this list will be a review. Because this chapter has presented a great deal of information and guidelines regarding both the pathophysiology of shock and the keen observations required to deal with it, this brief summary of nursing responsibilities should aid the reader in organizing and assimilating this vital information.

Prevention of shock is most important and can be achieved through competent care. For surgical patients, the gentle handling of tissue, maintenance of complete homeostasis, immediate replacement of large losses of blood, minimal exposure of the viscera to atmospheric air, minimalization of the patient's apprehension and pain, careful anesthetization, effective postoperative care, and prompt treatment are the means available to prevent surgical shock. The nurse can be a valuable aid in alleviating the patient's anxiety preoperatively and in supporting the patient through the postoperative period.

Controlling hemorrhage is important. Conservation of the patient's body heat with blankets is necessary, but not to the point of overheating the patient and thereby causing vasodilation. The nurse should remember to use shivering as a guide.

The patient should be moved as little as possible to prevent increased shock. By raising only the patient's feet, leaving the rest of the body in a supine position, approximately 500 mL of blood will be redirected to the circulatory system. The effect is one of giving an internal transfusion.

Because pain intensifies shock, pain management is crucial. Restlessness may indicate a need for oxygen or pain interventions. Maintaining adequate ventilation is vital to keep internal and external respiration functioning optimally. This can be done by proper positioning of the patient, correct breathing techniques and coughing, and active and passive exercises that will increase the venous flow and proper circulation.

Replacement of required fluids is essential in hypovolemic types of shock. Central venous pressure, atrial pressures, and pulmonary artery wedge pressure all indicate fluid volume status. Central venous pressure can also be estimated by observing the neck veins. Neck veins should normally fill to half of the column between the clavicle and the jaw.

Urine output should be closely monitored because renal perfusion is the best guide to tissue perfusion. This can be effectively accomplished by measuring hourly urine output.

The primary objective of the health team is the prevention of irreversible shock. The overall objective of treatment and nursing care, as well as of all the body's compensatory mechanisms, is to restore or maintain adequate tissue perfusion to maximize resistance of the patient to shock.

REFERENCES AND READINGS

Bell, T (1990). Disseminated intravascular coagulation and shock. Critical Care Clinics of North America 2(2), 255–262.

Burns, K (1990). Vasoactive drug therapy in shock. Critical Care Clinics of North America 2(2), 167–178.

Clochesy, J, et al (1993). Critical Care Nursing. Philadelphia: WB Saunders.

Colletti, R, Dew, R, & Goulart, A (1993). Antiendotoxin therapy in sepsis. Critical Care Clinics of North America 5(2), 345–354.

Daleiden, A (1993). Physiology and treatment of hemorrhagic shock during the early postoperative period. Critical Care Nursing Quarterly 16(1), 45–59.

Houston, M (1990). Pathophysiology of shock. Critical Care Clinics of North America 2(2), 143–149.

Klein, D (1990). Shock: Physiology, signs, symptoms. Nursing91 21(11), 74–76.

McCormac, M (1990). Managing hemorrhagic shock. American Journal of Nursing 90(12), 22–27.

McMorrow, M, & Cooney-Daniello, M (1991). When to suspect septic shock. RN 54(10), 32–37.

O'Neal, P (1994). How to spot early signs of cardiogenic shock. American Journal of Nursing 94(5), 36–40.

Rice, V (1991a). Shock, a clinical syndrome: An update. Part 1: An overview of shock. Critical Care Nurse 11(4), 20–24.

Rice, V (1991b). Shock, a clinical syndrome: An update. Part IV: Nursing care of the shock patient. Critical Care Nurse 11(7), 28–42.

Russell, S (1994). Hypovolemic shock: Is your patient at risk? Nursing94 24(4), 34–39.

Russell, S (1994). Septic shock: Can you recognize the clues? Nursing94 24(4), 40–48.

Singleton, KA, & Workman, ML (1995). Interventions for clients in shock. In DD Ignatavicius, ML Workman, & MA Mishler (eds), Medical-Surgical Nursing: A nursing process approach, ed 2. Philadelphia: WB Saunders, pp 965–984.

Stengle, J, & Dries, D (1994). Sepsis in the elderly. Critical Care Nursing Clinics of North America 6(2), 421–427.

CHAPTER 24

· · · · · · · ·
· · · · · · · ·
· · · · · · · ·
· · · · · · · ·
· · · · · · · ·
· · · · · · · ·
· · · · · · · ·
· · · · · · · ·
· · · · · · · ·
· · · · · · · ·

Burns

Burns are particularly critical in regard to fluid and electrolyte balance, because they present an immediate danger to life. The greatest threats to a patient with severe burns are shock and renal failure as a result of extensive fluid loss. Infections, particularly gram-negative respiratory infections, are an additional health hazard for the patient with burns.

TYPES OF BURNS

Burns may be caused by thermal, chemical, or electrical injury. Sunburn is the most common type of burn. Scalds represent about 70% of the self-inflicted burns in children. Direct contact with flame; strong acids, alkalis, or corrosive fluids; powder burns; high-voltage electricity; and inhalation of hot gases and smoke are other causes. Burns may also be indicative of child abuse or neglect. Excessive skin reaction to radiation therapy is also a type of burn (radiation burn).

EVALUATION OF BURNS

There is no system for evaluating the extent of burns that measures precisely the amount of fluid lost by a burn victim. Because it is so important to replace fluids immediately, however, a simple though somewhat imprecise method has been devised. Two factors must be considered when estimating the burn patient's need for fluid replacement. One is the amount of surface involvement and the other is the depth of the burn. The rule of nines and the Lund and Browder Chart provide approximate guides to determining surface involvement in percent of total body surface area.

The rule of nines divides the body on a rough visual basis into percentages that are multiples of 9 (Fig. 24–1).

The Lund and Browder Chart gives a more extensive breakdown of areas of the body and also distinguishes among different age groups (Fig. 24–2).

The depth as well as the surface area of a burn is important in determining fluid loss. Burn depth was formerly classified by degree, but is now more commonly classified by the extent of the burn or by the layers of tissue involved. All three systems are described here.

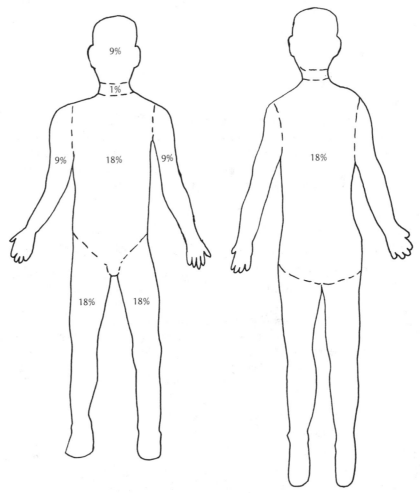

FIGURE 24–1
Rule of nines: head 9%; neck 1%; each arm 9% (front only, 4.5%); front 18%; back 18%; each leg 18% (front only, 9%).

LUND AND BROWDER CHART								
	Age -- Years					%	%	%
Area	0-1	1-4	5-9	10-15	Adult	2°	3°	Total
Head	19	17	13	10	7			
Neck	2	2	2	2	2			
Ant. trunk	13	17	13	13	13			
Post. trunk	13	13	13	13	13			
R. buttock	2½	2½	2½	2½	2½			
L. buttock	2½	2½	2½	2½	2½			
Genitalia	1	1	1	1	1			
R. U. Arm	4	4	4	4	4			
L. U. Arm	4	4	4	4	4			
R. L. Arm	3	3	3	3	3			
L. L. Arm	3	3	3	3	3			
R. Hand	2½	2½	2½	2½	2½			
L. Hand	2½	2½	2½	2½	2½			
R. Thigh	5½	6½	8½	8½	9½			
L. Thigh	5½	6½	8½	8½	9½			
R. Leg	5	5	5½	6	7			
L. Leg	5	5	5½	6	7			
R. Foot	3½	3½	3½	3½	3½			
L. Foot	3½	3½	3½	3½	3½			
					Total			

FIGURE 24–2
Lund and Browder chart. (Reprinted from Initial Burn Management, Flint Laboratories, Division of Baxter-Travenol Laboratories, Inc., Deerfield, IL, 1975, with permission.)

Degrees of Burns

First degree burns are indicated by dry, red skin. There are no blisters. This type of burn is minor unless age or infection causes it to be more critical. Although a first degree burn is painful, it usually heals quickly unless the patient is a young child or an adult over 65. The most common cause of first-degree burn is overexposure to the sun.

In *second degree* burns the skin is red, wet, and blistered. A second degree burn is very painful but will usually heal if not complicated by infection. Scalding with hot liquids is considered the primary cause of second degree burns.

Third degree burns are the most serious because they affect the deeper subcutaneous and muscle tissue. The skin is dry, frequently with a leathery eschar or slough. Charred blood vessels may be visible beneath the eschar. The skin color may be white or dark-colored and charred. There is little or no pain. Skin grafting is required for the burned area to heal. Third-degree burns are usually caused by direct contact with flame, chemicals, or electricity. *Fourth degree* burns are really severe third-degree injuries that can extend down to the bone.

Extent of Burns

In addition to being classified as first, second, or third degree, burns may also be described as minor or major. A burn is minor if it is a second degree burn, irrespective of appearance, that covers less than 15% of the total body surface area (TBSA) in adults and less than 10% of the TBSA in children or the elderly. With a third degree burn, it is minor if less than 2% of the TBSA is burned, and vital areas such as face or hands are not involved. The skin of a minor burn usually is reddened and unbroken and may have blisters.

A major burn is a second degree burn involving more than 15% of the TBSA in adults and more than 10% of the TBSA in children and adults over 40. Third degree burns covering more than 10% of the TBSA or significant burns involving the face, hands, feet, or perineum should also be considered major burns.

Thickness of Burns

Most clinicians classify burns by thickness of tissue involved. Burn wounds may be classified as:

- Superficial-thickness wounds
- Partial-thickness wounds (superficial and deep subgroups)
- Full-thickness wounds

Superficial-thickness burns affect only the epidermal layer of the skin. Erythema, mild edema, pain, and heat sensitivity result. The dead skin usually peels (desquamation) for several days after the burn and is replaced by new tissue.

A *partial-thickness* wound involves the epidermis and part of the dermis. Superficial partial-thickness burns result in blisters (vesicles) that heal in 10 to 14 days without scarring. Deeper wounds extend down into the dermis, have no blisters, and make the skin appear waxy or white. These burns take several weeks to heal and scarring occurs.

Full-thickness burns involve the entire epidermal and dermal layers of skin and are characterized by a hard, leathery covering, known as an eschar. Eschar must be removed or slough off to enable the area to heal. Wounds deeper than these levels are called fourth degree burns as described above. Fascia, tendons, muscle, and bone can be damaged.

PATHOPHYSIOLOGY AND ASSESSMENT

The pathophysiology of a severe burn involves rapid fluid and electrolyte changes. There is an apparent redistribution of fluid, protein, and minerals. The damaged capillary walls in the burn area permit plasma proteins to move into interstitial spaces. As the body responds to the burn or to a possible circulating burn toxin, capillaries in other parts of the body develop a permeability that allows plasma proteins to pass through this barrier. As a result of the decreased osmotic pressure within the blood vessels and the increased osmotic pressure in the interstitial spaces, fluid accumulates at the site of the thermally injured tissues, in interstitial spaces, and/or in blisters. The accumulation of fluid in blisters is a third space fluid loss.

Fluid is also lost as a result of evaporation. The skin, which normally restricts fluid and electrolyte loss and provides a barrier against infection, is damaged or destroyed. The loss of fluid by evaporation as a result of skin destruction is approximately 20 times greater than normal skin losses.

A redistribution or loss of other components of the body chemistry occurs along with the fluid loss. Large amounts of sodium move with the fluid from the intravascular to the interstitial fluid area during the first 24 hours (third spacing). The result of this type of third spacing is hypovolemia, which can lead to hypovolemic shock. For this reason, the first 24 hours or so after a burn is sometimes referred to as the "shock phase." (Also see Chap. 2 under Fluid Deficit for a more detailed discussion of third spacing.)

Sodium is normally present in approximately the same proportions in the intravascular and interstitial spaces; however, proportionately more sodium than water is drawn into interstitial spaces after a burn. Also, when a burn occurs, the sodium pump becomes less capable of holding saline in the intravascular space and more saline enters the intracellular space.

There is an increase in the release of aldosterone (renin-angiotensin-aldosterone mechanism) and antidiuretic hormone as a result of the general stress response and hypovolemia, which increases the amount of sodium and water retained. The amount of fluid retained in the interstitial and intracellular tissues because of a major burn may amount to several liters. At this time the patient usually becomes oliguric. In children an inappropriate secretion of ADH may cause dilutional hyponatremia accompanied by excretion of concentrated urine. Possible causes of the inappropriate ADH response include anxiety, prolonged pain, pyrexia, or a temporary hypothalamic dysfunction.

Potassium is elevated in the bloodstream because of the destruction of tissue and oliguria. The result of this shift is hyperkalemia.

Patients with burns also have a greater oxygen and caloric need in order to sustain the increased metabolic rate. The increased metabolic rate results partially from the body's need to replace the heat that is lost because of increased evaporation. It is this loss of body heat, as well as the stress response, that causes burn patients to complain of feeling chilly.

During the second 24 hours after the burn has occurred, the second fluid shift takes place. In what is sometimes called the diuretic phase, the sodium shifts back from the interstitial into the intravascular space. Diuresis occurs, and the patient may develop hypokalemia as a consequence of potassium lost via the urine. Other electrolytes, most significantly magnesium and serum proteins, are lost.

Vitamin deficiencies caused by lack of intake are also common in patients with severe burns. There is red blood cell loss from a microangiopathic anemia. The seriousness of this anemia is correlated to the severity and extent of the burn. Red blood cell destruction is caused initially by the heat of the burn and later by the hemolysis of heat-damaged cells. The result of the red blood cell destruction is critical, not only in terms of anemia but in terms of kidney damage as well. Normally, red blood cells do not pass into the kidney glomeruli. The damaged red blood cells release hemoglobin, which is filtered by the kidneys. A similar substance, myoglobin, is released by the damaged muscle tissue. Because both myoglobin and hemoglobin are filtered by the kidneys, and because this is an abnormal function, the kidney tubules may become plugged, possibly leading to acute tubular necrosis.

When body tissue is destroyed, potassium is released into the extracellular compartment, causing hyperkalemia. Further intensification of hyperkalemia is caused by oliguria. As the body attempts to compensate for increased potassium, additional hydrogen ions (acid) move from the intracellular fluid space to the intravascular space. The result is metabolic acidosis.

As kidney function continues to decrease from hypoperfusion, uremic waste products are retained (acute renal failure) and contribute to the acidosis. Age is a factor in recovery from the stress placed upon the kidneys. Patients over 60 or with a pre-existing renal disease are placed in a life-threatening position by these combined renal dysfunctions.

After 48 hours, capillary integrity is usually restored as the blood volume becomes expanded because of edema fluid mobilization; this may result in saline excess and circulatory overload. Fluid mobilization may cause clinical signs and laboratory values to be deceptive. Examples of such deceptive indications of improvement are high cardiac output and diuresis.

Lung injuries can be differentiated from hypovolemic shock by the presence of crackles, wheezes, and signs of hypoxia. There are three categories of respiratory injury: (1) smoke inhalation; (2) burns without smoke inhalation; and (3) a combination of burns and smoke inhalation. Respiratory failure is one of the major causes of death in patients with burns. The highest mortality is among those with respiratory damage caused by both burns and smoke inhalation.

Curling's ulcer or peptic ulcers may develop because of severe stress. Therefore gastric mucosal irritation or injury should be recognized and prevented by using nasogastric suction, antacids, H_2 histamine blockers (such as Zantac), and adequate nutritional support. Surgery is usually indicated if gastrointestinal hemorrhage cannot be controlled by medical procedures.

Morbidity in children includes shock and infection as well as thermal injuries. It also has been noted that hypertension in children is a significant complication of thermal injuries, sometimes leading to encephalopathy and convulsions.

The return of optimal function to the burned areas requires many months and frequently many years. This is especially true of the extremities, neck, and axillae. Immediate skilled attention by the health team, procedures such as surgical skin grafting, and the relief of contractures maximize the burn patient's chance of survival and return to optimal functioning.

COLLABORATIVE MANAGEMENT

Burn treatment must be discussed in terms of emergency treatment, and the more extensive treatment of fluid replacement and compensation for loss of life-sustaining functions.

Emergency Treatment

The *first priority* is to establish an adequate airway. When emergency personnel are available, more in-depth evaluation of the burn victim should be made. A tracheotomy is not necessary for every patient with a head or neck burn, but preparation should be made. Nasotracheal intubation should be attempted first, followed by endotracheal intubation if necessary. Later indications that a tracheotomy may be necessary are increased hoarseness, changes in respiratory pattern, and/or a sudden increase in laryngeal secretions or spasms.

An IV of lactated Ringer's solution is usually started if it requires more than 30 minutes to transport the patient to the hospital. The burned area is cleaned with cool water and bland soap.

Burn Unit Care

Burn therapy is complex because of the need to assess the interdependent factors of age, total body surface area (TBSA) of the patient, patient weight, percentage of body burned, renal output, and postburn time factor. The amount of fluid replacement should be calculated from the time of the burn. Because the maximal fluid loss occurs during the first 8 hours, the calculated loss should be replaced within the first 24 hours in burns covering 15% or more of the TBSA in adults and 10% or more of the TBSA in children and the elderly. This aggressive administration of fluid in the first 24 hours will usually sustain circulatory integrity, thereby maintaining renal function and preventing visceral failure and subsequent death. It is advisable to have several alternative methods of fluid replacement to suit the physiologic function of the individual (Table 24–1).

A number of fluid resuscitation formulas are used during the first 24 hours after a burn injury (Table 24–2). One regimen, the Parkland Regimen (Dallas Formula), reflects the more recent trend to avoid administration of colloid and hypotonic saline fluids during early treatment; however, there is some evidence that the leaking capillary bed heals earlier in children than in adults and that positive results may be achieved by administering colloids to children within the first 24 hours. According to the Parkland Regimen, a patient weighing 70 kg who has burns over 30% of the TBSA would receive 8400 mL of lactated Ringer's solution during the first 24 hours after the burn. The lactated Ringer's solution is administered at the rate of one half of the estimated fluid requirement during the first 8 hours, followed by one fourth the fluid requirement in the second 8 hours, and the remaining one fourth in the last 8 hours. This is the same distribution ratio as the Brooke Formula. Administration of lactated Ringer's solution is very successful in the treatment of children, providing an accurate guide for 99% of pediatric burns. Failure of the Parkland Regimen occurs for the most part in children in whom the burn is over 80% of the TBSA, or in adults over the age of 45.

These regimens are empirical guides, not hard-and-fast rules. The healthcare team must constantly reassess the individual patient's needs and adjust treatment accordingly. High-voltage electrical burns may be difficult

TABLE 24–1

THREE MAJOR REGIMENS FOR FLUID REPLACEMENT

The Brooke Formula	The Evans Formula	The Parkland Regimen (Dallas Formula)
During first 24 hours: Colloids (blood, dextran, or plasma): 0.5 mL per kg percent of body surface burned. Lactated Ringer's Solution: 1.5 mL per kg percent of body surface burned. Water requirement (dextrose in water): 2000 mL for adults; correspondingly less in children. If, in a burn of more than 50% in a large patient, fluid therapy based on a 50% restriction fails to prevent signs and symptoms of circulatory failure, therapy must be cautiously increased. One half of the estimated fluid requirements for the first 24 hours is usually given in the first 8 hours, one fourth in the second 8 hours, and one fourth in the third 8 hours.	During first 24 hours: Colloids: 1 mL per kg percent of body surface burned. Physiologic Saline Solution: 1 mL per kg percent of body surface burned. Nonelectrolytes: 2000 mL of 5% dextrose in water; correspondingly less in children.	During first 24 hours: Lactated Ringer's solution is given in the amount of 4 mL per kg percent of body surface burned.

NOTE: In adults, maintain urine output at 30 to 50 mL/h.

TABLE 24–2

COMMON FLUID RESUSCITATION FORMULAS FOR THE FIRST 24 HOURS AFTER A BURN INJURY

	Formula	Solution	Rate of Administration
Brooke	2 mL/kg per % BSA burn + 2000 mL/24 hr (maintenance fluid)	¾ crystalloid, ¼ colloid D_5W as maintenance	½ given in first 8 hr ½ given in next 16 hr
Parkland (Baxter)	4 mL/kg per % BSA burn for 24-hr period	Crystalloid only (lactated Ringer's)	½ given in first 8 hr ½ given in next 16 hr
Monafo		Crystalloid (hypertonic saline: sodium = 250 mEq/L)	Adjust to maintain urinary output of 30 mL/hr
Modified Parkland	4 mL/kg per % BSA burn + 15 mL/m² of BSA	Crystalloid only (lactated Ringer's)	½ given in first 8 hr ½ given in next 16 hr
Winski	2 mL/kg per % BSA burn + maintenance fluid	Crystalloid only (lactated Ringer's)	½ given in first 8 hr ½ given in next 16 hr

BSA, body surface area; D_5W, 5% dextrose in water.
Source: Ignatavicius, DD, Workman, ML, and Mishler, MA: Medical-Surgical Nursing: A Nursing Process Approach, ed 2. Philadelphia, WB Saunders, 1995, p 1989. Reprinted with permission.

to evaluate in terms of fluid need. The health team needs to be aware that large amounts of fluid are required in this category of burns.

Special consideration needs to be given to children under 36 months. Unique to this age group are temperature fluctuation, fluid loss through vomiting and diarrhea, fluctuation in basal metabolic rate, and variations in renal glomerular flow rate. Acidosis may occur suddenly, accompanied by hyperventilation and subsequent fluid loss. These factors, which are unique to children, indicate that a regimen based on adult norms may not provide accurate fluid replacement. Because body surface area is closely related to insensible water losses, surface area is a reliable guideline for calculation of fluid loss in children under 36 months.

Fluid therapy requires constant monitoring of urine values and output. Output for elderly patients should be maintained at approximately 50 mL/h. It should be noted that elderly patients have a low tolerance to fluid overload. Output in children should be maintained at 1 mL/kg of body weight per hour. Output for the average adult should be 30 to 50 mL/h. Monitoring urine output is extremely critical in light of the fact that the fluid sequestered at the burn site during the first 2 to 3 days is reabsorbed into the circulation in the following 3 to 6 days; consequently, fluid overload is a very real danger.

Once the immediate postburn need for massive intravenous fluid and electrolyte therapy has been met, it is imperative to fulfill the patient's increased nutritional requirements. Increased postburn morbidity is associated with a postburn weight loss of over 10%. Increase in protein synthesis and breakdown is proportional to the percentage of TBSA burned. If the increased caloric needs are not met, a weight loss occurs and the patient's defense mechanism against infection is reduced. Burned adults require about 60 calories/kg of body weight and about 3 g of protein/kg of body weight.

As with adults, the nutritional status of children must be evaluated upon admission. And as with adults, caloric requirements are increased in proportion to the percentage of TBSA burned. Children usually require approximately 85 calories/kg of body weight and 4 g of protein/kg of body weight. Most children maintain oral intake and do not require caloric supplementation or tube feeding; however, should enteral or parenteral feeding be required, it should be initiated by the third day of hospitalization. Both children and adults require large amounts of absorbic acid, the B vitamins, and the fat soluble vitamins A, D, and K; these should be administered as appropriate.

The potential for infection always exists with burns. A two-pronged attack is used to control infection. Surveillance of the wound, the burn unit equipment, and the surroundings for bacterial contamination reduces the introduction of harmful bacteria. Topical chemotherapy is used to prevent and control wound infection.

Topical agents that may be applied are mafenide acetate (Sulfamylon), silver sulfadiazine (Silvadene), povidone iodine (Betadine), and silver nitrate. A 2-year prospective study of 100 burn patients conducted by Vatche Ayvazian, MD, Director of the Burn Center, St. John's Mercy Medical Center, St. Louis, Missouri, indicated that more protection was received against gram-negative infections by using a combination of cerous nitrate with Silvadene. Gram-positive infections were better controlled with silver sulfadiazine (Silvadene) alone.

Pain is another important factor in planning the treatment procedure for the burn patient. Some pain reliever is usually administered during the initial emergency treatment. Pain medication should *never* be given intramuscularly; because of poor circulation, it may not be absorbed into the bloodstream, and buildup of pain medication administered intramuscularly could cause an overdose when normal circulation is restored.

Morphine may be administered intravenously, especially if the burn covers more than 20% of the TBSA. Large doses of narcotics should be avoided, especially in children. Patient restlessness in the acute phase is frequently caused by hypovolemia, hypoxia, or pain.

It is necessary to provide external heat to the burn patient. Various types of burn beds equipped with heat shields are available. Also, some burn units use permanent heat lamps installed in the ceiling as well as adjustable ceiling heat lamps. Other methods include increasing the room temperature.

The patient must be kept warm without direct contact of the bed linen with the exposed burn wound. An overbed cradle may be used to support blankets placed over the patient. Another alternative is a tent of aluminum foil, beneath which lamps are placed to reflect the heat off the interior of the tent and thereby warm the patient.

Nursing Interventions

Because burn patients are in a state of rapid fluctuation, nursing care is important throughout every stage of the burn patient's recovery. The nurse should:

- Monitor vital signs frequently.
- Maintain accurate intake and output records because urine output is the best guide to the patient's fluid status, the possibility of developing hypovolemic shock, and the development of acute tubular necrosis. Observe for dark-colored urine. A serial urine test will indicate whether or not the condition is improving or deteriorating. In the case of deterioration, mannitol, an osmotic diuretic, may be ordered to flush the tubules.
- Report anuria immediately to the physician.
- Prevent shock by administering replacement for calculated fluid loss intravenously. This is particularly important during the first 24 hours after the burn.
- Conserve the patient's body heat to minimize lactic acidosis that is due to inadequate metabolism of carbohydrates as a consequence of the body's increased need for oxygen.
- Administer pain medication intravenously as needed.
- Be alert to changes in respiratory pattern because changes may be indicative of inhalation injury or pulmonary edema.
- Maintain sterility of wounds and asepsis of environment.
- Meet the increased nutritional needs of the patient by increasing the amounts of protein, calories, and vitamins.
- Maintain a patent IV mainline at all times.

- Be alert to psychological and emotional needs, particularly grief, fear, and anxiety.

Early continuous psychosocial support by a nurse as well as a social worker, psychologist, psychiatrist, physical therapist, occupational therapist, and rehabilitation experts increases the patient's chances of returning to productive activity.

REFERENCES AND READINGS

Adams, LE, Purdue, GF, & Hunt, JL (1991). Tap-water scald burns. Journal of Burn Care and Rehabilitation 12, 91–95.

Bayley, E, et al (1992). Research priorities for burn nursing: Patient, nurse, and burn prevention education. Journal of Burn Care and Rehabilitation 12, 377–383.

Calistro, A (1993). Burn care basics and beyond. RN 56(3), 26–31.

Konigova, R (1992). The psychological problems of burned patients. Burns 18, 189–199.

Linares, AZ, & Linares, HA (1990). Burn prevention: The need for a comprehensive approach. Burns 18, 416–418.

Martyn, JA (1990). Acute Management of the Burned Patient. Philadelphia: WB Saunders.

Monofo, WW (1992). Then and now: 50 years of burn treatment. Burns 18, S7–S10.

Sadowski, DA (1992). Care of the child with burns. In MF Hazinski (ed), Nursing Care of the Critically Ill Child. St. Louis: CV Mosby.

Sadowski, DA (1995). Interventions for clients with burns. In DD Ignatavicius, ML Workman, & MA Mishler (eds), Medical-Surgical Nursing: A Nursing Process Approach. Philadelphia: WB Saunders.

Trofino, RB (1991). Nursing Care of the Burn-Injured Patient. Philadelphia: FA Davis.

Ward, RS (1991). Pressure therapy for the control of hypertrophic scar formation after burn injury. Journal of Burn Care and Rehabilitation 12, 257–262.

PART FOUR **QUESTIONS**

• • • • • • • • • • •

1. A patient with emphysema is admitted to the emergency department in a coma. What is the possible explanation for the coma in this type of patient?
   a. Carbon dioxide narcosis
   b. Hypoxia
   c. Hypokalemia
   d. Fluid deficit

2. What is a common laboratory finding in patients with congestive heart failure?
   a. Hyponatremia
   b. Metabolic alkalosis
   c. Hypokalemia
   d. Hyperchloremia

3. A patient is admitted with left-sided heart failure. Which of the following assessment findings is inconsistent with this diagnosis?
   a. Dyspnea
   b. Distended jugular veins
   c. Crackles
   d. Weakness

4. What type of dietary restriction is typically required for a patient with congestive heart failure?
   a. Potassium restriction
   b. Protein restriction
   c. Carbohydrate restriction
   d. Sodium restriction

5. Which of the following drugs is typically prescribed for the patient with congestive heart failure?
   a. Sodium bicarbonate
   b. Cipro
   c. Digoxin
   d. Aquamephyton

6. Which of the following clinical manifestations is usually present in patients with Addison's disease?
   a. Increased skin pigmentation
   b. Hypertension
   c. Hyperglycemia
   d. Edema

7. What major fluid or electrolyte imbalance occurs in patients with hyperglycemic, hyperosmolar nonketotic syndrome?
   a. Fluid overload
   b. Hypokalemia
   c. Hypercalcemia
   d. Fluid deficit

8. Which of the following signs and symptoms are not characteristic of diabetes mellitus?
   a. Polydipsia
   b. Polygraphia
   c. Polyphagia
   d. Polyuria

9. A type I diabetic complains of dizziness, shakiness, and hunger. What is the appropriate intervention for the patient?
   a. Take vital signs immediately.
   b. Give insulin as ordered.
   c. Provide a source of sugar.
   d. Test the urine for ketones.

10. What laboratory finding is expected in patients with oliguria related to renal failure?
    a. Increased creatinine level
    b. Decreased BUN
    c. Decreased potassium level
    d. Increased calcium level

11. What is the appropriate intervention for the patient in the oliguric phase of acute renal failure?
    a. Give potassium supplement as ordered.
    b. Provide high-protein foods.
    c. Encourage high-sodium foods.
    d. Limit fluids.

12. What acid-base imbalance is common in patients with chronic renal failure?
    a. Metabolic alkalosis
    b. Metabolic acidosis
    c. Respiratory alkalosis
    d. Respiratory acidosis

13. Which of the following assessment findings is common in patients experiencing hypovolemic shock?
    a. Increased blood pressure
    b. Decreased urinary output
    c. Decreased pulse
    d. Increased CVP reading

14. What is the most important nursing intervention related to shock?
    a. Prevent shock in high-risk patients.
    b. Give dopamine as ordered.
    c. Place patients in the Trendelenberg position.
    d. Use rotating tourniquets.

15. An unconscious person is taken from a burning house. What is the first priority for care of this patient?
    a. Provide IV fluids.
    b. Monitor for fluid and electrolyte imbalances.
    c. Establish an airway.
    d. Place the patient on a cardiac monitor.

APPENDIX

ANSWERS TO TEST QUESTIONS

PART ONE

1. b	5. a	9. a	13. b
2. d	6. b	10. d	14. c
3. d	7. b	11. c	15. c
4. c	8. c	12. a	

PART TWO

1. c	5. d	9. c	13. c
2. a	6. b	10. b	14. b
3. c	7. d	11. d	15. a
4. b	8. a	12. a	

PART THREE

1. b	5. c	9. d	13. a
2. d	6. b	10. c	14. c
3. a	7. a	11. b	15. b
4. c	8. d	12. a	

PART FOUR

1. a	5. c	9. c	13. b
2. c	6. a	10. a	14. a
3. b	7. d	11. d	15. c
4. d	8. b	12. b	

GLOSSARY

• • • • • • • • • •

acid. Any substance of a group that characteristically is sour in taste, neutralizes basic substances, makes litmus paper red, and produces hydrogen ions (protons) when reacting with certain substances.

acidosis. A disturbance of the acid-base balance of the body resulting in serum that is more acidic than normal (pH below 7.35).

active transport. The movement of substances, particularly electrolyte ions, across the cell membranes and epithelial layers—usually against a concentration gradient.

adenosine triphosphate (ATP). A chemical compound derived from nutrients which is used by the cell as a source of energy.

ADH. Abbreviation for antidiuretic hormone.

adrenocorticotropic hormone. An anterior pituitary hormone that stimulates the adrenal cortex to produce steroids.

aerobic. Living only in the presence of oxygen.

albumin. A simple protein found in most animal and vegetable tissue.

aldosterone. A hormone produced by the adrenal cortex that regulates the volume of blood and extracellular water through the reabsorption of sodium by the kidneys.

alkalemia. Excessive alkalinity of the blood because of a decrease in the hydrogen ion concentration or an increase in hydroxyl ions. The blood is normally slightly alkaline (pH 7.35 to 7.45).

alkalosis. Increase in body alkalinity resulting in a base excess in the serum (pH above 7.45).

alveolar-capillary diffusion. Exchange of oxygen and carbon dioxide gases across the alveolar-capillary membrane.

alveolar ventilation. The process by which atmospheric gases move into the alveoli.

alveoli. Microscopic air sacs in the lungs.

anabolism. Building up of the body tissue by combining simpler substances into more complex compounds.

anaerobic. Having the ability to live without air.

anasarca. Generalized massive edema.

anion. A negatively charged ion. Examples are Cl^-, HCO_3^-, HPO_4^{2-}, lactate.

anion gap. A concept used to estimate electrolyte (anion and cation) levels in the serum and conditions that influence them.

anorexia. Loss of appetite.

anoxia. Absence of oxygen in the body tissues.

antidiuretic hormone. A hormone produced by the hypothalamus and secreted by the pituitary gland that suppresses the secretion of urine and results in the retention of water. Abbreviation: ADH.

anuria. Urine output less than 10 mL/24 h.

apnea. Cessation of respiration in the resting expiratory position.

ascites. Serous fluid in the peritoneal cavity.

asterixis. A motor disturbance marked by inability to maintain an assumed posture as a result of intermittency of the sustained contraction of groups of muscles; flapping tremors of the extremities.

asthenia. Lack or loss of strength.

atom. The smallest particle of an element that can exist either alone or in combination.

azotemia. Presence of increased nitrogenous waste products in the blood (azo = nitrogen; emia = blood).

base. (1) The lower part of anything; (2) the principal substance in a mixture; (3) a substance that can bind a proton.

bicarbonate. A salt resulting from the incomplete neutralization of carbonic acid or from the passing of an excess of carbon dioxide into a solution of a base; HCO_3^- Synonym: base.

bicarbonate/carbonic acid ratio. A buffer mechanism formed by a balance of 20 parts of bicarbonate with 1 part carbonic acid in human blood.

Biot's respiration. Originally described in patients with meningitis; refers to sequences of uniformly deep gasps alternating with apnea.

blood-brain barrier. An anatomic-physiologic feature of the brain thought to consist of walls of capillaries in the central nervous system and surrounding glial membranes. The barrier separates the parenchyma of the central nervous system from blood. The blood-brain barrier functions in preventing or slowing the passage of various chemical compounds, radioactive ions, and disease-causing organisms, such as viruses, from the blood into the central nervous system.

borborygmus. A gurgling, splashing sound heard over the large intestine, caused by passage of gas through the intestines.

bound. Referring to substances that are joined to other substances, thus being held in chemical or physical combination temporarily or permanently.

buffer. A chemical substance that reacts to variations in the serum alkalinity/acidity in such a way that changes in the pH level are minimized; it also acts as a transport system that moves excess hydrogen ions to the lung.

buffer base. Any substance that will neutralize an acid substance: correlates with alkali.

BV. Abbreviation for blood volume.

cachectic. Malnourished.

calcium. A chemical element. The most abundant mineral in the body. In combination with phosphorous, it forms calcium phosphate, the dense hard material of the bones and teeth.

CAPD. Abbreviation for continuous ambulatory peritoneal dialysis.

carbonic acid. (H_2CO_3) Formed from water and carbon dioxide mixture.

cardiac insufficiency. Inadequate cardiac output due to failure of the heart to function properly; pump failure.

cardiogenic. Having origin in the heart.

catabolism. The breaking down of living cells or nutrients into simpler substances, most of which are excreted.

catecholamines. The hormones epinephrine and norepinephrine produced by the sympathetic nervous system.

cation. A positively charged ion. Examples are Na^-, K^+, CA^{2+}, Mg^{2+}.

CCPD. Abbreviation for continuous cycling peritoneal dialysis.

cellular. Pertaining, composed of, or derived from cells.

central nervous system (CNS). Brain and spinal cord, with their nerves and end-organs that control voluntary acts.

Cheyne-Stokes respiration. Cycles of gradually increasing tidal volume followed by gradually decreasing tidal volume and then by apnea.

chloride. A chemical element that is one of the major anions of the extracellular fluid; it functions to maintain the osmotic pressure of the blood.

CIPD. Abbreviation for chronic intermittent peritoneal dialysis.

cirrhosis. An interstitial inflammation with hardening, granulation, and contraction of the tissues of the liver.

clean technique. A procedure that renders a surface free, as nearly as possible, from pathogenic microorganisms.

CNS. Abbreviation for central nervous system.

colloid. A particle invisible to the naked eye, which instead of dissolving is held in a state of suspension.

colloid osmotic pressure. That component of the osmotic pressure supplied by large molecules, e.g., proteins.

compensation. The counterbalancing of any defect of structure or function.

compensation, metabolic. A body function in which lungs balance pH partially or completely. In metabolic acidosis, the lungs hyperventilate to blow off acid (decrease Pco_2); in metabolic alkalosis, the lungs hypoventilate to retain acid.

compensation, respiratory. A body function in which kidneys balance the pH partially or completely. In respiratory acidosis, the kidneys retain bicarbonate (HCO_3^-) and excrete excess hydrogen ions; in respiratory alkalosis the kidneys retain hydrogen ions.

COP. Abbreviation for colloid osmotic pressure; that component of the osmotic pressure supplied by large protein molecules, especially albumin.

COPD. Abbreviation for chronic obstructive pulmonary disease.

cor pulmonale. Heart disease due to lung disease.

crenation. The reduction in size of a red blood cell due to mixture with a hypertonic solution.

CRF. Abbreviation for corticotropin releasing factor and chronic renal failure.

crystalloid. Clear like a crystal; a substance capable of crystallization that when dissolved separates into individual small molecules that can be diffused through animal membranes, e.g., salt, sugar.

cubital fossa. The furrow in the elbow area.

dialysis. Process used to separate substances (solutes) in solution by means of their unequal diffusion through a selectively permeable membrane, natural or synthetic.

dialyzer. An apparatus for effecting dialysis. The porous septum or diaphragm of such an apparatus.

diaphoresis. Profuse sweating, causing loss of fluid and electrolytes.

diffusion. Movement of solutes or gases from an area of higher concentration to an area of lower concentration; a passive transport system.

diuresis. Secretion and passage of large quantities of urine.

diuretic. An agent that increases the excretion of urine.

dysphagia. Difficulty in swallowing.

dyspnea. Labored or difficult breathing.

dysrhythmia. Abnormal rhythm of the heart.

ECBV. Abbreviation for effective circulating blood volume.

ecchymosis. Blood under the skin; a bruise.

ECF. Abbreviation for extracellular fluid.

ECW. Abbreviation for extracellular water.

edema. A condition in which the interstitial (tissue) spaces contain an excessive amount of extracellular fluid.

edema, dependent. The collection of extracellular fluid in interstitial spaces secondary to gravitational flow.

edema, pitting. The collection of extracellular fluid in tissue spaces without gravity flow. Finger impressions remain when applied pressure is released.

electrolyte. A substance capable of dissociating into ions and which in solution will conduct an electric current.

-emia (acidemia, alkalemia). A word ending denoting the presence of a substance in the blood.

endocrine. Pertaining to glands that release internal secretions.

endotracheal tube, cuffed. Tube used to provide an airway through the trachea. After insertion, the cuff, which surrounds the tube, is inflated to prevent aspiration of foreign substances into the bronchus.

erythropoietin. A hormone produced by the kidney that stimulates red blood cell production in the bone marrow.

extracellular. Any area inside the body, but outside the cell; interstitial and intravascular areas.

extracellular fluid. A chemical solution, primarily saline, which occupies the areas outside the cells. Abbreviation: ECF.

FE. Abbreviation for fluid and electrolyte.

Fe. Chemical symbol for iron.

febrile. Feverish; having a high body temperature.

fibrillation. Tremor or rapid action of the heart without effective cardiac output.

fibrin. A part of the blood clotting system. Fibrin is a fine mesh formed from fibrinogen that traps red blood cells, white blood cells, and platelets to form a clot.

filtration. The passage of a solution through a filter.

fistula. Any abnormal, tubelike passage between normal body cavities or a normal body cavity and a free surface.

fixed acid. An acid that is not converted to carbon dioxide and water and must be excreted by the kidneys.

forced expiratory reserve. The residual air left in the lungs after strenuous exhalation.

gavage. A feeding administered into the stomach via a nasogastric or orogastric tube.

GFR. Abbreviation for glomerular filtration rate.

globulin. A type of simple animal protein found in body fluids and cells.

glomerular. Pertaining to, produced by, or involving a glomerulus, a tuft of capillaries within the nephron, the functional unit of the kidney.

glomerular filtration. The passage of the ultrafiltrate of plasma (equivalent to plasma, minus cells and protein) that passes across the membranes of the malpighian corpuscles of the kidney to the lumen of Bowman's capsule.

glomerular filtration rate. The speed of clearance of a certain volume of plasma (including waste products) through the glomerulus; e.g., inulin clearance and GFR in man is 120 mL/min.

glomerulonephritis. Inflammation of the capillary loops in the glomeruli of the kidney. This condition has acute, subacute, and chronic forms and is usually secondary to an infection—especially hemolytic streptococcus.

glucocorticoids. Hormones of the steroid class secreted by the adrenal cortex.

glucogenesis. Production of glycogen from glucose or other sugars in the body.

gluconeogenesis. Transformation of noncarbohydrates into glycogen.

gonads. Reproductive gland of males or females.

Hct. Abbreviation for hematocrit.

hematemesis. Vomiting of blood.

hematocrit. A laboratory test evaluating the relationship between erythrocytes and plasma. Abbreviation: Hct.

hematogenic. Pertaining to the formation of blood; produced by or derived from blood.

hemoconcentration. An increase in the concentration of red blood cells resulting from a decrease in the volume of plasma.

hemodialysis. Removal of endogenous metabolites or exogenous poisons from the blood through a semipermeable membrane while the blood is being circulated outside the body.

hemodilution. An increase in the volume of blood plasma resulting in reduced relative concentration of red blood cells.

hemoglobin. An iron-containing pigment of the red blood cells that carries oxygen from the lungs to the tissues.

hemolysis. The enlargement of a red blood cell to a swollen globular state in hypotonic saline solution or distilled water.

hemoptysis. Blood-tinged sputum arising from hemorrhage of larynx, trachea, bronchi, or lungs.

Henderson-Hasselbalch Equation. An equation giving the pH of a buffer solution: $pH = 6.1 + \log \frac{[HCO_3^-]}{[H_2CO_3]}$

hepatic. Pertaining to the liver.

Hgb. Abbreviation for hemoglobin.

homeostasis. The integrated balance of all body systems.

hormone. A substance originating in an organ, gland, or body part that is conveyed through the blood to another part of the body, stimulating it by chemical action to increased functional activity and increased secretion.

hydrogen ions. See: ions, hydrogen.

hydrostatic pressure. Pressure exerted on liquids by the weight of the liquids above it.

hyperalimentation. The intake of large amounts of basic nutrients sufficient to achieve tissue synthesis and growth; may be either oral ingestion or intravenous infusion.

hypercalcemia. Excessive calcium in the blood.

hypercapnia. Excessive carbon dioxide in the blood.

hyperchloremia. Excessive chloride in the blood.

hyperglycemia. Excess of glucose in the blood.

hyperkalemia. Excessive potassium in the blood due to a deficit of water.

hypernatremia. Excessive sodium in the blood due to a deficit of water.

hyperosmolality. Abnormally increased osmolal concentration.

hyperosmolar. Having an osmotic pressure greater than that of normal plasma.

hyperosmolarity. Abnormally increased osmolar concentration.

hyperphosphatemia. Abnormal amount of phosphorus in the blood.

hyperplasia. Abnormal or unusual increase in components of tissue; can be benign or malignant growth.

hyperpnea. Increased breathing; usually refers to increased tidal volume with or without increased frequency.

hyperpotassemia. Abnormal amounts of potassium in the blood.

hyperreflexia. Increased action of the reflexes.

hypertonic. A solution that has a higher osmotic pressure than the one with which it is compared.

hypertonicity. (1) Excess muscular tonus; (2) Osmotic pressure or concentration of solutes greater than that of blood.

hyperventilation. Respirations that are rapid and increased both in depth and rate causing a deficit of Pa_{CO_2}. Synonym: hyperpnea.

hypervolemia. Abnormal increase in the volume of circulating fluid in the body.

hypocalcemia. Decreased calcium in the blood.

hypochloremia. Decreased chloride in the blood.

hypokalemia. Decreased potassium in the blood.

hyponatremia. Decreased sodium in the blood due to an excess of water.

hypo-osmolality. Abnormally decreased osmolal concentration.

hypo-osmolarity. Abnormally decreased osmolar concentration.

hypophosphoremia. Decreased amount of phosphorous compounds in the blood.

hypopotassemia. Decreased potassium in the blood.

hypoproteinemia. Decrease in the normal quantity of protein (usually albumin and globulin) in the blood.

hyporeflexia. Diminished function of the reflexes.

hypothalamic. Pertaining to the hypothalamus.

hypotonic. Having an osmotic pressure lower than the one with which it is compared.

hypoventilation. Decreased alveolar ventilation in relation to metabolic rate.

hypovolemia. Diminished circulating blood volume.

hypoxemia. Inadequate oxygenation of the blood.

hypoxia. Diminished availability of oxygen to the body tissues.

iatrogenic. Adverse condition resulting from treatment by the physician.

ICF. Abbreviation for intracellular fluid.

ICW. Abbreviation for intracellular water.

idiopathic. Pertaining to conditions without clear pathogenesis, or disease without recognizable cause, as of spontaneous origin.

infarction. Cessation of the blood supply to the tissue of an organ with resultant death of that tissue; usually results from vascular occlusion or stenosis.

inspiratory reserve. The volume of air that can be inhaled after normal exhalation.

inspiratory reserve volume. The maximum volume of gas that can be inspired from the end-tidal inspiratory level.

inspiratory stridor. A harsh, high-pitched sound usually due to obstruction in the air passages.

insensible loss. Loss of body fluids that occurs without the individual's awareness.

integumentary. Related to integument. Synonym: skin.

interstitial. The area between and around the cells.

interstitial fluid. Extracellular fluid composed of water and electrolytes which circulates between and around the cells.

intracellular. Area within the cell.

intracellular fluid. A solution of water and electrolytes which circulates within the cell. Abbreviation: ICF.

intravascular. Within the blood vessels.

intubation. Insertion of a tube into any hollow organ, as into the larynx or trachea through the glottis for entrance of air, or to dilate a stricture.

ion. An electrically charged particle.

ion gap. Differences in NA^+ and Cl^- and HCO_3^- in metabolic conditions.

ion, hydrogen. An electrically charged hydrogen molecule.

ionize. The process by which an electrically charged chemical moves from one chemical compound to another.

IPD. Abbreviation for intermittent peritoneal dialysis.

IPPB. Abbreviation for intermittent positive pressure breathing.

ischemia. Localized tissue anemia due to obstruction of the inflow of arterial blood.

isosthenuria. Poor urine; little variation in the urinary specific gravity of a person with nephritis.

isotonic. A solution having the same osmotic pressure as the one with which it is compared.

kg, 75. The weight of an average-sized man (165 pounds). See kilogram.

kilogram. A weight measurement equal to 2.2 pounds. Abbreviation: kg.

Kreb's cycle. Citric acid cycle. A complicated series of reactions in the body involving the oxidative metabolism of pyruvic acid and liberation of energy. It is the main pathway of terminal oxidation in the process of which not only carbohydrates but proteins and fats are used.

Kussmaul's respiration. A pattern of deep and gasping breaths, as seen in severe cases of ketoacidosis.

KVO. Abbreviation for keep vein open.

lactic acid. A colorless syrupy liquid ($C_3H_6O_3$) formed in muscle during activity by the breakdown of glycogen (glycolysis) during the anaerobic cycle.

lactic dehydrogenase. An enzyme released after injury in selected organs, especially the heart and liver. Abbreviation: LDH.

laryngectomy. Removal of the larynx.

lavage. Washing out of the stomach via intubation.

LDH. Abbreviation for lactic dehydrogenase.

marasmus. Emaciation, wasting. Infantile atrophy that occurs almost wholly as a sequel to acute diseases, especially diarrheic disease of infancy.

matrix. A structure in which something else permeates; the intracellular substance of a tissue such as cartilage.

melena. Blood in the feces.

mEq. Milliequivalent; measure of the amount of ion charge or of a reactant.

metabolic water. See water of combustion.

metabolism. The sum total of the physical and chemical processes and reactions taking place among the ions, atoms, and molecules of the body.

metamorphosis. Transformation.

milliequivalent. A measurement of solutes in terms of combining power; the weight of a substance contained in 1 mL of an IN (normal) saline solution: 1/1000 molar concentration of the ion × charge.

milligram (mg). A metric unit of weight equal to one thousandth (10^{-3}) of a gram.

milligrams percent. In biochemistry, the number of milligrams of a substance per 100 mL of blood; symbol mg%, mg.%, mg/dL.

milliosmol. Measure of solute concentration expressed as 1/1000 of the molar (or molal) concentration of an equivalent perfect solute.

mineralocorticoid. A hormone of the steroid group, produced by the adrenal cortex.

molecule. The smallest division of a chemical compound that maintains its original characteristics.

morbidity. Diseased state.

mortality. Death rate.

mOsm. Abbreviation for milliosmol. A measure of solute concentration expressed as 1/1000 of the molar (or molal) concentration of an equivalent perfect solute.

narcosis. A reversible condition characterized by stupor or insensibility.

nasopharynx. Part of the pharynx situated above the soft palate (postnasal space).

nephron. The functional unit of the kidney.

nephrosis. Condition in which there are degenerative changes in the kidneys without the occurrence of inflammation.

neurogenic. Originating from nervous tissue.

neutralize. To convert the entire amount of an acid or a base into a salt by the addition of exactly sufficient quantity of a base or an acid, respectively; to render ineffective.

nonelectrolyte. Substance such as sugar or urea that, in solution, does not dissociate into ions and is thus electrically neutral.

nonvolatile. Does not easily evaporate.

NPO. An abbreviation meaning nothing by mouth (Latin: *nihil per os*).

oliguria. Diminished amount of urination: less than 20 mL/h, 160 mL in 8 hours, 480 mL in 24 hours.

oncotic pressure. Colloid osmotic pressure.

organic acid. An acid containing one or more carboxyl (COOH) groups.

oropharynx. Central portion of the pharynx lying between the soft palate and upper portion of the epiglottis.

orthopnea. Difficulty in breathing except in an upright position.

-osis (acidosis, alkalosis). A word ending denoting a process, especially a diseased or morbid process.

osmolality. Molal concentration of osmotic particles (mOsm/kg).

osmolarity. Molar concentration of osmotic particles (mOsm/L).

osmoreceptors. Structures in the hypothalamus that respond to changes in osmotic pressure of the blood by regulating the secretion of the neurohypophyseal antidiuretic hormone.

osmoregulation. Maintenance of osmolarity by a simple organism or body cell with respect to the surrounding medium.

osmosis. The passage of water (solvent) through a semipermeable membrane from an area of lesser concentration to an area of greater concentration of solute; a passive transport system.

osmotic pressure. The pressure developed when two solutions of different concentrations are separated by a semipermeable membrane; partial pressure of dissolved solutes (from displacement of water molecules).

paralytic ileus. Paralysis of the wall of the intestine, involving distention and symptoms of obstruction.

parenchyma. The essential parts of an organ that are concerned with its function in contradistinction to its framework.

paresthesia. A numbness or a prickling sensation.

parotitis. Inflammation of the salivary glands.

paroxysmal nocturnal dyspnea. A periodically occurring attack of shortness of breath during the night or when in a supine position. Abbreviation: PND.

particles. An extremely small mass or portion of substance or matter.

pathophysiologic. Related to pathophysiology, the study of an abnormal function.

Pco_2. Partial pressure of carbon dioxide (gas).

PE. Abbreviation for physical examination.

perfusion. Flow of blood in capillary bed.

pericarditis. Inflammation of the outside layer of the heart.

periphery. Outer part or a surface of a body.

peristalsis. The progressive action caused by the contraction and relaxation of the muscles of the gastrointestinal tract.

peritonitis. Inflammation of the peritoneum.

pH. A symbol used to express the degree of acidity or alkalinity of the blood, that is, the hydrogen ion concentration; potential of hydrogen, the negative logarithm of the hydrogen ion concentration.

phlebothrombosis. Condition of clots in the veins without concurrent inflammation.

phosphate. A salt of phosphoric acid.

phosphorus. A chemical element. In combination with calcium, oxygen, and hydrogen, it forms the substance of bones.

plasma. The liquid part of lymph and blood composed of serum and protein substances in solution.

Po_2. Partial pressure of oxygen.

polycythemia. Excess number of red corpuscles in the blood.

polydipsia. Excessive thirst.

polyuria. Excessive excretion of urine.

postural blood pressure. Blood pressure as affected by the posture of the body. A decrease of 10 mmHg or more in the systolic pressure from a supine to a standing position can indicate an extracellular fluid deficit.

potassemia. Presence of excessive quantity of potassium in the blood.

potassium. A chemical element. As the ion K^+ it plays an essential role in body processes, including maintenance of the acid-base and water balance.

pressure, osmotic. See osmotic pressure.

rale. An abnormal sound heard on auscultation of the chest and produced by the passage of air through bronchi that contain secretion or exudate or are constricted by spasms.

renin. A protein formed in an ischemic kidney that converts an alpha globulin (hypertensinogen) of the blood into hypertensin (angiotensin), a powerful vasoconstrictor.

respiration. The process of O_2 and CO_2 exchange as carried out by the lungs.

rhonchus (pl., rhonchi). A coarse, dry rale or rattling in the bronchial tubes.

RPS. Abbreviation for renal pressor substance.

saline. Combination of sodium ion, chloride ion, and water.

saline, normal. A 0.9% salt solution.

salt. A crystalloid chemical compound composed of a cation and an anion equally balanced by valence; most frequently thought of as sodium chloride.

semipermeable membrane. A membrane that only permits passage of certain molecules.

sensible loss. Perceptible loss of body fluid and electrolyte, for example, urine.

sensorium. Degree of mental alertness.

serum. Plasma minus blood protein fibrinogen.

shock. A state of collapse resulting from acute peripheral circulatory failure. It may occur following hemorrhage, severe trauma, surgery, burns, extracellular fluid deficit, infections, or drug toxicity.

singultus. Hiccoughs.

skin turgor. The degree of tension of the skin.

SMA. Abbreviation for sequential multiple analysis; a form of grouping results of several laboratory tests at once. SMA 6 equals six different tests.

SOB. Abbreviation for shortness of breath.

sodium. Sodium constitutes approximately 0.15% of elements of the body. Sodium (Na^+) is the principal cation found in extracellular fluids.

sodium pump. A theoretical mechanism for transporting sodium ions across cell membranes against an opposing concentration gradient.

solute. A substance that is dissolved in a solution.

solution. Liquid containing a dissolved substance.

solvent. The liquid in which another substance (the solute) is dissolved to form a solution.

somnolence. Prolonged drowsiness or a condition resembling a trance, which may continue for a number of days; sleepiness.

specific gravity. The weight of a substance compared with an equal volume of water. The weight of water is taken as 1.000.

steatorrhea. Fatty stools.

sternal retraction. Strong contraction of the chest muscles in an attempt to pull in more air.

stress mechanism. The body's automatic defenses against stress.

sulfate. A salt of sulfuric acid.

supine. Lying with the face upward, or on the dorsal surface.

syncope. A transient form of unconsciousness due to cerebral hypoxia: fainting.

tachypnea. Increased frequency of breathing.

tamponade, cardiac. Compression of the heart due to collection of fluid or blood in the pericardium.

TBSA. Abbreviation for total body surface area.

TBW. Abbreviation for total body water.

TCDB. Abbreviation for turn, cough, deep breathe.

tetany. Intermittent or steady contraction of a muscle.

third space. An area in which extravasated fluid collects and becomes physiologically unavailable to the body.

thrombophlebitis. The development of venous thrombi in the presence of inflammatory changes in the vessel wall.

thromboplastin. A substance found in the tissues or plasma that accelerates clotting of the blood.

tidal volume. The amount of air in one quiet respiration.

tonicity. Measurement of osmotic pressure of a solution; property of possessing tone, especially muscular tone.

tortuous. Marked by repeated twists.

toxic. Poisonous or caused by poison.

tracheostomy. Operation of cutting into the trachea through the neck, usually for insertion of a tube to overcome tracheal obstruction and provide an airway.

tracheotomy. Incision of the trachea through the skin and muscles of the neck overlying the trachea; see: tracheostomy.

transport systems. The movement or transference of biochemical substances. There are two types of transport systems: active and passive.

ultrafiltration. Filtration of a colloidal substance in which the dispersed particles, but not the liquid, are held back.

urea. The chief nitrogenous constituent of urine and the final product of protein metabolism in the body.

uric acid. A crystalline acid occurring as an end-product of purine metabolism.

valence. Property of an atom or group of atoms causing them to combine in definite proportion with other atoms or groups of atoms: the degree of combining power of an element or radical.

vasogenic. Related to the vascular system.

ventilation. The constant supplying of oxygen through the lungs.

villi. Small vascular processes of protrusions as from the free surface of a membrane.

vital capacity. The greatest volume of air that can be expressed from the lungs after a maximum inspiration.

volatile. Easily vaporized or evaporated.

volatile acid. An acid that is reduced to carbon dioxide (CO_2) and water (H_2O) and is expelled by the lungs.

water of combustion (metabolic water). Water in the body derived from metabolism of a food element such as starch, glucose, or fat; called also metabolic water or water of oxidation.

INDEX

• • • • • • • • • • •

Numbers in italics indicate figures; numbers followed by "t" indicate tables.